Excited State Dynamics in DNA and RNA

Excited State Dynamics in DNA and RNA

Experimental and Computational Approaches on Triplet State

Edited by

Jinquan Chen

WILEY-VCH

Editor

Prof. Jinquan Chen
State Key Lab. Precision Spectroscopy
East China Normal University
Shanghai 200241
China

Cover Design: Wiley
Cover Image: © Yuichiro Chino/Getty Images

Typesetting: Straive, Chennai, India

Library of Congress Card No.: applied for

British Library Cataloguing-in-Publication Data A catalogue record for this book is available from the British Library.

Bibliographic information published by the Deutsche Nationalbibliothek
The Deutsche Nationalbibliothek lists this publication in the Deutsche Nationalbibliografie; detailed bibliographic data are available on the Internet at <http://dnb.d-nb.de>.

Print ISBN: 978-3-527-35467-2
ePDF ISBN: 978-3-527-85036-5
ePub ISBN: 978-3-527-85035-8
oBook ISBN: 978-3-527-85037-2

Printed and bound by CPI Group (UK) Ltd, Croydon, CR0 4YY

C97835273554672_050226

Contents

Preface

The integrity of nucleic acids is jeopardized by the initial assault of highly absorbent ultraviolet (UV) solar photons, posing a significant risk to genomic stability. However, the unique chemical composition of DNA nucleobases and duplexes facilitates the deactivation of excited singlet electronic states within nanoseconds, effectively dissipating the absorbed photon energy and reducing the likelihood of DNA photodamage. In contrast, the longer-lived triplet excited states have garnered interdisciplinary attention due to their role in initiating various detrimental processes, including one-electron oxidation, H-abstraction, and crosslinking within DNA/RNA. Indeed, the existence of a triplet channel for DNA/RNA photodamage formation after direct UV irradiation of DNA/RNA has been confirmed since the 1960s, and it is the research hotspot during 1960s–1970s. The findings are summarized in the book entitled *Excited States of Proteins and Nucleic Acids* edited by R. F. Steiner (1971, Plenum Press).

However, studies on the DNA/RNA triplet states from a dynamic aspect are rare due to their low extinction coefficients, their small quantum yields, and the time resolution of instruments at that time. Thanks to the newly developed highly sensitive spectroscopic technological tools, advanced high-precision quantum chemical calculations, and the well-designed DNA/RNA modeling systems, the study of the photoactivated dynamics of DNA/RNA has once again become one of the hottest topics in photochemistry.

Precise detection of DNA/RNA triplet states was achieved in the past decade, enabling significant advances in our knowledge of their photo-activated dynamics. Now, there is more intuitive evidence for the mechanistic understanding of the triple state-mediated DNA photodamage response, such as cyclobutane pyrimidine dimers (CPDs), pyrimidine (6-4) pyrimidone photoproducts, one-electron oxidation, H-abstraction, and cross-link in DNA/RNA. Furthermore, triplet states have also been identified as playing a crucial role in the effectiveness of DNA-intercalated photosensitizing drugs. Photosensitized DNA/RNA damage can occur through a variety of processes, including electron transfer, hydrogen abstraction, triplet–triplet energy transfer, or generation of reactive oxygen species.

We witnessed the evolution of the dynamics studies on DNA/RNA triplet states and propose this book in order to systematically summarize the progress made in the study of triplet excited state dynamics in DNA/RNA over the past years.

This comprehensive work would provide valuable insights into the field, enabling researchers to better understand the behavior of triplet excited states and their implications for DNA-targeted photosensitizing drugs and other applications. This book aims to identify the most important research trends within the field and to synthesize their major findings. Furthermore, the authors have strived to present this substantial body of knowledge in a pedagogical manner, ensuring accessibility for all scientists active in the field, beyond specialists alone.

The first chapter is devoted to providing the significance of triplet state in DNA/RNA and a review of the development of history of such field. Chapters 2 and 3 describe the main discoveries of triplet state dynamics in canonical and important noncanonical nucleobases. Chapters 4 and 5 focus on the triplet-mediated photochemistry in nucleobases and DNA from their intrinsic excited state properties and drug-sensitized aspect, respectively. Chapters 6–8 review recent advances in computational photobiology, concerning the triplet excited state dynamics in DNA/RNA from nucleobases to poly-oligomers.

We are deeply grateful to all the authors for their outstanding work and highly informative chapters. We hope that the readers of this book will find its content insightful and a valuable contribution to the field.

22 August 2025
Jinquan Chen
East China Normal University Campus
Shanghai, China

List of Abbreviations

Preface

UV	ultraviolet
CPDs	cyclobutane pyrimidine dimers

Chapter 1

ISC	intersystem crossing
CPD	cyclobutane pyrimidine dimer
UV	ultraviolet
VR	vibrational relaxation
IC	internal conversion
AMP	adenosine monophosphate
GMP	guanosine monophosphate
TMP	thymidine monophosphate
CMP	cytidine monophosphate
UMP	uridine monophosphate
TTET	triplet–triplet energy transfer
T–T	triplet–triplet
k_{et}	electron transfer rate constant
TRIR	time-resolved infrared
FC	Franck–Condon
SOC	spin–orbit coupling

Chapter 2

ISC	intersystem crossing
A	adenine
G	guanine
C	cytosine
T	thymine
U	uracil
Thy	thymine

Thd	thymidine
TMP	thymidine-5′-monophosphate
Urd	Uridine
UMP	uridine-5′-monophosphate
MeCN	acetonitrile
TRIR	time-resolved infrared
EADS	evolution-associated difference spectra
dCyd	2′-deoxycytidine
5mdCyd	2′-deoxy-5-methylcytidine
FC	Franck–Condon
TRTL	time-resolved thermal lens
2-tUra	2-thiouracil
3-tThd	2-thiothymidine
2,4-dtThy	2,4-dithiothymine
TRPES	time-resolved photoelectron spectroscopy
4-tUra	4-thiouracil
2,4-dtUra	2,4-dithiouracil
GS	ground state
2-tCyt	2-thiocytosine
PES	potential energy surface
3-tCyd	2-thiocytidine
$^3O_2$	triplet oxygen
REMPI	resonance-enhanced multiphoton ionization
5-Me-tGua	6-methylthioguanine
2,6-dtPur	2,6-dithiopurine
6-SeGua	6-selenoguanine
4-aCyt	5-azacytosine
5-aUra	6-azauracil
6-aThy	6-azathymine
7-aAde	7-azaadenine
8-aGua	8-azaguanine
TRPD	time-resolved phosphorescence detection
7-aThd	6-azathymidine
6-aUrd	6-azauridine

Chapter 3

ISC	intersystem crossing
5mC	5-methylcytosine
5hmC	5-hydroxymethylcytosine
5fC	5-formylcytosine
5caC	5-carboxycytosine
5FC	5-fluorocytosine
5ClC	5-chlorocytosine
5BrC	5-bromocytosine

5AC	5-azacytidine
CIs	conical intersections
CT	charge-transfer
5mCyd	5-methylcytidine
5mdCyd	5-2′-deoxy-5-methylcytidine
PES	Potential energy surface
5hmdCyd	2′-deoxy-5-hydroxymethylcytidine
TA	transient absorption
5fdCyd	2′-deoxy-5-formylcytidine
5cadCyd	2′-deoxy-5-carboxylcytidine
ICT	intramolecular charge transfer
AC	5-azacytosine
5ACyd	5-azacytidine
5AdCyd	2′-deoxy-5-azacytidine
5fU	5-formyluraci
N6-mA	N6-methyladenine
N6-hmA	N6-hydroxymethyladenine
N6-fA	N6-formyladenine
N6-mAdo	N6-methyladenosine
D_2O	deuterated aqueous
LE	local excited
N6-hmAdo	N6-hydroxymethyladenosine
N6-fAdo	6-formyladenosine
fs-TA	femtosecond transient absorption
ns-TA	nanosecond transient absorption
oop	out-of-plane
SOC	spin–orbit coupling
SOCT-ISC	spin–orbit charge transfer ISC
m7G	N7-methylguanosine
FUPC	fluorescence up-conversion
5mC	5-methylcytosine
C	cytosine
ESA	excited state absorption
3Intra-CT	intrastrand triplet CT state
HPLC	high-performance liquid chromatography
PAGE	polyacrylamide gel electrophoresis

Chapter 4

CPDs	cyclobutane pyrimidine dimers
SPs	spore photoproducts
G^{SO3}	guanosine-6-sulfonate
DPCs	DNA–protein crosslinks
TiO_2	titanium dioxide
6-TG	6-thioguanine

AuNPs	gold nanoparticles
$^1O_2$	singlet oxygen
RNA	ribonucleic acid
DNA	deoxyribonucleic acid
UV	ultraviolet
ISC	intersystem crossing
CI	conical intersection
IC	internal conversion
SP	spore photoproduct
SPL	spore photoproduct lyase
SACs	single-atom catalysts
PDT	photodynamic therapy
7-MP	6-mercaptopurine
6-TG	6-thioguanine
ROS	reactive oxygen species
G^{SO2}	guanosine-6-sulfinate
G	guanosine
G^{SO}	guanine sulfenic acid
MCSCF	multiconfigurational self-consistent field
TD-DFT	time-dependent density functional theory
UVA	ultraviolet A
TR-FTIR	infrared transient absorption
DFT	density functional theory
4-TU	4-thiouracil
U	uracil
U^{SO3}	uracil-6-sulfonate
GSH	glutathione
Trp	tryptophan
ΔG	electron-transfer driving force
TrpH	tryptophan

Chapter 5

5-MOP	5-methoxypsolaren
6-4PPs	pyrimidine 6-4 pyrimidone photoproducts
8-MOP	8-methoxypsolaren
AB	avobenzone
AB(E)	enol form of avobenzone
AB(K)	diketone form of avobenzone
AMP	adenosine 5′-monophosphate
ANFX	N(4′)-acetyl norfloxacin
AP	apyrimidinic-apurinic
BNP	benoxaprofen
BP	benzophenone
CMP	cytidine 5′-monophosphate

CP	carprofen
CPDs	cyclobutane pyrimidine dimers
CPX	ciprofloxacin
Cyd	2′-deoxycytidine
Cyt (^{3}Cyt*)	cytosine (triplet excited state of cytosine)
Cyt<>Cyt	cyclobutane cytosine dimer
DPA	dipicolinic acid
DSB	double-stranded break
E. coli	*Escherichia coli*
ΔE	energy gap
ESIPT	excited state intramolecular proton transfer
E_T	triplet energy
ENX	enoxacin
FA	fenofibric acid
FLX	fleroxacin
FQ	fluoroquinolone
GMP	guanosine 5′-monophosphate
GV	gilvocarcin V
HaCaT	human keratinocytes cells
HCT	hydrochlorothiazide
HPLC	high-performance liquid chromatography
INP (^{3}INP*)	indoprofen (triplet excited state of indoprofen)
ISC	intersystem crossing
KP (^{3}KP*)	ketoprofen (triplet excited state of ketoprofen)
k_{TTET}	triplet–triplet energy transfer rate constant
k_D	diffusion-controlled rate constant
LFX	lomefloxacin
LPF	laser flash photolysis
MFQ	mefloquine
MS	mass detector
MS/MS	tandem mass detector
N (^{3}N*)	nucleobase (triplet excited state of nucleobase)
NSAID	nonsteroidal anti-inflammatory drug
NFX	norlfoxacin
O_2	molecular oxygen
$^1O_2$	singlet oxygen
OFX	ofloxacin
PAGE	polyacrylamide gel electrophoresis
PDT	photodynamic therapy
PFX	pefloxacin
ppRSV	photoproduct of rosuvastatin
Ps	psolaren
Phs (^{3}Phs*)	photosensitizer (triplet excited state of photosensitizer)
Pur	purine
PUVA	psolaren-mediated therapy

Pyr (^{1}Pyr*) (^{3}Pyr*)	pyrimidine (singlet excited state of pyrimidine) (triplet excited state of pyrimidine)
PyPs (^{3}PyPs*)	pyridopsolaren (triplet excited state of pyridopsolaren)
S_1	singlet excited state
SCC	squamous cell carcinoma
SP	spore photoproduct
RFX	rufloxacin
ROS	reactive oxygen species
T_1	triplet excited state
T4 endo V	T4 endonuclease V
T–T transition	triplet–triplet transition
TMP (^{3}TMP*)	thymidine 5′-monophosphate (triplet excited state of thymidine 5′-monophosphate)
TPA	tiaprofenic acid
TTET	triplet–triplet energy transfer
Thd<>Thd	cyclobutane thymidine dimer
Thy	thymidine
Thy (^{3}Thy*)	thymine (triplet excited state of thymine)
Thy<>Thy	cyclobutane thymine dimer
TCS	triclosan
UV	ultraviolet
SSB	single-stranded break
SCGE	single-cell gel electrophoresis assay
XP	xeroderma pigmentosum

Chapter 6

UV	ultraviolet
TD-DFT	time-dependent density functional theory
ISC	intersystem crossing
SOC	spin–orbit coupling
IC	internal conversion
CIs	conical intersections
2tThd	2-thiothymidine
6tGuo	6-thioguanosine
PDT	photodynamic therapy
2tThd	2-thiothymidine
2tThy	2-thiothymine
DFT	density functional theory
6n-2tThy	6-aza-2-thiothymine
op-S	out-of-plane sulfur displacement
2TU	2-thiouracil
6tGua	6-thioguanine
6SeGua	6-selenoguanine
2tThd	2-thiothymidine

PCM	polarizable continuum model
ROS	reactive oxygen species
QM/MM	quantum mechanics/molecular mechanics
ML	machine learning
OLEDs	organic light-emitting diodes

Chapter 7

ISC	intersystem crossing
CPD	cyclobutane pyrimidine dimer
TSH	Tully's surface hopping approach
PES	potential energy surface
AS	active space
MEP	minimum energy path
FC	Franck–Condon
CI	conical intersection
SOC	spin–orbit coupling
MECP	minimum energy crossing point
SH	surface-hopping
AIMS	ab initio multiple spawning

Chapter 8

TR	time-resolved
PES	potential energy surface
FC	Franck–Condon
CI	conical intersection
5hmC	5-hydroxymethylcytosine
5fC	5-formyl-cytosine
5caC	5-carboxyl-cytosine
εdC	N4-etheno-2′-deoxycytidine
εdG	N2-etheno-2′-deoxyguanosine
CPDs	cyclobutane pyrimidine dimers
CT	charge transfer
5hmc	5-hydroxymethyl-cytosine
5caC	5-carboxyl-cytosine
7mG	7methyl-guanine
6hmA	6-hydroxymethyl-adenosine
6fA	N6-formyladenosine
ESIPT	excited state intramolecular proton transfer

1

Introduce of Triplet Excited State in DNA and RNA

Yanyan Jia[1] *and Jinquan Chen*[2]

[1] *East China University of Science and Technology, School of Chemistry and Molecular Engineering, 130 Meilong Rd, Shanghai 200030, China*
[2] *East China Normal University, State Key Laboratory of Precision Spectroscopy, 500 Dongchuan Rd, Shanghai 200241, China*

1.1 Introduction of the Significance of Triplet Excited State in Nucleic Acids

Although the sunlight is essential for life on earth, it contains a large amount of ultraviolet (UV) photons, and the integrity of nucleic acids is jeopardized by the initial assault of those highly absorbent UV solar photons, posing a significant risk to genomic stability of life on Earth [1]. Fortunately, the unique chemical composition of DNA/RNA nucleobases as well as their second- and third-order structures facilitates the deactivation of excited singlet electronic states typically on the picosecond timescale, effectively dissipating the absorbed photon energy and reducing the likelihood of DNA/RNA photodamage. Indeed, excited state dynamics of nucleic acids has been extensively studied due to the development of state-of-the-art ultrafast time-resolved spectroscopy, more accurate computational methods, more efficient software, and the continuously increasing computing power. The research progress achieved in the past two decades has been timely reviewed from both experimental and theoretical aspects [2–11]. However, the main focus of those studies usually starts with excited singlet states because absorption of a UV photon in nucleic acids can only produce excited states with singlet multiplicity, and their decay pathways have to be fully understood before one can further study the intersystem crossing (ISC) process that generates triplet excited state as well as their dynamics.

Triplet excited states represent a distinct electronic configuration characterized by two electrons occupying different orbitals with parallel spin orientations. This spin alignment endows triplet states with specific quantum number features in spin configuration systems. The total spin angular momentum quantum number (S) equals 1 for triplet states, compared to $S = 0$ for singlet states. According to quantum mechanical principles, when $S = 1$, the magnetic quantum number (m_s) can take three

Excited State Dynamics in DNA and RNA: Experimental and Computational Approaches on Triplet State, First Edition. Edited by Jinquan Chen.

projection values: (−1), (0), and (+1), corresponding to three quantized orientations of spin angular momentum in a magnetic field. Under external magnetic fields, this spin degeneracy interacts with the field through the Zeeman effect, splitting the originally degenerate triplet energy levels into three nondegenerate sublevels—a phenomenon that directly accounts for the "triplet" nomenclature (Scheme 1.1).

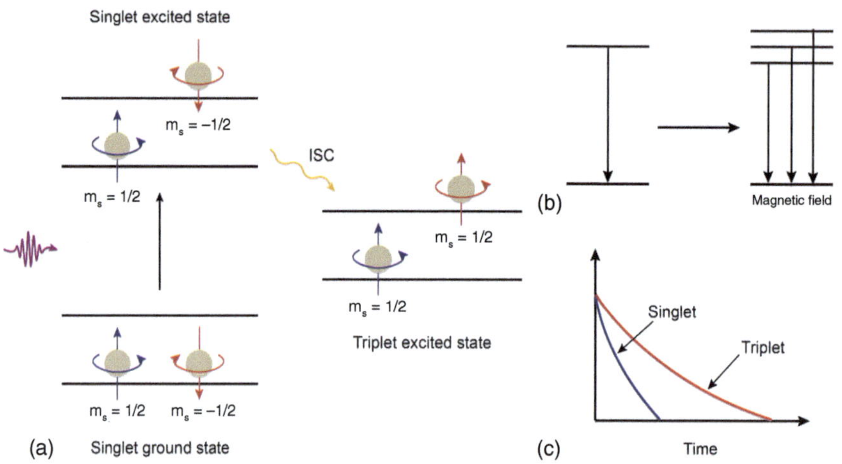

Scheme 1.1 (a) Illustration of singlet and triplet excited states. (b) Illustration of Zeeman effect. (c) Comparison of lifetimes between singlet and triplet excited states.

Triplet states exhibit fundamental differences from singlet states in both spin configuration and spectroscopic transitions. Transitions between triplet and singlet states involve electron spin state alterations, making them subject to strict selection rules with distinct transition probabilities and rates. According to selection rules, transitions preserving spin multiplicity ($\Delta S = 0$) are allowed, while those altering multiplicity ($\Delta S \neq 0$) are forbidden. In ground states (S_0), electrons will usually occupy lower energy molecular orbitals with opposite spins (singlet configuration). Upon photon absorption, molecules undergo vertical transitions to singlet excited states (S_n), where electrons maintain identical spin multiplicity. Subsequent processes follow Kasha's rule: rapid vibrational relaxation (VR) and internal conversion (IC) lead to the lowest energy conformation of the first excited singlet state (S_1), which may differ geometrically from S_0. Spin-flipping through ISC then generates triplet excited states. Direct triplet excitation via photons is virtually impossible due to the spin–orbit coupling (SOC) prohibition, rendering triplet states typically "dark" in nature. It is worth noting that the transition from triplet state to the ground state ($\Delta S \neq 0$) is also prohibited, leading to much longer lifetimes (up to microseconds or even milliseconds) compared with singlet states (usually in nanoseconds).

Triplet excited state of nucleic acids and their participation in DNA photodamage after direct UV irradiation have discovered in the 1960s with the help of low-temperature spectroscopy techniques [12–16]. On the other hand, experimental spectroscopy study on the triplet excited state dynamics in nucleic acids at room temperature is not straightforward because of its low extinction coefficient (~1000 M^{-1} cm^{-1}) and small quantum yields (~0.02) [17–20]. Therefore, a simplifying experimental approach, in which a triplet sensitizer is excited in the presence of nucleic acids at a wavelength where the latter one's absorption is negligible, is commonly used during the studies on the photophysics and photochemistry of nucleic acids when triplet excited states are involved [21–26]. Thanks to the development of advanced spectroscopic techniques and high-precision quantum chemical calculations, there is more intuitive evidence for the mechanistic understanding of the triple excited state dynamics as well as triple excited state-mediated photodamage of nucleic acids in the past decade [27–39]. Furthermore, triplet states have also been identified as playing a crucial role in the effectiveness of DNA-intercalated photosensitizing medicines [40–42]. Therefore, a comprehensive review focusing on the topic of triple excited state dynamics in DNA/RNA and the related photophysics and photochemistry is timely. In this book, we focus on those works that provide insight into the triplet excited states in DNA/RNA, which is a choice that represents our own research interests and also the limited perspective on the vast area of photophysics and photochemistry of nucleic acids [8, 11]. We review research progress in both experimental and theoretical studies on the triplet excited state dynamics in DNA/RNA and try to integrate the seemingly disparate results into a comprehensive model for explaining the remarkable triplet excited state-related photophysical properties.

1.2 Triplet Excited State Studies in the 1960–1970s with Low Temperature Spectroscopy

The study of triplet excited state in nucleotides by phosphorescence spectroscopy has been a subject of fascination within the fields of biochemistry and physical chemistry and nucleic acids. In this section, we briefly recall the main discoveries on DNA/RNA triplet excited state from the early 1960s through the 1970s, which is the first Golden Age of DNA photophysics. The first report on the phosphorescence in nucleic acids, which laid the foundation for understanding the origins of phosphorescence, was registered by Steele and Szent-Györgyi [43]. Then, the first results on isolated DNA/RNA monomers were obtained by Longworth in 1962 [44] and Bersohn in 1964 [45]. Phosphorescence from DNA/RNA at 77 K is typically observed as the sum of emissions from individual nucleotides, reflecting the triplet states of adenosine monophosphate (AMP), guanosine monophosphate (GMP), thymidine monophosphate (TMP), cytidine monophosphate (CMP), and uridine monophosphate (UMP), as shown in Figure 1.1 [16]. The phosphorescence quantum yields and lifetimes of the abovementioned monomers at different conditions are summarized in Table 1.1 [46]. It is worth pointing that the data in Table 1.1 are under the conditions of direct UV excitation of those nucleotides, which is the limitation of content in this chapter, and these values may vary under sensitization conditions. Indeed, triplet sensitization is an extremely important technique since it not only

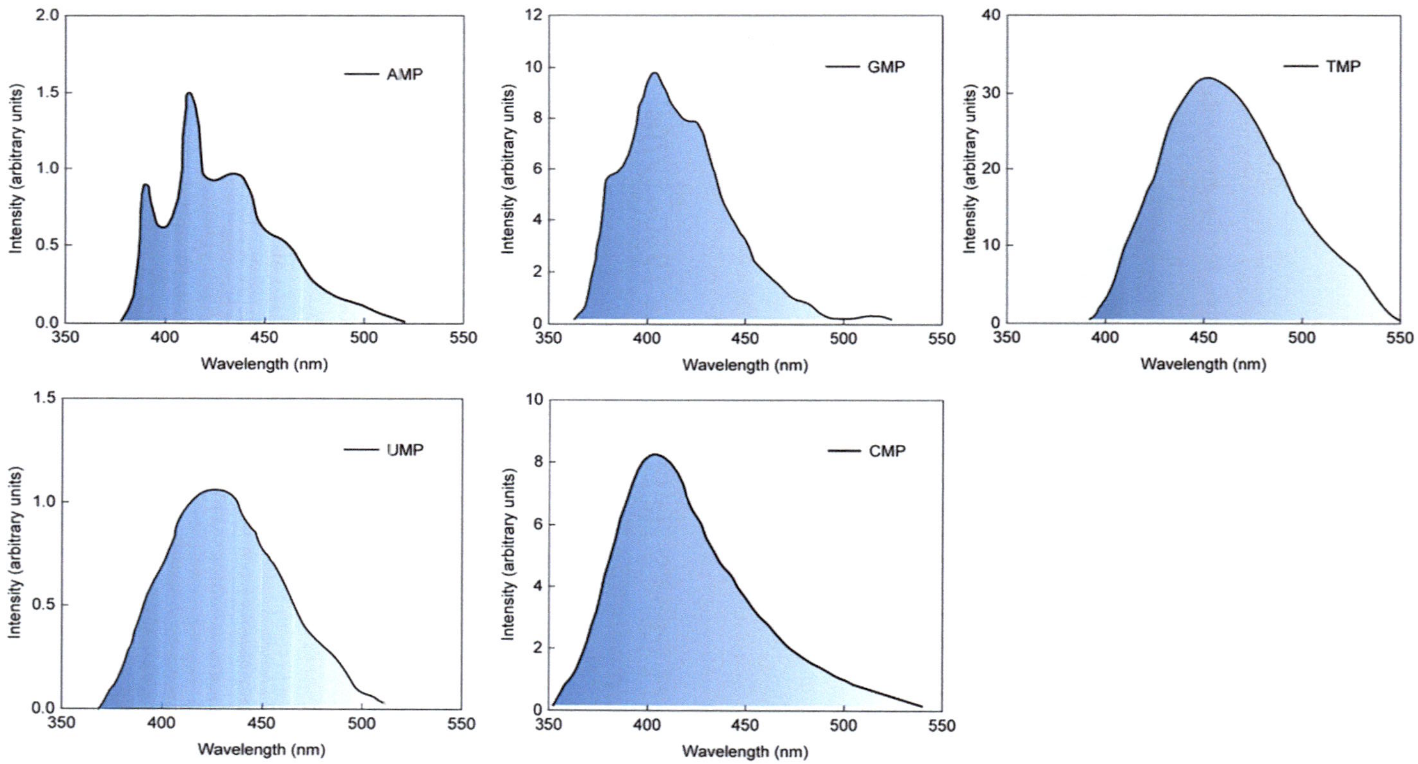

Figure 1.1 Phosphorescence of DNA/RNA nucleotides at 77 K under direct UV excitation.

Table 1.1 Phosphorescence quantum yields and lifetimes of DNA/RNA nucleotides at 80 K under direct UV excitation.

	pH	φ_p	φ_f	$^3\tau$(s)	Estimated triplet state energy (cm^{-1})
	pH = 2	0.005	—	1.75	27 350 (pH = 2)
AMP	pH = 7	0.025	0.01	2.4	26 700 (pH = 7)
	pH = 11	0.028	—	—	
	pH = 1	≤0.02	—	—	
GMP	pH = 7	0.095	0.13	1.3	27 200 (pH = 7)
	pH = 11	0.06	—	1.3	
	pH = 1	0.005	—	—	27 900 (pH = 7)
CMP	pH = 7	0.015	0.05	0.34	
	pH = 11	0.018	—	—	
	pH = 1	≤0.02	—	1.2	26 300 (pH = 7)
TMP	pH = 7	≤0.01	0.16	0.35	27 000 (pH = 12)
	pH = 12	0.03	—	0.45	
	pH = 1	≤0.006	—	—	
UMP	pH = 7	0.006	0.01	0.55	27 400 (pH = 7)
	pH = 12	0.006	—	—	28 400 (pH = 12)

makes it possible to study triplet states that could not be populated by direct excitation but also could selectively produce certain photolesions in DNA (see details in Chapter 5).

As shown in Figure 1.1 and Table 1.1, pyrimidine and purine nucleotides exhibit distinct phosphorescence behaviors. In general, the triplet states of pyrimidine nucleotides are less stable and exhibit lower quantum yields and shorter phosphorescence [16, 46, 47]. The main factor that affects the phosphorescence is the pH value. For example, the phosphorescence quantum yield of thymine base increases sharply in acidic conditions (pH = 1–2) while it remains low in neutral and basic environments. This is because deprotonation at the N3 position of the thymine ring under alkaline conditions disrupts the conjugated π-electron system and reduces molecular rigidity, which increases nonradiative decay rate (e.g. rotational/vibrational motions). In this case of TMP, the phosphorescence is strongly suppressed in acidic pH (pH < 4) while it intensifies near-neutral to mildly alkaline pH (pH = 6–9). Such observation is due to the protonation of the phosphate group in acidic media, leading to promoting collisional quenching because of the increased molecular solubility and flexibility. On the other hand, the negatively charged phosphate stabilizes the triplet state through electrostatic interactions and rigidifies the structure at neutral or mildly alkaline pH. The pH can also induce tautomerization of thymine. In acidic conditions, the keto form of thymine is more stable, leading to a longer phosphorescence lifetime. In neutral and basic conditions, the enol form predominates, resulting in shorter lifetimes. The keto form of thymine is more stable in acidic conditions due to the stabilization of the thymine triplet state, which is facilitated by the protonation of the thymine ring [48, 49]. Furthermore, the presence of substituents on the pyrimidine ring can further reduce the phosphorescence quantum yield and lifetime by disrupting the electronic structure and the hydrogen bonding interactions [49]. Additionally, the phosphorescence of pyrimidine bases is more sensitive to the solvent environment, with polar solvents like ethanol and ethylene glycol enhancing the phosphorescence quantum yield [15, 46].

On the other hand, purine nucleotides such as AMP and GMP are known to show higher quantum yields and longer phosphorescence lifetimes, compared to pyrimidine nucleotides [45, 46, 49]. This difference is attributed to the electronic structure and the hydrogen bonding properties of the nucleotides. Specifically, the triplet state of AMP (adenine) is more stable due to its planar structure and the ability to form hydrogen bonds with neighboring bases, which enhances the phosphorescence quantum yield [16, 45, 47, 48]. Similar to pyrimidines, the pH value could affect the phosphorescence of purine nucleotides. For adenine base itself, the phosphorescence intensity decreases sharply in alkaline pH (e.g. $pH > 7$) but remains relatively stable in acidic to neutral conditions [45, 48]. This is because the deprotonation of the N9 position under alkaline conditions disrupts the rigid planar structure of adenine, enhancing nonradiative decay pathways (e.g. vibrational/rotational motions) and reducing triplet state lifetimes. Meanwhile, the phosphorescence of AMP is strongly quenched in acidic pH (e.g. $pH < 4$) but enhances significantly near-neutral pH (pH 6–8). In this case, protonation of the phosphate group in acidic media increases solubility and molecular flexibility, promoting collisional quenching. At neutral pH, the negatively charged phosphate group again stabilizes the triplet state through electrostatic interactions and rigidifies the structure, suppressing nonradiative decay [20, 49, 50].

1.2.1 Phosphorescence in Dinucleotides and Polynucleotides in 1960–1970s

The results of DNA/RNA monomers set the foundation for further research on phosphorescence in dinucleotides and polynucleotides. In dinucleotides, the observed phosphorescence spectra and decay kinetics are dominated by the monomer with the lowest triplet energy, reflecting efficient triplet–triplet energy transfer (TTET). For example, ApG (3′–5′) produces adenine-like phosphorescence with a lifetime of ~1.3 s, thus matching monomeric AMP. On the other hand, TpA (3′–5′) exhibits thymine-like phosphorescence with a lifetime of ~0.35 s, which again matches that of TMP. The TTET mechanism is proven by the observation that ApT (3′–5′) exhibits thymine-like phosphorescence at neutral pH while showing phosphorescence from both adenine and thymine anions (T^-) at pH 11.5, indicating that the deprotonation of thymine ($T \rightarrow T^-$) at high pH could disrupt base stacking and reduce TTET [49].

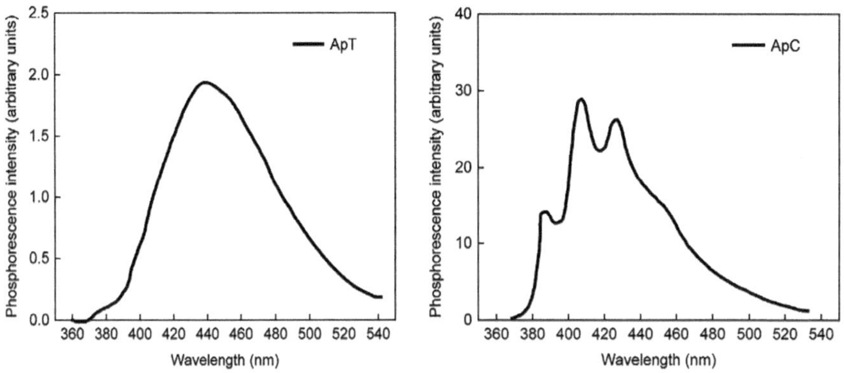

Figure 1.2 Phosphorescence spectra of representative dinucleotides (ApT and ApC) at 77 K.

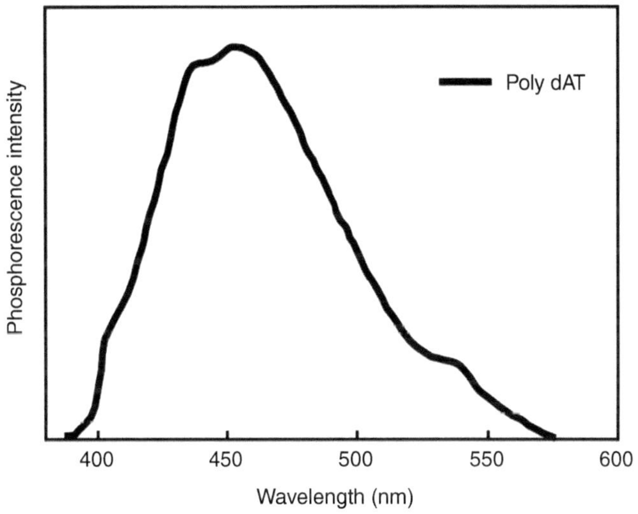

Figure 1.3 Phosphorescence spectra of representative polynucleotides at 80 K.

We have summarized the phosphorescence spectra of representative dinucleotides at 77 K in Figure 1.2 and the lowest triplet energies of DNA/RNA nucleotides in Table 1.1.

The phosphorescence spectra of representative polynucleotides are shown in Figure 1.3. For native DNAs and polynucleotides containing T (such as poly dAT:dAT) [51], they exhibit unstructured phosphorescence with a maximum at 450 nm and a lifetime of ~0.3 s at 80 K. The phosphorescence is attributed to T triplets, and it is confirmed by the sensitized phosphorescence of native DNAs using acetone or acetophenone as triplet energy donors. The underlying mechanism is similar to that found in dinucleotides, where triplet energy migrates from higher energy bases (A, G, C) to T via exchange interactions between stacked bases. For model polynucleotides that do not contain T base (such as poly rAU and poly rA), they showed phosphorescence spectra resemble that of adenine, yet the lifetime could vary depending on the sequence [49, 51]. A special case is noted for poly rG:rC as it exhibits negligible phosphorescence due to complete quenching of G and C fluorescence at the singlet state level, which is proven due to efficient excited state proton-coupled electron transfer between G and C in more recent studies [52, 53].

1.2.2 The Understanding of Temperature Effect on Triplet Excited State Formation and Decay During 1960s–1970s

UV excitation of DNA/RNA nucleotides populates that their singlet states and triplet excited states are generated by ISC from the singlet states. However, the low phosphorescence quantum yields of DNA/RNA nucleotides summarized in Table 1.1 indicate that ISC is almost negligible in isolated nucleotide monomers. Meanwhile, fluorescence quantum yields of DNA/RNA nucleotides are also at the level of $\sim 10^{-2}$. Therefore, it is concluded that nonradiative decay dominates the excited state dynamics of DNA/RNA nucleotides even at 80 K (the values of $1\text{-}\varphi_f\text{-}\varphi_p$ are in the range of 0.72–0.99). Since there are only negligible changes in the absorption spectra of DNA/RNA nucleotides from 80 K to room temperature, it suggests that the energies necessary to populate the singlet states and the low-lying triplet states are essentially temperature independent [49]. Table 1.2 compares the ISC yields and the nonradiative decay rate constants of the singlet states of DNA/RNA nucleotides from the available data in the 1960s–1970s, revealing that the ratio of nonradiative decay rates of the singlet states increases by one to three orders of magnitude from 80 K to room temperature. Meanwhile, the ISC rates for GMP and CMP are in the same order as GMP and CMP from 80 K to room temperature. For AMP, UMP, and TMP, the ISC rates are two to three orders of magnitude higher at room temperature. Therefore, scientists have foreseen that the IC of DNA/RNA nucleotides is extremely fast at room temperature and that this is the key reason for their observed low fluorescence and phosphorescence at room temperature. However, this hypothesis was verified until the year 2000 using femtosecond time-resolved spectroscopy [54, 55].

Table 1.2 Comparison of ISC and nonradiative decay rate constants of the singlet states of DNA/RNA nucleotides with the data available in the 1960s–1970s.

	φ_{ISC} (80 K)	φ_{ISC} (298 K)	$\frac{k_{ISC}(298\ K)}{k_{ISC}(80\ K)}$	$\frac{k_{nr}(298\ K)}{k_{nr}(80\ K)}$
AMP	2×10^{-2}	4×10^{-4}	<30	<1500
TMP	$<3\times10^{-3}$	8×10^{-3}	≥530	300
CMP	3×10^{-2}	1.5×10^{-3}	5	100
GMP	1.5×10^{-1}	4.6×10^{-4}	1.3	450
UMP	$<3\times10^{-3}$	7×10^{-3}	≥30	13

1.3 Breakthroughs of Time-Resolved Spectroscopy Techniques and Progress of Triplet Excited State Study of DNA/RNA in 1980–2000

In 1980s, both time-resolved transient absorption and time-resolved emission experiments were carried out on DNA/RNA monomers at room temperature, even though most of the studies focus on singlet excited state dynamics. This is possible due to the development of picosecond lasers in the early 1980s. For example, Nikogosyan et al. performed the first transient absorption measurements on DNA/RNA bases and nucleosides at room temperature in 1981 with picosecond UV pulses and reported that the obtained S1 lifetimes are equal to or somewhat shorter than their instrumental time resolution [56]. The first time-resolved emission experiments on DNA/RNA nucleosides at room temperature were also reported by Ballini et al. in 1982 [57]. However, they were unable to distinguish emission decays from the instrumental response function and only estimated an upper limit of 100 ps for the monomer fluorescence lifetimes. Later, Kobayashi reported that the fluorescence lifetime for adenine is 6 ps in a room temperature aqueous solution with a streak camera in 1984. The abovementioned two experimental approaches (transient absorption and time-resolved emission by a streak camera) continued into the 1990s. Nikogosyan et al. extended their experiments to the femtosecond timescale [58, 59]. Although the instrumental response time is greatly improved by femtosecond excitation pulses, high pump intensities (20–200 GW cm^{-2}) used in the experiments could lead to substantial multiphoton absorption and ionization, resulting in errors in the detected singlet state lifetimes. To avoid such systematic uncertainty, Reuther et al. used visible wavelengths and estimated ~1 ps lifetimes for adenine, cytosine, thymine, and uracil [60]. The use of femtosecond laser pulses did not improve the time resolution for streak camera detection in the 1990s. The measured lifetimes of DNA/RNA bases were estimated to be ~4 ps, which is around the instrumental response time of the streak camera [61]. Therefore, the authors claimed that 1 ps is not a lower limit for IC in DNA/RNA bases.

Despite the advances in the time-resolved spectroscopy techniques and the progress in the studies of the singlet excited state in DNA/RNA monomers, studies on their triplet excited state are rarely reported during the 1980s–1990s. Kemp et al. used laser flash photolysis with 249 nm excitation to generate triplet excited states of thymine and uracil in acetonitrile and examined the quenching rates by various electron acceptors [62]. They found that the lifetimes of both triplet thymine and uracil were reduced upon addition of electron-accepting molecules and the correlations between rate constants and thermodynamics of electron-transfer quenching followed the Rehm–Weller relation, indicating an electron-transfer mechanism. Because of the low ISC yields of DNA/RNA monomers under direct UV excitation condition at room temperature, acetone sensitization emerged as a breakthrough. Acetone's high triplet energy (337 kJ mol^{-1}) and high Φ_{ISC} (~1) efficient energy transfer to all DNA/RNA nucleotides, enhancing their triplet yields by two to three orders of magnitude [21, 22].

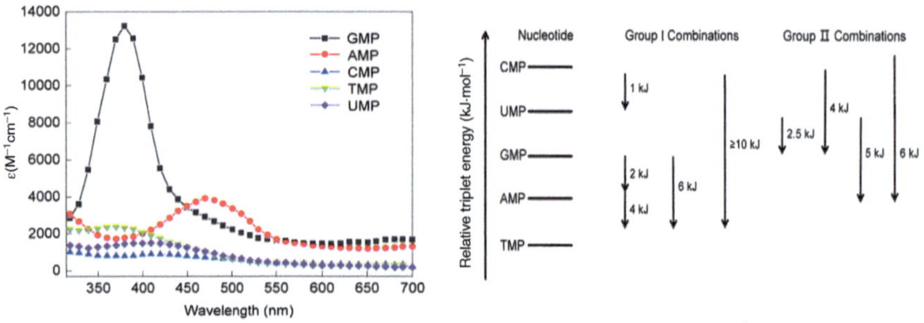

Figure 1.4 Triplet–triplet (T–T) absorption spectra of DNA/RNA nucleotides and scheme of their triplet state energy level.

For example, Song et al. demonstrated that acetone-sensitized triplet states of thymine components (Thy, Thd, TMP) in aqueous solution exhibited identical kinetic parameters to those from direct excitation, validating the method's reliability [23]. Enhanced signal amplitudes facilitated accurate spectral and kinetic measurements. Triplet–triplet (T–T) absorption spectra of all DNA/RNA bases and nucleotides were resolved using acetone sensitization [21–23, 63]. Notably, Zuo et al. first reported cytosine triplet spectra ($\varepsilon_T \approx 200$–$400\,M^{-1}\,cm^{-1}$), which are weaker than thymine ($\varepsilon_T \approx 4000\,M^{-1}\,cm^{-1}$) due to lower Φ_{ISC} and rapid self-quenching. Self-quenching rate constants (k_{sq}) for triplet excited states vary by base: thymine ($4.2\times10^8\,M^{-1}\,s^{-1}$) > cytosine ($2.4\times10^8\,M^{-1}\,s^{-1}$) > adenine ($1.9\times10^8\,M^{-1}\,s^{-1}$) ($k_{sq} \approx 1.8$–$4.2\times10^8\,M^{-1}\,s^{-1}$) [23, 63]. Redmond and coworkers established the triplet energy hierarchy for DNA/RNA nucleotides in aqueous solution at room temperature: CMP ($321\,kJ\,mol^{-1}$) > UMP ($320\,kJ\,mol^{-1}$) > GMP ($317\,kJ\,mol^{-1}$) > AMP ($314\,kJ\,mol^{-1}$) > TMP ($310\,kJ\,mol^{-1}$) [21, 22]. Triplet energy transfer equilibria were modeled using the Sandros equation, revealing that thymine acts as a partial energy sink, with an equilibrium ratio favoring triplet localization at lower energy bases. Moreover, in purine/pyrimidine pairs (e.g. GMP–UMP, GMP–CMP), triplet-mediated electron transfer occurs, producing purine radical cations (e.g. $G^{\bullet+}$) that deprotonate to neutral radicals ($G(-H)^{\bullet}$) [21, 22]. Electron transfer rate constant (k_{et}) for GMP–UMP reached $3.5\times10^8\,M^{-1}\,s^{-1}$, with radical yields doubling in 1:1 mixtures compared to pure GMP. This mechanism is absent in thymine-containing pairs due to its low triplet energy (Figure 1.4).

1.4 Current Research Frontiers (2000s–Present)

1.4.1 Ultrafast Spectroscopy Study

The development of femtosecond time-resolved spectroscopy in the 1990s has facilitated the research on the triplet state of DNA/RNA bases and oligonucleotides, and direct observations of the excited state dynamics of the initially populated singlet states have been achieved since 2000 [54, 55, 64]. The experimental research progress on the excited state dynamics of the singlet states in nucleic acids has been well documented in the recent two decades [2–4, 6–8, 10]. Thanks to those studies, it is now possible to distinguish the ISC process in DNA/RNA bases and oligonucleotides. In this section, we go through the major discoveries after the year 2000 in order of time, and the detailed mechanism will be discussed in the next chapter.

As there is barely any triplet state observed after direct excitation of purine bases at room temperature, the studies are focused on the triplet excited state dynamics of pyrimidines. For thymine and its derivatives in solution, a low-lying $^1n\pi^*$ state is directly observed using UV pump and UV probe femtosecond transient absorption spectroscopy and such state is proposed to be a gateway to its lowest triplet state [65]. Such hypothesis was supported by a study on thymidine monomer and a single-stranded thymine oligonucleotide in solution, in which a doorway state with singlet nature is believed to facilitate bifurcation of the initially populated singlet state to either the ground state or the lowest triplet state [66]. Nevertheless, triplet states of thymine and thymidine are shown to be fully formed within the first 10 ps in acetonitrile-d_3 solution after UV excitation by time-resolved infrared (TRIR) spectroscopy[67] and such observation ruled out the possibility that the relaxed $^1n\pi^*$ state is the triplet precursor. In this case, the precursor of the triplet excited state is either the initially populated $^1\pi\pi^*$ state or the hot $^1n\pi^*$ state. Orrewing and coworkers revisit thymidine in chloroform solution and found that the triplet state is generated after 5–10 ps [68]. However, due to spectral overlap and anharmonic spreading of the hot band, they could not identify its precursor. Later, Pilles et al. revisited the decay pathways of thymine and TMP by time-resolved UV/vis and IR spectroscopy. They reported that the formation time of the lowest triplet state of thymine and TMP is ~10 ps, but the quantum yield is solvent dependent, ranging from 0.01 in D_2O to 0.1 in acetonitrile-d_3/methanol-d_4 [29]. Recently, our group reported that varying the excitation wavelength can significantly alter the branching of the excited state population at the Franck–Condon (FC) region, leading to different fluorescence and triplet state quantum yields of thymidine [38]. On the other hand, one study of thymine and its derivatives in the gas phase found dark states with sub-nanosecond to hundreds of nanosecond lifetimes and the nature of these states is proposed to be either the lower-lying singlet $n\pi^*$ state or triplet states generated through ultrafast ISC [69–72]. Although direct observation of the ISC process in thymine and its derivatives has not been successful yet, current evidence suggests that the ISC should be ultrafast and that the rate constant could reach the level of ~10^{11} s^{-1}.

As for uracil and its derivatives, the low-lying $^1n\pi^*$ state is also directly observed using UV pump and UV probe femtosecond transient absorption spectroscopy [65, 73]. Meanwhile, triplet excited states were observed in a few picoseconds after excitation, and they are reported not to form from the thermalized $^1n\pi^*$ state. Therefore, efficient ISC prior to vibrational cooling of the dark $^1n\pi^*$ state is proposed to be the reason for the wavelength- and solvent-dependent triplet yields seen in uracil and its derivatives. Brister and Crespo-Hernández used broadband transient absorption spectroscopy to study the triplet state population dynamics of 1-cyclohexyluracil, finding that SIC to the triplet state occurs in less than 1 ps [74]. They also investigated the excited state dynamics of UMP in aqueous solution using the same method, revealing that excitation at 267 nm leads to the simultaneous population of the hot ground state, a long-lived $^1n\pi^*$ state, and a triplet state within 200 fs [75]. The quantum yield of the triplet in uridine was also reported to be significantly affected by the excitation wavelength, and ultrafast ISC ($k_{ISC} > 10^{11}\ s^{-1}$) is the key reason [38].

Compared to thymine/uracil and their derivatives, there is less experimental study on the triplet excited state of cytosine and its derivatives. Gas-phase studies by Lobsiger and Leutwyler revealed that the triplet energy of the T1 state of keto-amino cytosine is 3.26 eV above the ground state and the ISC rate is $\sim 10^{10}\ s^{-1}$ [76]. However, a more detailed study by the same group showed that if the S1 state excess energy is below 550 cm^{-1}, the ISC rate of keto-amino cytosine reduces to $0.4–1.5 \times 10^9\ s^{-1}$, which is at least one order of magnitude smaller than that of IC from S1 to S0 [77]. Such observation suggests that previously reported ultrafast ISC of keto-amino cytosine has to occur at the hot S1 state rather than at the energy minimum of the S1 state. Our group has pioneered in the study of ISC in cytidine in solution and directly observed its ultrafast ISC ($k_{ISC} > 10^{10}\ s^{-1}$) using femtosecond transient absorption spectroscopy. It is shown that the ISC in cytidine is sensitive to the excitation wavelength, and a spin-vibronic ISC mechanism at the FC region of the initial excited S1 state is proposed [37].

1.4.2 Theoretical Study

Theoretical studies on the photophysics and photochemistry of nucleic acids and nucleobases have also been extensively performed since 2000. Among those works, triplet states of pyrimidines are commonly involved. We also go through the major discoveries in order of time in this section, and the detailed mechanism can be found in Chapters 6 and 7.

For thymine/uracil and their derivatives, the most likely pathway for ISC is from the dark $^1n\pi^*$ state because a region of singlet/triplet degeneracy has been located near the minimum of this state according to the theoretical results at the CASPT2 (complete active space second-order perturbation)/CASSCF (complete active space self-consistent field) level [78, 79]. The SOC strength between the $^1n\pi^*$ and $^3\pi\pi^*$ states is reported to be $\sim 60\ cm^{-1}$, suggesting the high probability of ISC in the crossing region even though the experimental results suggest ISC should take from the hot $^1n\pi^*$ state rather than at its energy minimum [67]. Meanwhile, the coupling of the T1 ($^3\pi\pi^*$) state with S0 is only 2 cm^{-1}, which explains the long lifetime of the T1 state. The abovementioned mechanism has been corroborated with ISC rate calculations for the $^1n\pi^*$ to $^3\pi\pi^*$ transition using Fermi's golden rule approach [80]. Other pathways for the triplet state population of thymine to S0 have been found to be less favorable [78].

The general picture obtained for thymine has been confirmed for uracil [81]. Furthermore, a dynamic calculation using trajectory surface hopping at the CASSCF level confirms that ISC from the $^1n\pi^*$ to the $^3\pi\pi^*$ state is the favored mechanism for uracil [82]. They also found that the $^1n\pi^*$ state is formed after decay through the $^1\pi_O\pi^*$-min intermediate and further decay through the S2/S1 crossing. The triplet population of uracil is 20–25% after 1 ps of trajectory simulation time, depending on the initial conditions. These results are in line with the solvent-dependent triplet quantum yields reported by Hare et al. [73]

For cytosine and its derivatives, the first computational study on the possible ISC mechanism was provided by CASPT2 calculations [83]. An energy region of S1/T1 degeneracy with substantial SOC (20 cm^{-1}) was located along the decay pathways to the S0. The large SOC values are explained by the mixed $\pi\pi^*/n_O\pi^*$ character induced by C6 pyramidalization and a small singlet–triplet energy gap at the S1/T1 crossing region. ISC from the singlet to the triplet states has been confirmed by simulations at the CASSCF(14,10) electronic structure level. Richter et al. reported that the ultrafast ISC should follow S1 ($^1n\pi^*$) → T2 ($^3n\pi^*$) → T1 ($^3\pi\pi^*$) path due to a small energy difference (<0.05 eV) between S1 and T2 states induced by pyramidalization of the NH2 group or the out-of-plane oscillation of the hydrogen in the NH group [84]. Mai et al. confirmed the above mentioned mechanism in their study and proposed that efficient ISC in keto-cytosine is achieved due to large SOC at a three-state (S1/T2/T1) energy near-degeneracy point [85]. The S1 → T2 ISC is enhanced by the mixed $\pi\pi^*/n_O\pi^*$ character of the states. However, these results seem to contradict the abovementioned experimental study of Lobsiger et al., in which ISC is not ultrafast when the excess energy of the S1 state is below ~550 cm^{-1} [77]. Pepino et al. pointed out that there are two efficient ISC regions in cytidine. One is $^1\pi\pi^*$ to $^3\pi^2\pi^*$ transition (S1 → T2) that occurs close to FC region with SOC value of ~12 cm^{-1}. The other one is close to the S1/S0 conical intersection, and it is $^1\pi\pi^*$ to $^3\pi\pi^*$ transition (S1 → T1) with SOC value of ~24 cm^{-1}. Meanwhile, they found efficient $^1n_O\pi^*$ to $^3\pi\pi^*$ ISC path with a larger SOC value (~42 cm^{-1}), which is believed to be the main ISC channel in all pyrimidine nucleosides [33]. Such mechanism is further supported by our experimental study of cytidine in solution [37]

1.4.3 Study on the Reaction Mechanism of Cyclobutane Pyrimidine Dimer (CPD) Photolesions in DNA/RNA

DNA/RNA photodamage can lead to a variety of photolesions, including cyclobutane pyrimidine dimer (CPD), the (6-4) lesion, and its Dewar valence isomer. Among them, CPD is the photoproducts that cause the most DNA damage under UVB and UVA irradiation in cellular DNA, as well as in DNA model systems [86]. Due to its relatively high abundance, CPDs between thymine bases became the most studied system in scientific research, and the reaction is expected to occur through either excited singlet or triplet states. However, the underlying reaction mechanism of CPD formation, especially those involving triplet excited states, has been under debate since the pioneering works in the 1960s [87–89]. This is because the importance of a triplet channel for CPD formation after the direct UV irradiation of DNA/RNA has been overlooked due to the explanation of the singlet excitation state reaction channel for CPD formation of thymine bases or nucleotides in frozen matrices and aggregates [15, 90–92]. In this section, we briefly recall the advances of the studies on the triplet excited state reaction mechanism of CPD formation by direct UV excitation, and the details of reaction mechanism by triplet photosensitization will be discussed in Chapters 4 and 5 (Figure 1.5).

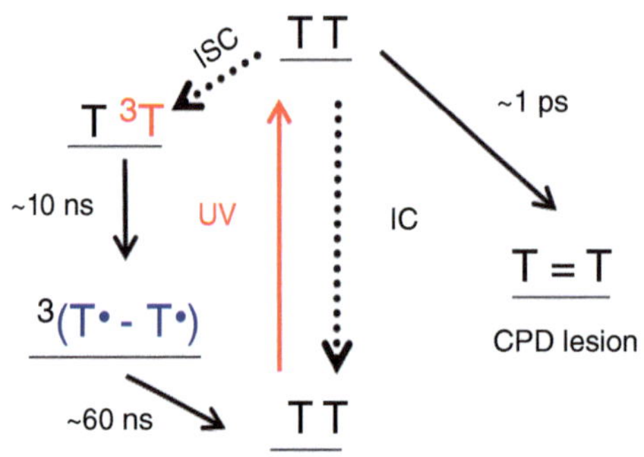

Figure 1.5 Schemes for cyclobutane pyrimidine dimer (CPD) formation in triplet states.

Through photosensitization method, Bosca et al. determined the triplet energy of thymine in DNA is 270 kJ mol^{-1} [25]. Meanwhile, the first transient absorption experiments that try to address CPD formation on the femtosecond to nanosecond timescale were performed in the UV–vis probe range. However, no triplet state was observed in all-thymine oligonucleotides [93]. Later, Banyasz et al. revisit this problem and reported wavelength-dependent quantum yields for CPD and triplet state formation [94]. They estimated that the contribution of triplet states to CPD formation in single-stranded DNA is below 10%. Pilles and coworkers measured decay of the triplet states in $(dT)_{18}$ and showed that triplet states predominantly decay to the electronic ground state via a biradical intermediate on a 10-ns timescale. Such assignment was supported by theoretical studies [95, 96], explaining the low efficiencies for CPD formation from triplet states as well as the low propensity of thymine triplets to form CPDs in sensitization experiments [97–99]. Very similar mechanism was also reported for the formation of CPDs in TpC and CpT dimers [31] as well as in Tp^{5m}C dimer (^{5m}C refers to 5-methyl-cytosine) [100, 101].

1.5 Book Structure Overview

This book is organized as follows: Chapter 1, "Introduce of Triplet Excited State in DNA and RNA," Chapter 2, "Triplet Excited State in Canonical Nucleobases and Their Sulfur Substituted Derivatives," Chapter 3, "Triplet States in Epigenetic Modified Nucleobases and DNA/RNA," Chapter 4, "Triplet-Mediated Photochemistry of Nucleic Acid Bases," Chapter 5, "Drug-Sensitized Triplet States Generation in DNA," Chapter 6, "Triplet Decay in Canonical and Noncanonical DNA/RNA Nucleobases," Chapter 7, "Triplet State Dynamics in DNA/RNA Nucleobases and DNA by Ab Initio Static and Surface-Hopping Dynamics Simulations," and Chapter 8, "The Photophysics and Photochemistry of Epigenetics and Natural Noncanonical Bases: Insights from Quantum Mechanical Calculations." Although comprehensive understanding of triplet excited state in DNA/RNA is not yet at hand, particularly for excitations in systems containing two or more nucleobases bases, it is our hope that this book will communicate the current achievement in this field and inspire future experimental and theoretical work.

Acknowledgments

Jinquan Chen thanks Lina Wang for help with the figures and Dr. Xueli Wang for helpful discussion.

References

1 Alexandrov, L.B., Kim, J., Haradhvala, N.J. et al. (2020). The repertoire of mutational signatures in human cancer. *Nature* 578: 94–101.
2 Crespo-Hernández, C.E., Cohen, B., Hare, P.M., and Kohler, B. (2004). Ultrafast excited-state dynamics in nucleic acids. *Chemical Reviews* 104: 1977–2019.
3 Vries, M.S.D. and Hobza, P. (2007). Gas-phase spectroscopy of biomolecular building blocks. *Annual Review of Physical Chemistry* 58: 585–612.
4 Middleton, C.T., de La Harpe, K., Su, C. et al. (2009). DNA excited-state dynamics: from single bases to the double helix. *Annual Review of Physical Chemistry* 60: 217–239.
5 Serrano-Andrés, L. and Merchán, M. (2009). Are the five natural DNA/RNA base monomers a good choice from natural selection? A photochemical perspective. *Journal of Photochemistry and Photobiology, C: Photochemistry Reviews* 10: 21–32.
6 Kohler, B. (2010). Nonradiative decay mechanisms in DNA model systems. *Journal of Physical Chemistry Letters* 1: 2047–2053.
7 Kleinermanns, K., Nachtigallová, D., and de Vries, M.S. (2013). Excited state dynamics of DNA bases. *International Reviews in Physical Chemistry* 32: 308–342.
8 Barbatti, M., Borin, A.C., and Ullrich, S. (2015). Photoinduced processes in nucleic acids. In: *Photoinduced Phenomena in Nucleic Acids I: Nucleobases in the Gas Phase and in Solvents*, vol. 355 (ed. M. Barbatti, A.C. Borin, and S. Ullrich), 1–32. Springer.
9 Improta, R., Santoro, F., and Blancafort, L. (2016). Quantum mechanical studies on the photophysics and the photochemistry of nucleic acids and nucleobases. *Chemical Reviews* 116: 3540–3593.
10 Boldissar, S. and de Vries, M.S. (2018). How nature covers its bases. *Physical Chemistry Chemical Physics* 20: 9701–9716.
11 Improta, R. and Douki, T. (2021). *DNA Photodamage: From Light Absorption to Cellular Responses and Skin Cancer*. Cambridge: Royal Society of Chemistry.
12 Rahn, R.O., Longworth, J.W., Eisinger, J., and Shulman, R.G. (1964). Electron spin resonance and luminescence studies of excited states of nucleic acids. *Proceedings of the National Academy of Sciences of the United States of America* 51: 1299–1303.
13 Eisinger, J. and Shulman, R.G. (1966). Triplet excitons in polyadenylic acid. *Proceedings of the National Academy of Sciences of the United States of America* 55: 1387–1391.

14 Lamola, A.A. and Yamane, T. (1967). Sensitized photodimerization of thymine in DNA. *Proceedings of the National Academy of Sciences of the United States of America* 58: 443–446.

15 Eisinger, J. and Shulman, R.G. (1967). Precursor of the thymine dimer in ice. *Proceedings of the National Academy of Sciences of the United States of America* 58: 895–900.

16 Eisinger, J. and Shulman, R.G. (1968). Excited electronic states of DNA. *Science* 161: 1311–1319.

17 Johns, H.E. and Whillans, D.W. (1971). Properties of the triplet states of thymine and uracil in aqueous solution. *Journal of the American Chemical Society* 93: 1358–1362.

18 Salet, C. and Bensasson, R. (1975). Studies on thymine and uracil triplet excited state in acetonitrile and water. *Photochemistry and Photobiology* 22: 231–235.

19 Salet, C., Bensasson, R., and Becker, R.S. (1979). Triplet excited states of pyrimidine nucleosides and nucleotides. *Photochemistry and Photobiology* 30: 325–329.

20 Görner, H. (1990). Transients of uracil and thymine derivatives and the quantum yields of electron ejection and intersystem crossing upon 20 ns photolysis at 248 nm. *Photochemistry and Photobiology* 52: 935–948.

21 Gut, I.G., Wood, P.D., and Redmond, R.W. (1996). Interaction of triplet photosensitizers with nucleotides and DNA in aqueous solution at room temperature. *Journal of the American Chemical Society* 118: 2366–2373.

22 Wood, P.D. and Redmond, R.W. (1996). Triplet state interactions between nucleic acid bases in solution at room temperature: intermolecular energy and electron transfer. *Journal of the American Chemical Society* 118: 4256–4263.

23 Song, Q., Lin, W., Yao, S., and Lin, N. (1998). Comparative studies of triplet states of thymine components by acetone sensitization and direct excitation in aqueous solution at room temperature. *Journal of Photochemistry and Photobiology, A: Chemistry* 114: 181–184.

24 Lhiaubet-Vallet, V., Trzcionka, J., Encinas, S. et al. (2004). The triplet state of a n-phenylphthalimidine with high intersystem crossing efficiency: characterization by transient absorption spectroscopy and DNA sensitization properties. *The Journal of Physical Chemistry. B* 108: 14148–14153.

25 Bosca, F., Lhiaubet-Vallet, V., Cuquerella, M.C. et al. (2006). The triplet energy of thymine in DNA. *Journal of the American Chemical Society* 128: 6318–6319.

26 Cuquerella, M.C., Lhiaubet-Vallet, V., Cadet, J., and Miranda, M.A. (2012). Benzophenone photosensitized DNA damage. *Accounts of Chemical Research* 45: 1558–1570.

27 Schreier, W.J., Gilch, P., and Zinth, W. (2015). Early events of DNA photodamage. *Annual Review of Physical Chemistry* 66: 497–519.

28 Mai, S., Pollum, M., Martínez-Fernández, L. et al. (2016). The origin of efficient triplet state population in sulfur-substituted nucleobases. *Nature Communications* 7: 13077.

29 Pilles, B.M., Maerz, B., Chen, J. et al. (2018). Decay pathways of thymine revisited. *The Journal of Physical Chemistry. A* 122: 4819–4828.

30 Francés-Monerris, A., Hognon, C., Miranda, M.A. et al. (2018). Triplet photosensitization mechanism of thymine by an oxidized nucleobase: from a dimeric model to DNA environment. *Physical Chemistry Chemical Physics* 20: 25666–25675.

31 Gontcharov, J., Liu, L., Pilles, B.M. et al. (2019). Triplet-induced lesion formation at CpT and TpC sites in DNA. *Chemistry - A European Journal* 25: 15164–15172.

32 Aparici-Espert, I., Garcia-Lainez, G., Andreu, I. et al. (2018). Oxidatively generated lesions as internal photosensitizers for pyrimidine dimerization in DNA. *ACS Chemical Biology* 13: 542–547.

33 Pepino, A.J., Segarra-Marti, J., Nenov, A. et al. (2018). Uv-induced long-lived decays in solvated pyrimidine nucleosides resolved at the MS-CASPT2/MM level. *Physical Chemistry Chemical Physics* 20: 6877–6890.

34 Francés-Monerris, A., Lineros-Rosa, M., Miranda, M.A. et al. (2020). Photoinduced intersystem crossing in DNA oxidative lesions and epigenetic intermediates. *Chemical Communications* 56: 4404–4407.

35 Lineros-Rosa, M., Francés-Monerris, A., Monari, A. et al. (2020). Experimental and theoretical studies on thymine photodimerization mediated by oxidatively generated DNA lesions and epigenetic intermediates. *Physical Chemistry Chemical Physics* 22: 25661–25668.

36 Wang, X., Martínez-Fernández, L., Zhang, Y. et al. (2021). Solvent-dependent stabilization of a charge transfer state is the key to ultrafast triplet state formation in an epigenetic DNA nucleoside. *Chemistry - A European Journal* 27: 10932–10940.

37 Wu, P., Wang, X., Pan, H., and Chen, J. (2022). Direct observation of excitation wavelength-dependent ultrafast intersystem crossing in cytosine nucleoside solution. *The Journal of Physical Chemistry. B* 126: 7975–7980.

38 Wu, P., Wang, X., Pan, H., and Chen, J. (2023). Wavelength dependent excited state dynamics observed in canonical pyrimidine nucleosides. *Journal of Photochemistry and Photobiology* 18: 100211.

39 Wang, X., Martínez-Fernández, L., Zhang, Y. et al. (2024). Ultrafast formation of a delocalized triplet-excited state in an epigenetically modified DNA duplex under direct uv excitation. *Journal of the American Chemical Society* 146: 1839–1848.

40 Vendrell-Criado, V., Rodríguez-Muñiz, G.M., Cuquerella, M.C. et al. (2013). Photosensitization of DNA by 5-methyl-2-pyrimidone deoxyribonucleoside: (6-4) photoproduct as a possible trojan horse. *Angewandte Chemie, International Edition* 52: 6476–6479.

41 Diekmann, J., Gontcharov, J., Fröbel, S. et al. (2019). The photoaddition of a psoralen to DNA proceeds via the triplet state. *Journal of the American Chemical Society* 141: 13643–13653.

42 Jin, P., Wang, X., Jia, M. et al. (2024). Ultrafast and multichannel generation of the triplet state in DNA-intercalated gilvocarcin V for an enhanced photoaddition reaction. *Journal of Physical Chemistry Letters* 15: 2765–2771.

43 Steele, R.H. and Szent-Gyorgyi, A. (1957). On excitation of biological substances. *Proceedings of the National Academy of Sciences of the United States of America* 43: 477–491.

44 Longworth, J.W. (1962). Luminescence of purines and pyrimidines. *Biochemical Journal* 84: 104.

45 Bersohn, R. and Isenberg, I. (1964). Phosphorescence in nucleotides and nucleic acids. *The Journal of Chemical Physics* 40: 3175–3180.

46 Görner, H. (1990). Phosphorescence of nucleic acids and DNA components at 77 k. *Journal of Photochemistry and Photobiology B: Biology* 5: 359–377.

47 Eisinger, J., Guéron, M., and Shulman, R.G. (1968). Excited states of DNA. *Advances in Biological and Medical Physics* 12: 219–238.

48 Rahn, R.O., Shulman, R.G., and Longworth, J.W. (1966). Phosphorescence and electron spin resonance studies of the UV-excited triplet state of DNA. *The Journal of Chemical Physics* 45: 2955–2965.

49 Eisinger, J., Lamola, A.A., Steiner, R.F., and Weinryb, I. (1971). The excited states of nucleic acids. In: *Excited States of Proteins and Nucleic Acids* (ed. R.F. Steiner and I. Weinryb), 107–198. New York: Plenum Press.

50 Longworth, J.W., Rahn, R.O., and Shulman, R.G. (1966). Luminescence of pyrimidines, purines, nucleosides, and nucleotides at 77 k. The effect of ionization and tautomerization. *The Journal of Chemical Physics* 45: 2930–2939.

51 Lamola, A.A., Guéron, M., Yamane, T. et al. (1967). Triplet state of DNA. *The Journal of Chemical Physics* 47: 2210–2217.

52 Zhang, Y., de La Harpe, K., Beckstead, A.A. et al. (2015). UV-induced proton transfer between DNA strands. *Journal of the American Chemical Society* 137: 7059–7062.

53 Zhang, Y., Li, X.B., Fleming, A.M. et al. (2016). UV-induced proton-coupled electron transfer in cyclic DNA miniduplexes. *Journal of the American Chemical Society* 138: 7395–7401.

54 Pecourt, J.-M.L., Peon, J., and Kohler, B. (2000). Ultrafast internal conversion of electronically excited RNA and DNA nucleosides in water. *Journal of the American Chemical Society* 122: 9348–9349.

55 Peon, J. and Zewail, A.H. (2001). DNA/RNA nucleotides and nucleosides: direct measurement of excited-state lifetimes by femtosecond fluorescence up-conversion. *Chemical Physics Letters* 348: 255–262.

56 Oraevsky, A.A., Sharkov, A.V., and Nikogosyan, D.N. (1981). Picosecond study of electronically excited singlet states of nucleic acid components. *Chemical Physics Letters* 83: 276–280.

57 Ballini, J.P., Daniels, M., and Vigny, P. (1982). Wavelength-resolved lifetime measurements of emissions from DNA components and poly rA at room temperature excited with synchrotron radiation. *Journal of Luminescence* 27: 389–400.

58 Nikogosyan, D.N., Angelov, D., Soep, B., and Lindqvist, L. (1996). Direct measurement of excited singlet-state lifetime in the homologous sequence adenine, adenosine, adenosine 5′-monophosphate and in calf thymus DNA. *Chemical Physics Letters* 252: 322–326.

59 Reuther, A., Nikogosyan, D.N., and Laubereau, A. (1996). Primary photochemical processes in thymine in concentrated aqueous solution studied by femtosecond UV spectroscopy. *The Journal of Physical Chemistry* 100: 5570–5577.

60 Reuther, A., Iglev, H., Laenen, R., and Laubereau, A. (2000). Femtosecond photo-ionization of nucleic acid bases: electronic lifetimes and electron yields. *Chemical Physics Letters* 325: 360–368.

61 Häupl, T., Windolph, C., Jochum, T. et al. (1997). Picosecond fluorescence of nucleic acid bases. *Chemical Physics Letters* 280: 520–524.

62 Kemp, T.J., Parker, A.W., and Wardman, P. (1985). Electron-transfer quenching of triplet state thymine and uracil. *Journal of the Chemical Society, Chemical Communications* 1377–1378.

63 Zuo, Z.-h., Yao, S.-d., Luo, J. et al. (1992). Laser photolysis of cytosine, cytidine and DCMP in aqueous solution. *Journal of Photochemistry and Photobiology B: Biology* 15: 215–222.

64 Pecourt, J.-M.L., Peon, J., and Kohler, B. (2001). DNA excited-state dynamics: ultrafast internal conversion and vibrational cooling in a series of nucleosides. *Journal of the American Chemical Society* 123: 10370–10378.

65 Hare, P.M., Crespo-Hernández, C.E., and Kohler, B. (2007). Internal conversion to the electronic ground state occurs via two distinct pathways for pyrimidine bases in aqueous solution. *Proceedings of the National Academy of Sciences of the United States of America* 104: 435–440.

66 Kwok, W.M., Ma, C., and Phillips, D.L. (2008). A doorway state leads to photostability or triplet photodamage in thymine DNA. *Journal of the American Chemical Society* 130: 5131–5139.

67 Hare, P.M., Middleton, C.T., Mertel, K.I. et al. (2008). Time-resolved infrared spectroscopy of the lowest triplet state of thymine and thymidine. *Chemical Physics* 347: 383–392.

68 Röttger, K., Marroux, H., Böhnke, H. et al. (2016). Probing the excited state relaxation dynamics of pyrimidine nucleosides in chloroform solution. *Faraday Discussions* 194: 683–708.

69 He, Y., Wu, C., and Kong, W. (2003). Decay pathways of thymine and methyl-substituted uracil and thymine in the gas phase. *The Journal of Physical Chemistry. A* 107: 5145–5148.

70 He, Y., Wu, C., and Kong, W. (2004). Photophysics of methyl-substituted uracils and thymines and their water complexes in the gas phase. *The Journal of Physical Chemistry. A* 108: 943–949.

71 Gonzalez-Vazquez, J., Gonzalez, L., Samoylova, E., and Schultz, T. (2009). Thymine relaxation after UV irradiation: the role of tautomerization and $\pi\sigma^*$ states. *Physical Chemistry Chemical Physics* 11: 3927–3934.

72 Ligare, M., Siouri, F., Bludsky, O. et al. (2015). Characterizing the dark state in thymine and uracil by double resonant spectroscopy and quantum computation. *Physical Chemistry Chemical Physics* 17: 24336–24341.

73 Hare, P.M., Crespo-Hernández, C.E., and Kohler, B. (2006). Solvent-dependent photophysics of 1-cyclohexyluracil: ultrafast branching in the initial bright state leads nonradiatively to the electronic ground state and a long-lived 1 nπ* state. *The Journal of Physical Chemistry. B* 110: 18641–18650.

74 Brister, M.M. and Crespo-Hernández, C.E. (2015). Direct observation of triplet-state population dynamics in the RNA uracil derivative 1-cyclohexyluracil. *Journal of Physical Chemistry Letters* 6: 4404–4409.

75 Brister, M.M. and Crespo-Hernández, C.E. (2019). Excited-state dynamics in the RNA nucleotide uridine 5′-monophosphate investigated using femtosecond broadband transient absorption spectroscopy. *Journal of Physical Chemistry Letters* 10: 2156–2161.

76 Lobsiger, S. and Leutwyler, S. (2012). The adiabatic ionization energy and triplet T1 energy of jet-cooled keto-amino cytosine. *Journal of Physical Chemistry Letters* 3: 3576–3580.

77 Lobsiger, S., Etinski, M., Blaser, S. et al. (2015). Intersystem crossing rates of S-1 state keto-amino cytosine at low excess energy. *The Journal of Chemical Physics* 143: 234301.

78 Serrano-Perez, J.J., Gonzalez-Luque, R., Merchán, M., and Serrano-Andrés, L. (2007). On the intrinsic population of the lowest triplet state of thymine. *The Journal of Physical Chemistry. B* 111: 11880–11883.

79 González-Luque, R., Climent, T., González-Ramírez, I. et al. (2010). Singlet–triplet states interaction regions in DNA/RNA nucleobase hypersurfaces. *Journal of Chemical Theory and Computation* 6: 2103–2114.

80 Etinski, M., Fleig, T., and Marian, C.A. (2009). Intersystem crossing and characterization of dark states in the pyrimidine nucleobases uracil, thymine, and 1-methylthymine. *The Journal of Physical Chemistry. A* 113: 11809–11816.

81 Climent, T., González-Luque, R., Merchán, M., and Serrano-Andrés, L. (2007). On the intrinsic population of the lowest triplet state of uracil. *Chemical Physics Letters* 441: 327–331.

82 Richter, M., Mai, S., Marquetand, P., and González, L. (2014). Ultrafast intersystem crossing dynamics in uracil unravelled by ab initio molecular dynamics. *Physical Chemistry Chemical Physics* 16: 24423–24436.

83 Merchán, M., Serrano-Andrés, L., Robb, M.A., and Blancafort, L. (2005). Triplet-state formation along the ultrafast decay of excited singlet cytosine. *Journal of the American Chemical Society* 127: 1820–1825.

84 Richter, M., Marquetand, P., Gonzalez-Vazquez, J. et al. (2012). Femtosecond intersystem crossing in the DNA nucleobase cytosine. *Journal of Physical Chemistry Letters* 3: 3090–3095.

85 Mai, S., Marquetand, P., Richter, M. et al. (2013). Singlet and triplet excited-state dynamics study of the keto and enol tautomers of cytosine. *ChemPhysChem* 14: 2920–2931.

86 Douki, T., Court, M., Sauvaigo, S. et al. (2000). Formation of the main UV-induced thymine dimeric lesions within isolated and cellular DNA as measured by high performance liquid chromatography-tandem mass spectrometry. *The Journal of Biological Chemistry* 275: 11678–11685.

87 Beukers, R. and Berends, W. (1960). Isolation and identification of the irradiation product of thymine. *Biochimica et Biophysica Acta* 41: 550–551.

88 Johns, H.E., Rapaport, S.A., and Delbruck, M. (1962). Photochemistry of thymine dimers. *Journal of Molecular Biology* 4: 104–114.

89 Fisher, G.J., Johns, H.E., and Wang, S.Y. (1976). Pyrimidine photodimers. In: *Photochemistry and Photobiology of Nucleic Acids*, vol. 1 (ed. S.Y. Wang), 225–294. New York: Academic Press.

90 Eisinger, J. and Lamola, A.A. (1967). Excited-state precursor for the thymine dimer. *Biochemical and Biophysical Research Communications* 28: 558–565.

91 Lamola, A.A. (1968). Excited state precursors of thymine photodimers. *Photochemistry and Photobiology* 7: 619–632.

92 Fisher, G.J. and Johns, H.E. (1970). Ultraviolet photochemistry of thymine in aqueous solution. *Photochemistry and Photobiology* 11: 429–444.

93 Marguet, S. and Markovitsi, D. (2005). Time-resolved study of thymine photodimer formation. *Journal of the American Chemical Society* 127: 5780–5781.

94 Banyasz, A., Douki, T., Improta, R. et al. (2012). Electronic excited states responsible for dimer formation upon UV absorption directly by thymine strands: joint experimental and theoretical study. *Journal of the American Chemical Society* 134: 14834–14845.

95 Zhang, R.B. and Eriksson, L.A. (2006). A triplet mechanism for the formation of cyclobutane pyrimidine dimers in UV-irradiated DNA. *The Journal of Physical Chemistry. B* 110: 7556–7562.

96 Climent, T., González-Ramírez, I., González-Luque, R. et al. (2010). Cyclobutane pyrimidine photodimerization of DNA/RNA nucleobases in the triplet state. *Journal of Physical Chemistry Letters* 1: 2072–2076.

97 Lhiaubet-Vallet, V., Cuquerella, M.C., Castell, J.V. et al. (2007). Triplet excited fluoroquinolones as mediators for thymine cyclobutane dimer formation in DNA. *The Journal of Physical Chemistry. B* 111: 7409–7414.

98 Cuquerella, M.C., Lhiaubetvallet, V., Bosca, F., and Miranda, M.A. (2011). Photosensitised pyrimidine dimerisation in DNA. *Chemical Science* 2: 1219–1232.

99 Liu, L., Pilles, B.M., Gontcharov, J. et al. (2016). Quantum yield of cyclobutane pyrimidine dimer formation via the triplet channel determined by photosensitization. *The Journal of Physical Chemistry. B* 120: 292–298.

100 Jin, P., Wang, J.-N., Wang, X. et al. (2023). Tracking the early stage of triplet-induced photodamage in a DNA dimer and oligomer containing 5-methylcytosine. *The Journal of Physical Chemistry. B* 127: 6878–6886.

101 Lee, W. and Matsika, S. (2020). Stabilization of the triplet biradical intermediate of 5-methylcytosine enhances cyclobutane pyrimidine dimer (CPD) formation in DNA. *Chemistry - A European Journal* 26: 14181–14186.

2

Triplet Excited State in Canonical Nucleobases and Their Derivatives

Jing Li[1], *Yanyan Jia*[2], *and Jinquan Chen*[1]

[1]*East China Normal University, State Key Laboratory of Precision Spectroscopy, 500 Dongchuan Rd, Shanghai 200241, China*
[2]*East China University of Science and Technology, School of Chemistry and Molecular Engineering, 130 Meilong Rd, Shanghai 200030, China*

As the fundamental building blocks of DNA and RNA—carriers of genetic information—the excited state behaviors of adenine (A), guanine (G), cytosine (C), thymine (T), and uracil (U) are deeply associated with key biological activities such as DNA photodamage repair and light-regulated gene expression. Studies on the triplet excited state dynamics of canonical nucleobases are of crucial significance for revealing the intrinsic mechanisms underlying the photochemical processes of biomolecules. The continuous innovation of time-resolved spectroscopy techniques has provided high-resolution observational tools for accurately capturing the transient evolution processes of nucleobases in their triplet excited states, enabling researchers to resolve their dynamic mechanisms at the molecular level.

2.1 Study on the Triplet Excited States Dynamics of Canonical Nucleobases by Laser Flash Photolysis and Femtosecond Transient Absorption Spectroscopy

2.1.1 Early Research Foundation

From the 1960s to the 1990s, the application of laser flash photolysis technology laid a crucial foundation for research on the triplet excited states of nucleobases. Lamola [1, 2] and Johns and Whillans [3], respectively, employed specific triplet quenchers combined with flash photolysis techniques to investigate the photodimerization mechanisms of thymine (Thy) and uracil in different solvents, revealing significant differences in their photodimerization pathways. Johns and co-workers utilized flash photolysis to study the triplet state properties of Thy and uracil in aqueous

Excited State Dynamics in DNA and RNA: Experimental and Computational Approaches on Triplet State, First Edition. Edited by Jinquan Chen.

solutions and found that approximately half of the triplet states in uracil were quenched by the parent molecule to form dimers, compared to one-fifth in Thy [3]. They also determined that the ratio of intersystem crossing (ISC) efficiency between uracil and Thy was 3.8. Subsequently, Szabo et al. measured the lifetime of the Thy triplet state as $14 \pm 1\,\mu s$ by laser flash excitation of deoxythymidine solutions in acetonitrile [4]. Salet and Bensasson observed that the ratio of triplet quantum yields between uracil and Thy was numerically similar in acetonitrile and water; however, the triplet quantum yields of both uracil and Thy in water were reduced by a factor of 10 compared to those in acetonitrile [5]. Similar to the findings for uracil and Thy, Salet et al. determined the quantum yields of the lowest triplet states for thymidine (Thd), thymidine-5′-monophosphate (TMP) dipotassium salt, uridine (Urd), and uridine-5′-monophosphate (UMP) disodium salt in various solvents [6]. They found that the quantum yields of these four compounds in water were significantly lower than in any other solvent. Additionally, their experiments on Ura in different solvents demonstrated that the quantum yield of Ura's lowest triplet state gradually decreases with increasing hydrogen-bonding capacity of the solvent. Further studies quantified the triplet lifetimes and quantum yields of Thy and Ura in acetonitrile and water [4, 6], clarifying the significant impact of solvent effects on ISC efficiency. Song et al. investigated the triplet states of Thy, Thd, and TMP in aqueous solutions using both acetone sensitization and direct excitation methods [7]. The consistent kinetic parameters obtained from the two methods validated the acetone sensitization approach, which also enhanced signal amplitude, enabling more reliable detection of triplet states. Wardman generated triplet states of Thy and Ura in acetonitrile (MeCN) solutions via 249-nm laser flash photolysis and studied their quenching by various electron acceptors [8]. The quenching rates were found to conform to the Rehm–Weller relationship, leading to the conclusion that an electron transfer mechanism is involved.

Zuo et al. employed laser photolysis technology to investigate the triplet states of cytosine and its derivatives in aqueous solutions [9]. They revealed the characteristics of their absorption spectra, clarified the reaction mechanism involving triplet state acetone excitation transfer, and obtained relevant kinetic parameters. The Redmond group utilized laser flash photolysis to successfully measure the triplet–triplet absorption spectra of nucleobases in complex solution systems [10, 11]. Through quantitative analysis, they determined the triplet energy order as C > U > G > A > T and found that the triplet quantum yield under sensitized conditions was significantly higher than that under direct ultraviolet (UV) irradiation. Later, Markovitsi studied TMP and polythymidylic acid $(dT)_{20}$ using laser flash photolysis and observed that TMP exhibited a significant triplet state with an absorption peak at 360 nm and a decay constant of $5\,\mu s$, whereas the triplet absorption of $(dT)_{20}$ was negligible [12].

These early studies established the fundamental theoretical framework for research on the triplet excited states of nucleobases, providing crucial data support and methodological references for subsequent in-depth explorations based on time-resolved spectroscopy techniques.

2.1.2 In-Depth Research Promoted by Time-Resolved Spectroscopy Techniques

Since its inception, time-resolved spectroscopy technology has undergone a leapfrog development in terms of time resolution, progressing from the nanosecond scale to the femtosecond scale and even the attosecond scale. By integrating ultrafast laser pulses with spectral detection technology, this technique enables the real-time capture of transient spectral changes in molecular excited states across different timescales. Leveraging its unique advantages of high time resolution and spectral resolution, it can precisely analyze dynamic processes such as energy transfer, ISC, and state-to-state transitions in excited state molecules. This provides an effective means to gain microscopic insights into the dynamic behavior of molecules for the study of triplet excited states of nucleic acid bases, significantly propelling research in this field to a deeper level.

2.1.2.1 Thy Monomer

Kohler group innovatively proposed the theoretical hypothesis that the $^1n\pi^*$ state serves as the gateway to the lowest triplet state [13]. Subsequently, they investigated the lowest triplet state of Thy using time-resolved infrared (TRIR) spectroscopy, revealing that the carbonyl stretching vibration in the triplet state undergoes a significant redshift and confirming that the triplet state is fully formed within 10 ps after excitation [14]. Similarly, Kwok et al. conducted studies using femtosecond broadband time-resolved fluorescence spectroscopy and transient absorption spectroscopy and observed that the singlet-state gateway state S_n' in Thy monomers can further populate the triplet state [15]. They proposed the relaxation mechanism of Thd: upon excitation to the S_π ($\pi\pi^*$) state, Thd populates the S_n' state; from the S_n' state, molecules can either decay nonradiatively to the ground state (GS) (S_0) via rapid internal conversion or populate the first triplet state T_1 ($^3\pi\pi^*$) through ISC. Finally, molecules populating the T_1 ($^3\pi\pi^*$) state primarily relax back to the S_0 state. It is suggested that vibronic coupling between the S_π ($\pi\pi^*$) state and the lowest-energy S_n ($n\pi^*$) state may endow the S_n state with mixed properties, and the gateway state S_n' is attributed to such a mixed S_n state.

Previous studies generally held that the ISC process of natural pyrimidine bases and their derivatives occurs from the $^1n\pi^*$ state to the $^3\pi\pi^*$ triplet state. However, Pilles et al. studied the electronic excited state decay of Thy and TMP using time-resolved UV/vis and IR spectroscopy and found that ultrafast ISC may also occur directly from the bright $^1\pi\pi^*$ state to the $^3n\pi^*$ triplet state [16]. Specifically, starting from the $^1\pi\pi^*$ state, molecules can populate the lowest triplet state ($^3\pi\pi^*$) either directly via ISC or indirectly through an intermediate $^1n\pi^*$ state. Notably, the quantum yield of the lowest $^3\pi\pi^*$ triplet state of Thy varies significantly with the type of solvent.

Building upon the aforementioned research, Chen group has conducted a series of pioneering studies on the excited state dynamics of Thy. To clarify the newly observed ISC channel from the bright $^1\pi\pi^*$ state directly to the triplet $^3n\pi^*$ state,

Pilles et al. selected three pyrimidone molecules (2-pyrimidone deoxynucleoside, 1,5-dimethylpyrimidin-2(1*H*)-one, and 1,4,5-trimethylpyrimidin-2(1*H*)-one) as model compounds for investigating this mechanism, as they can amplify the triplet state signals of the pyrimidine ring [17]. Systematic studies on their excited state dynamics were carried out in various typical aprotic and protic solvents using femtosecond transient absorption spectroscopy combined with quantum chemical calculations. The research revealed that in protic solvents, after being excited to the hot S_1 ($^1\pi\pi^*$) state, the pyrimidone molecules undergo vibrational cooling to populate the S_1 ($^1\pi\pi^*$) state, followed by ISC to populate the T_3 ($^3n\pi^*$) state. In aprotic solvents, the S_1 ($^1\pi\pi^*$) and S_2 ($^1n\pi^*$) states of the pyrimidone molecules, due to their close energy levels, undergo mixing, which further promotes ISC to the T_3 ($^3\pi\pi^*$) state. The study confirmed that the $n\pi^*$ state of the triplet can facilitate efficient ISC. Additionally, solvent-related experiments demonstrated that hydrogen-bonding interactions between pyrimidone molecules and protic solvents enhance the vibrational cooling efficiency of the $^1\pi\pi^*$ state. These findings not only validate the previously proposed ISC mechanism from the bright $^1\pi\pi^*$ state directly to the triplet $^3n\pi^*$ state but also provide a crucial basis for understanding the ISC processes in pyrimidines and their derivatives (Figure 2.1).

Banyasz et al. observed that the triplet quantum yield of TMP in water decreases significantly with the red shift of the excitation wavelength [18]. Previous studies have indicated that the excitation wavelength affects the initially populated excited states and subsequent relaxation pathways in Thd. However, how the excitation wavelength regulates the formation of long-lived triplet states in Thd and its excited state relaxation pathways remains unclear. To address this, Chen group employed femtosecond broadband transient absorption spectroscopy to real-time monitor the excited state dynamics of 2′-deoxy-thymidine (Thd) under different excitation wavelengths and constructed an excited state relaxation model [19]. The experimental results demonstrated that Thd, upon excitation, may immediately

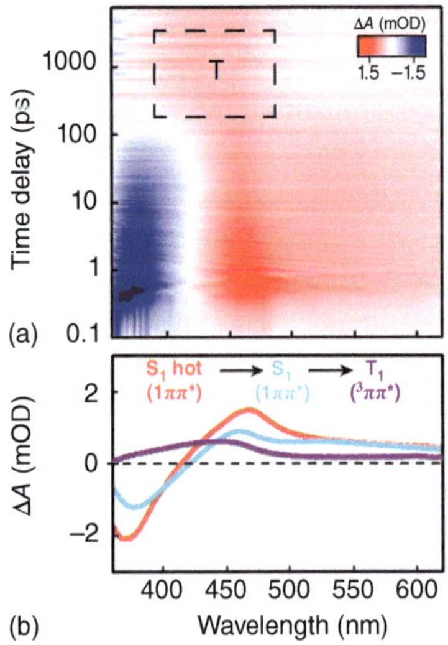

Figure 2.1 (a) Femtosecond broadband transient absorption spectra and (b) evolution-associated difference spectra (EADS) of 2-Pyrimidone deoxynucleoside under 320 nm excitation in phosphate-buffered solution (PBS) ([17] / with permission of American Chemical Society).

relax to the hot GS, $^1n_O\pi^*$ state, and $^1\pi\pi^*$ state. From the $^1n_O\pi^*$ state, it is likely to populate the $^3\pi_H\pi^*$ state via ISC. The evolution-associated difference spectra (EADS) obtained under 267 and 285 nm excitation exhibit high similarity in shape and lifetime, but different excitation wavelengths significantly affect the excited state population and quantum yield of Thd: the triplet quantum yield under 267-nm excitation (0.16 ± 0.03) is significantly higher than that under 285-nm excitation (0.07 ± 0.02). This indicates that changing the excitation wavelength affects the population entering different channels, with more populations relaxing to the GS through the $^1\pi_H\pi^*$ state rather than decaying to the dark $n_O\pi^*$ state. This study reveals the regulatory mechanism of excitation wavelength on the excited state relaxation pathways and triplet quantum yield of Thd, which contributes to understanding the wavelength-dependent photodamage of these nucleosides.

It is now clear that Thy, Thd, and TMP can all populate the $^3\pi\pi^*$ triplet state. Among them, Thy and TMP can populate the $^3\pi\pi^*$ state via ISC from the $^1n\pi^*$ state [15, 16, 19], while Thd can populate the $^3\pi\pi^*$ state via ISC from the gateway state S_n' with mixed properties of the $^1n\pi^*$ state and $^1\pi\pi^*$ state [15]. In addition, Thy and TMP can also populate the $^3n\pi^*$ state via ISC from the $^1\pi\pi^*$ state[16].

2.1.2.2 Uracil Monomer

Brown and Johns conducted research on Ura molecules in the past century and found that the quantum yield of ISC decreases significantly with the red shift of the excitation wavelength [20]. In the 21st century, Brister and Crespo-Hernández, in their study on the excited state dynamics of UMP in aqueous solution, utilized broadband transient absorption spectroscopy to reveal that after UMP is excited to the $^1\pi\pi^*$ state, it simultaneously populates the vibrationally excited GS, the long-lived $^1n\pi^*$ state, and the "receiver triplet state" within 200 fs. This receiver triplet state is most likely a $^3n_O\pi^*$ state, and it can populate the long-lived $^3\pi\pi^*$ state via internal conversion on an ultrafast timescale [21]. According to previous studies, the excitation wavelength should affect the initially populated excited states and subsequent relaxation pathways in oxy-Urd, but the detailed mechanism remains unclear.

To address this, Chen research group further expanded the depth and breadth of research on the excited states of Ura. They took Urd as the research object, used femtosecond transient absorption spectroscopy to construct an excited state relaxation model, and explored in depth the influence of excitation wavelength on the excited state relaxation pathways of Urd molecules [19]. The study revealed that under excitation at 267 and 285 nm, Urd is initially populated to the $^1\pi_H\pi^*$ state and then can either relax back to the GS or relax to the $^1n_O\pi^*$ state or $^3n_O\pi^*$ state. Starting from the $^3n_O\pi^*$ state, the molecule may relax to the $^3\pi_H\pi^*$ state on an ultrafast timescale; however, the yield of the $^3\pi_H\pi^*$ state mainly comes from the pathway where the $^1n_O\pi^*$ state populates this state via ISC. When the excitation wavelength is red-shifted from 267 to 285 nm, the spectral shape of Urd is roughly the same and the lifetime is similar, but the stimulated emission signal is enhanced and the triplet quantum yield is reduced, indicating that changing the excitation wavelength can regulate the population distribution of different relaxation pathways. Specifically, when the excitation wavelength is red-shifted to 285 nm, the number of particles returning to the GS from the $^1\pi_H\pi^*$ state via radiative transition increases, while the number of particles decaying to the $^1n_O\pi^*$ state decreases. These findings provide important data for understanding the photobiological processes of Ura in RNA and lay a crucial foundation for exploring the mechanism of RNA photoregulation.

Both UMP and Urd can relax to the $^3\pi\pi^*$ state upon excitation, but their relaxation pathways for populating the $^3\pi\pi^*$ state differ. Specifically, after being excited to the $^1\pi\pi^*$ state, UMP relaxes to a "receiver triplet state" (which is highly likely a $^3n_O\pi^*$ state) and then populates the $^3\pi\pi^*$ state via ultrafast internal conversion [21]. In contrast, after being excited to the $^1\pi\pi^*$ state, Urd relaxes to the $^1n_O\pi^*$ state first, followed by populating the $^3\pi\pi^*$ state through ISC. Moreover, the number of particles populating the $^3\pi\pi^*$ state can be regulated by changing the excitation wavelength, thereby controlling the triplet quantum yield [19] (Figure 2.2).

2.1.2.3 Cytosine Monomer

Among the five nucleoside monomers, cytidine molecules exhibit the most complex excited state dynamics. Additionally, methylation at the C5 position of cytosine plays a crucial role in gene epigenetics, yet the excited state relaxation processes of cytidine and 5-methylcytidine remain controversial. To clarify the excited state relaxation mechanisms of 2′-deoxycytidine (dCyd) and 2′-deoxy-5-methylcytidine (5mdCyd), Chen research group employed femtosecond transient absorption spectroscopy and conducted systematic studies on the excited state relaxation of dCyd and 5mdCyd through experiments under different excitation wavelengths, excitation powers, solvent environments, and pH conditions [22]. The study revealed that varying excitation wavelengths leads to differences in the initially populated excited states of dCyd and 5mdCyd, thereby resulting in two distinct decay pathways for the two molecules. Specifically, when dCyd is excited at 240 nm, it is initially populated to the $\pi\pi^*_2$ state, followed by relaxation to the $\pi\pi^*_1$ state. In the $\pi\pi^*_1$ state, approximately 70% of the population relaxes back to the GS within ~1 ps, while the remaining 30% decays to the $n_O\pi^*$ state. However, when dCyd is excited at 290 nm,

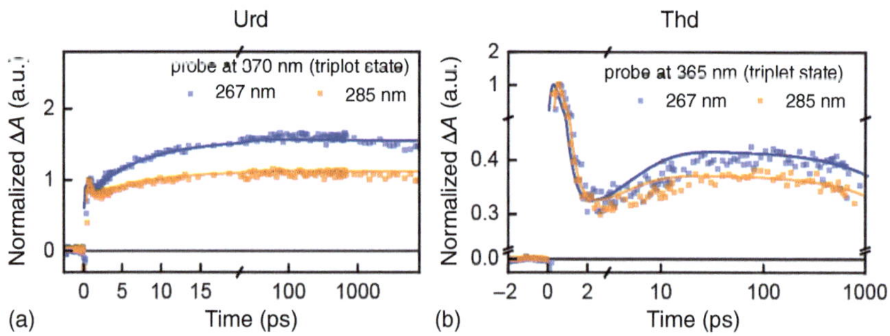

Figure 2.2 Representative decay traces recording the dynamics of the $^3\pi_H\pi^*$ state for (a) oxy-uridine (Urd) and (b) 2′-deoxy-thymidine (Thd) under 267 and 285 nm excitation in acetonitrile ([19] / with permission of Elsevier).

it is initially populated to the $\pi\pi^*_1$ state. Subsequently, approximately 60% of the population in the $\pi\pi^*_1$ state relax back to the GS within ~1 ps, ~30% relax to the $n_O\pi^*$ state, and ~10% decay to a long-lived state. For 5mdCyd, when excited to the $\pi\pi^*_2$ state at 240 nm, it populates the $\pi\pi^*_1$ state and then relaxes to the minimum energy of the $\pi\pi^*_1$ state. Population in the $\pi\pi^*_1$ state then almost completely decay back to the GS via internal conversion. Meanwhile, when 5mdCyd is excited to the $\pi\pi^*_1$ state by 290 nm pulses, relaxation from the Franck–Condon (FC) region to the minimum of the $\pi\pi^*_1$ state occurs, accompanied by relaxation back to the GS via internal conversion while it can also populate a long-lived state. Experiments showed that the long-lived components only emerge when excited at 266 and 290 nm in both dCyd and 5mdCyd. To explore the origin of this long-lived component, further experiments were conducted. By varying excitation power and solvent environments, the possibility that the long-lived component is a cationic state or a hydrate was ruled out. By adjusting the pH environment of dCyd and 5mdCyd, it was hypothesized that the long-lived component observed under 266 and 290 nm excitation might originate from the relaxation of the $n_N\pi^*$ state, but this inference contradicts theoretical calculation results. This study by the Chen group provides important clues for investigating the properties of cytidine monomers, but the definitive identification of the long-lived component requires further experimental exploration.

To clearly and robustly identify the long-lived components in the excited state relaxation pathways of cytidine monomers, Chen research group further investigated the long-lived components of dCyd under excitation at different wavelengths by precisely tuning the excitation wavelength and combining femtosecond broadband transient absorption spectroscopy [23]. Fitting and analysis of the spectral data revealed that the first observed long-lived component is the $^1n_N\pi^*$ state of dCyd, which can be detected under excitation at 250, 255, 260, 265, 270, and 275 nm. The second long-lived component is the triplet state of dCyd, which was more clearly resolved under excitation at 260 and 265 nm. This is the first experimental confirmation that the triplet state of dCyd can be generated via ISC from the FC region. The study experimentally demonstrated that the triplet yield of dCyd is dependent on the excitation wavelength, as the relative contributions of several ISC pathways change when the excess energy is varied within the spectral range of 250–275 nm. In particular, under excitation at 260 and 265 nm, the spin-vibronic coupled ISC pathway may account for the increased triplet yield. Elucidating the ISC mechanism in dCyd is of great significance, as long-lived triplet excited states may lead to the formation of harmful photoproducts in DNA and RNA. These findings significantly deepen the understanding of dCyd's excited state dynamics, lay a foundation for further studies on the physical and photochemical properties of the dCyd triplet state in DNA and RNA, and provide an important theoretical and experimental basis for exploring the role of cytosine in DNA photochemical processes (Figure 2.3).

The abovementioned experiments show that exciting dCyd at different wavelengths can alter its relaxation pathways. At longer wavelengths (260, 290 nm),

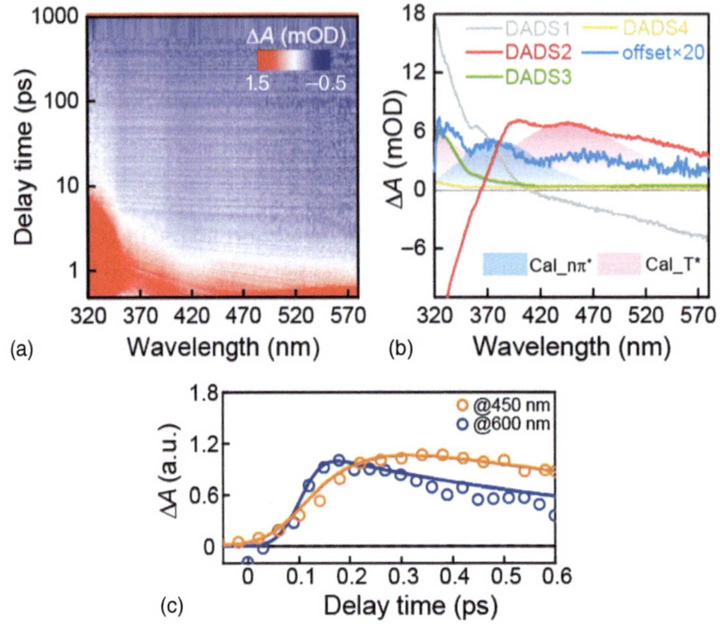

Figure 2.3 (a) Femtosecond broadband transient absorption spectra of 2′-deoxycytidine (dCyd) under 265-nm excitation in acetonitrile. (b) The decay-associated difference spectra (DADS) derived from global fitting of Figure (a) (gradient blue box and pink box extracted from calculated results for comparison). (c) Kinetic traces at 450 and 600 nm probe wavelengths within the first 0.6 ps time delay of dCyd under 265-nm excitation in acetonitrile ([23] / with permission of American Chemical Society).

dCyd can relax to a long-lived state [22], whereas no such component appears in its relaxation pathways when excited at a shorter wavelength (240 nm) [22]. Further experimental investigations revealed that the long-lived components in dCyd's relaxation pathways include the $^1n_N\pi^*$ state and the triplet state, with the triplet state being more clearly observable under excitation at 260 and 265 nm [23].

2.1.3 Summary

Driven by time-resolved spectroscopy techniques, significant progress has been made in the research on the triplet excited state dynamics of canonical nucleobases, and the key parameters are summarized in Table 2.1. Through systematic and innovative research, Chen research group has achieved a series of breakthrough results in such directions as the solvent effect and excitation wavelength effect of Thy, the excited state behavior of uracil in simulated biological environments, and the excited state decay mechanism of cytosine. These achievements have greatly enriched people's understanding of the triplet excited state dynamics of nucleobases and provided important theoretical support and experimental basis for the development of this field.

Table 2.1 Triplet state properties of canonical nucleobases.

References		Φ_{ISC}	τ_{ISC}	Φ_T	τ_T	Experimental conditions
[9]	Cytosine				10 μs	Water, N_2
	Cytidine				11.2 μs	
	2′-Deoxycytidine-5′-monophosphate (dCMP)				10.6 μs	
[5]	Thymine (Thy)			0.06 ± 20%	2.5 μs	Acetonitrile, Ar
	Uracil (Ura)			0.2 ± 20%	0.6 μs	
	Thy			0.0062	600 ns	Water, Ar
	Ura			0.023	350 ns	
[6]	Thymidine (Thd)			1.4×10^{-2}		Water, Ar
				6.9×10^{-2}		Acetonitrile, Ar
	Thymidine-5′-monophosphate (TMP) dipotassium salt			1.5×10^{-2}		Water, Ar
				5.5×10^{-2}		Ethanol, Ar
	Uridine (Urd)			1.5×10^{-2}		Water, Ar
				7.8×10^{-2}		Acetonitrile, Ar
	Uridine-5′-monophosphate (UMP) dipotassium salt			1.1×10^{-2}		Water, Ar
				4.4×10^{-2}		Ethanol, Ar
	Uracil (Ura)			6×10^{-2}		Methanol, Ar
				10×10^{-2}		Ethanol, Ar
[4]	Thy				14 ± 1 μs	Degassed acetonitrile
[12]	TMP	$(1.4 \pm 0.1) \times 10^{-2}$				Phosphate buffer, N_2
[14]	Thy				0.56 ± 0.03 μs	Acentonitrile-d_3, Ar

(Continued)

Table 2.1 (Continued)

References		Φ_{ISC}	τ_{ISC}	Φ_T	τ_T	Experimental conditions
[15]	Thd				4000 ps	Buffered aqueous solution
[16]	Thy			0.13		Acetonitrile-d_3
	Thd TMP			0.018		Deuterium oxide
	Thymidine TMP			0.11		Methanol-d_4
[17]	2-Pyrimidone deoxynucleoside (2-dP)		87 ± 4 ps			Phosphate-buffered solution (pH 7.4)
			92 ± 3 ps			Methanol
			69 ± 1 ps			Acetonitrile
	1,5-Dimethylpyrimidin-2(1*H*)-one (1,5-Dim2P)		$(2.9 \pm 0.3) \times 10^3$ ps			Phosphate-buffered solution (pH 7.4)
			$(1.9 \pm 0.1) \times 10^3$ ps			Methanol
			$(1.5 \pm 0.1) \times 10^3$ ps			Acetonitrile
			$(1.6 \pm 0.1) \times 10^3$ ps			Dichloromethane
	1,4,5-Trimethylpyrimidin-2(1*H*)-one (1,4,5-Trim2P)		$(2.6 \pm 0.1) \times 10^3$ ps			Phosphate-buffered solution (pH 7.4)
			$(1.7 \pm 0.1) \times 10^3$ ps			Methanol
			510 ± 5 ps			Acetonitrile
			580 ± 5 ps			Dichloromethane

[19]	Oxy-uridine (Urd)	7.0 ± 0.2 ps ($^1n_O\pi^* \rightarrow {}^3\pi_H\pi^*$) 0.12 ± 0.10 ps ($^1\pi_H\pi^* \rightarrow {}^3n_O\pi^*$) ($\lambda_{ex} = 267$ nm)	0.26 ± 0.02 ($\lambda_{ex} = 267$ nm)	Acetonitrile
		7.8 ± 0.3 ps ($^1n_O\pi^* \rightarrow {}^3\pi_H\pi^*$) 0.12 ± 0.10 ps ($^1\pi_H\pi^* \rightarrow {}^3n_O\pi^*$) ($\lambda_{ex} = 285$ nm)	0.13 ± 0.01 ($\lambda_{ex} = 285$ nm)	
	2′-Deoxy-thymidine (Thd)	5.1 ± 0.2 ps ($\lambda_{ex} = 267$ nm)	0.16 ± 0.03 ($\lambda_{ex} = 267$ nm)	
		3.9 ± 0.3 ps ($\lambda_{ex} = 285$ nm)	0.07 ± 0.02 ($\lambda_{ex} = 285$ nm)	
[23]	2′-Deoxycytidine (dCyd)		2.08%	Acetonitrile

2.2 Study on Triplet Excited State Dynamics of Thio- and Selenium-Substituted Nucleobases

Research on thio- and selenium-substituted nucleobases holds significant scientific significance and application value. In practical applications, thio-substituted nucleobases exhibit great potential in fields such as photodynamic therapy and drug development. From the perspective of scientific research, as fundamental building blocks of genetic material, thio- and selenium-substituted modifications of nucleobases provide a unique insight into exploring the intrinsic relationship between molecular structure and photophysical/photochemical properties. Due to their unique electronic structure and chemical properties, sulfur and selenium atoms can alter the energy level distribution and electron cloud density characteristics of nucleobases. Investigating the excited state dynamics of thio- and selenium-substituted nucleobases facilitates in-depth analysis of the mechanisms underlying key processes such as excited state ISC, thereby enriching and improving the basic theoretical system of nucleic acid chemistry and photobiology. This section focuses on the triplet state properties of thio- and selenium-substituted nucleobases, with a review of relevant previous studies as follows.

2.2.1 Thio-Substituted Pyrimidine Bases

Harada et al. conducted in-depth studies on the excited state dynamics of 4-thiothymidine (4-tThd) using time-resolved thermal lens (TRTL) technology. They found that UVA photolysis of 4-tThd induces internal conversion from S_2 ($\pi\pi^*$) to S_1 ($n\pi^*$), followed by population of the T_1 ($\pi\pi^*$) state from S_1 ($n\pi^*$) [24]. The Harada group also investigated the triplet state of 4-tThd in aqueous solution [25] and observed that upon excitation at 263 nm, the excited singlet state of 4-tThd primarily relaxes to the triplet state via ISC, with the triplet formation time being less than 10 ps.

Crespo-Hernández group excited 4-tThd in phosphate-buffered saline with UVA and observed room temperature phosphorescence [26]. Combining results from femtosecond transient absorption experiments and theoretical calculations, they found that the initially excited singlet state can populate the triplet state with nearly unit yield on an extremely short timescale (~240 fs). They also proposed a simplified kinetic model for 4-tThd upon excitation with 340 or 360 nm light: after being excited to the S_2 state, only a negligible fraction of particles relaxes back to the GS via radiative transition, while most of the initial excited state population populates the T_1 state through ISC. However, whether the S_1 state is involved in the excited state dynamics of 4-tThd remains unclear. Lara Martínez-Fernández et al. studied 4-tThd and proposed that it has a complex relaxation pathway involving two S_2 minima and three quasi-degenerate regions, with the main relaxation pathway being $S_2/S_1/T_1$ and a minor contribution from the $S_2/T_2/T_1$ pathway [27].

Taras-Goślińska et al. employed experimental techniques such as nanosecond laser flash photolysis to study 2-thiothymine (2-tThy), the riboside 5-methyl-2-thiouridine (5-Me-2tUrd), and the deoxyriboside 2-thiothymidine (2-tThd) in acetonitrile [28]. They found that ISC in these three compounds occurs on an ultrafast timescale, with triplet quantum yields approaching 1.

Pollum and Crespo-Hernández performed femtosecond broadband transient absorption experiments on 2-tThy and 2-thiouracil (2-tUra) in pH 7.4 phosphate-buffered saline and acetonitrile and proposed a continuous excited state relaxation model for both: $S_2(\pi\pi^*) \rightarrow S_1(n\pi^*) \rightarrow T_1(\pi\pi^*)$ [29].

They also investigated the triplet states of 2-tThy, 2-thiothymidine (2-tThd), 4-tThd, and 2,4-dithiothymine (2,4-dtThy) in phosphate-buffered saline using femtosecond transient absorption and nanosecond luminescence spectroscopy [30]. They found that sulfur substitution significantly enhances ISC rates and triplet yields. Among them, 2,4-dtThy exhibits the fastest ISC rate (180 ± 40 fs) and a nearly 100% triplet yield, with a singlet oxygen quantum yield of 0.46 ± 0.02. This indicates that 2,4-dtThy has potential applications as a UVA chemotherapeutic agent, possibly more promising than 4-tThd.

Recently, Ullrich group combined time-resolved photoelectron spectroscopy (TRPES) with ab initio calculations to further compare the triplet dynamics of 2-tUra and 2-tThy in the gas phase [31]. The results showed that methylation in 2-tThy slows down its ISC dynamics compared to 2-tUra. Upon excitation to the S_2 ($\pi_S\pi_6^*$) state, 2-tUra and 2-tThy populate the lowest singlet excited state S_1 ($n_S\pi_2^*$) via internal conversion, followed by ISC to the lowest triplet $\pi\pi^*$ state, which forms a double-well potential energy surface (PES). The minima of this potential well correspond to the out-of-plane geometry of the sulfur atom ($T_{1\ op\text{-}S}$) and a slightly boat-like ring geometry ($T_{1\ boat}$), with $T_{1\ op\text{-}S}$ initially populated via ISC from the S_1 ($n_S\pi_2^*$) state. The system achieves equilibrium between the two T_1 minima through vibrational energy transfer. The $T_{1\ boat}$ minimum may decay via phosphorescence emission, ISC back to the GS, or cross the barrier to return to $T_{1\ op\text{-}S}$, with depopulation occurring through the $T_{1\ op\text{-}S}/S_0$ crossing point. However, ISC from the $T_{1\ boat}$ minimum back to the GS is less favorable.

Based on previous studies, both 4-tThd and 2-tThd can populate the T_1 ($\pi\pi^*$) state via the relaxation pathway S_2 ($\pi\pi^*$) $\rightarrow$ S_1 ($n\pi^*$) $\rightarrow$ T_1 ($\pi\pi^*$) [24, 27, 31], while 4-tThd can also populate the T_1 ($\pi\pi^*$) state via the minor pathway S_2 ($\pi\pi^*$) $\rightarrow$ T_2 ($n\pi^*$) $\rightarrow$ T_1 ($\pi\pi^*$) [27].

Crespo-Hernández and co-workers further conducted spectroscopic and photochemical studies on 2-tUra, 4-thiouracil (4-tUra), and 2,4-dithiouracil (2,4-dtUra) in pH 7.4 phosphate-buffered saline solution in 2015, with a focus on investigating the photoreactivity of 2,4-dtUra [32]. It was found that upon excitation, 2,4-dtUra can populate the triplet state via ISC within 220 ± 40 fs, with a triplet quantum yield approaching unity ($\Phi_T = 0.90 \pm 0.15$) and a singlet oxygen quantum yield of approximately 50%.

Ullrich group applied TRPES to study the photodynamics of sulfur-substituted nucleobases such as 2-tUra [33] and 4-tUra [34] under ultraviolet excitation. The group first performed gas-phase photodynamics studies on unsubstituted uracil, Thy, and 2-tUra [33]. It was observed that when uracil and Thy are excited to the S_2 ($\pi\pi^*$) state, only a small fraction populates the lowest-energy triplet state: this fraction of uracil populates the S_1 ($^1n\pi^*$) state via internal conversion within 170 fs, followed by ISC to the triplet state T_1 ($^3\pi\pi^*$) within 2.35 ps. For Thy, it populates the S_1 ($^1n\pi^*$) state within 175 fs and then reaches T_1 ($^3\pi\pi^*$) via ISC within 6.13 ps, with both triplet states having lifetimes exceeding 1 ns. In contrast, the internal conversion from S_2 ($^1\pi\pi^*$) to S_1 ($^1n\pi^*$) in 2-tUra is more rapid, completing within less than 100 fs, followed by ISC to the T_1 ($^3\pi\pi^*$) state within 775 fs, with the decay lifetime of the T_1 ($^3\pi\pi^*$) state being approximately 203 ps.

Subsequently, they excited 2-tUra in the gas phase and condensed phase using different wavelengths of light [35]. For 2-tUra in the gas phase, under the various excitation wavelengths used in the experiments, relaxation primarily proceeds via the proposed pathway $^1\pi_S\pi_6^* \rightarrow {}^1n_S\pi_2^* \rightarrow {}^3\pi\pi^*$. However, when 2-tUra in the gas phase is excited with shorter wavelengths, it may be initially populated to the $^1\pi_S\pi_2^*$ state, which internally converts to the $^1\pi_S\pi_6^*$ state and further converts to the $^1n_S\pi_2^*$ state. They hypothesized that with shorter pump wavelengths, a small fraction of population might populate the triplet state via the pathway $^1\pi_S\pi_2^* \rightarrow {}^3n_S\pi_2^* \rightarrow {}^3\pi_S\pi_2^*$. When the excitation wavelength is 249 nm, the time constants for triplet population via ISC and repopulation of the GS from the triplet state in gas-phase 2-tUra decrease to a quarter of their original values. They postulated that higher excitation energies increase the probability of molecular populations overcoming energy barriers to access the vicinity of the $^1n_S\pi_2^*/^3\pi_S\pi_2$ and $^3\pi_S\pi_2/S_0$ crossing seams. For 2-tUra in acetonitrile, after being excited to the $^1\pi_S\pi_6^*$ or $^1\pi_S\pi_2^*$ state, it populates the $^1n_S\pi_2^*$(hot) state via internal conversion and further populates the triplet state from the $^1n_S\pi_2^*$ state via ISC. Studies in the condensed phase showed that when the excitation wavelength is reduced from 316 to 268 nm, the time required for 2-tUra to populate the triplet state from the singlet state increases slightly. They postulated that this arises from dissipation of excess vibrational energy in the $^1n_S\pi_2^*$ state, with the additional possibility that initial excitation to the higher lying $^1\pi_S\pi_2^*$ state requires surmounting a modest energy barrier prior to internal conversion to the $^1n_S\pi_2^*$ state. After excitation at 268 or 290 nm, the decay of the triplet state of 2-tUra in acetonitrile is five times faster than that upon excitation at 316 nm. They proposed that excitation at 316 nm traps 2-tUra at the $^3\pi_5\pi_6^*$ minimum, while excitation wavelengths shorter than 308 nm provide sufficient excess vibrational energy to stabilize it at the $^3\pi_S\pi_2^*$ minimum. Since the $^3\pi_S\pi_2^*$ minimum is predicted to reach the T_1/S_0 crossing point more efficiently, shorter excitation wave lengths lead to faster decay of the triplet population. The lifetime of the 2-tUra triplet state increases by nearly three orders of magnitude from the gas phase to acetonitrile solvent, which results from the synergistic effect of electronic and vibrational effects.

In the case of 4-tUra, when gas-phase 4-tUra is excited at 347.2, 330.16, and 309.02 nm, it first populates the S_2 ($^1\pi\pi^*$) state, then populates the S_1 ($^1n\pi^*$) state via ultrafast internal conversion, and subsequently fills into a long-lived state on a picosecond timescale[34]. This long-lived state is assigned to the triplet states T_2 ($^3\pi\pi^*$) and T_1 ($^3n\pi^*$), with a triplet lifetime exceeding 1.5 ns. The relaxation mechanisms for populating triplet states in gas-phase 4-tUra and 2-tUra are similar, but the triplet lifetime of 4-tUra is much longer than that of 2-tUra. This is due to the difference in the topological structures of their triplet states: the T_1 state of 2-tUra is a $^3\pi\pi^*$ double-well potential, while the T_1 state of 4-tUra has a $^3n\pi^*$ minimum.

Ullrich has also investigated 2,4-dtUra in the gas phase using TRPES and proposed a relaxation model [36]. Upon excitation to the $^1\pi_{S42}\pi_{S4}{}^*$ state, gas-phase 2,4-dtUra enters the minimum of the singlet excited state $^1n_{S24}\pi_{S2}{}^*$ via ultrafast internal conversion. It then undergoes internal conversion to the $^1n_{S42}\pi_{S4}{}^*$ state through a four-state crossing point formed by this minimum and three degenerate states ($^3\pi_{S42}\pi_{S2}{}^*$, $^3n_{S42}\pi_{S4}{}^*$, $^1n_{S42}\pi_{S4}{}^*$). Subsequently, it populates the lowest triplet state $^3\pi_{S24}\pi_{S4}{}^*$ via ISC on a sub-picosecond timescale. Overall, the observed time constants decrease with decreasing photoexcitation wavelength, indicating the presence of energy barriers in the relaxation pathway, and increasing vibrational excitation energy facilitates overcoming these barriers.

Recently, Mohamadzade et al. combined gas-phase TRPES experiments with theoretical quantum chemical methods to investigate 2,4-dtUra in the gas phase under excitation at different wavelengths, identifying four relaxation pathways for the molecule [37]. **Pathway 1**: When 2,4-dtUra is excited by 376 and 361 nm light, it populates the FC $^1\pi_{S4}\pi_{S4}{}^*$ state. After reaching the $^1\pi_{S4}\pi_{S4}{}^*/^1n_{S4}\pi_{S4}{}^*$ crossing region (CI-$^1\pi_{S4}\pi_{S4}{}^*/^1n_{S4}\pi_{S4}{}^*$), it populates the minimum of the $^1n_{S4}\pi_{S4}{}^*$ state (Min-$^1n_{S4}\pi_{S4}{}^*$). Subsequently, it populates the minimum of the $^3\pi_{S4}\pi_{S4}{}^*$ state (Min-$^3\pi_{S4}\pi_{S4}{}^*$) via ISC. In this relaxation pathway, the 2,4-dtUra molecule remains mostly planar. 2,4-dtUra excited by 376 and 361 nm light exhibits characteristics similar to those of 4-tUra, with almost barrierless ISC associated with Min-$^1n_{S4}\pi_{S4}{}^*$ and $^3\pi_{S4}\pi_{S4}{}^*$. The faster ISC time under 361-nm excitation can be attributed to sufficient excitation energy activating random out-of-plane vibrational modes. The energy of 344 and 329 nm pump light lies between the calculated vertical excitation energies of the $^1\pi_{S4}\pi_{S4}{}^*$ and $^1\pi_{S2}\pi_{S4}{}^*$ states. When 2,4-dtUra is excited at these two wavelengths, **Pathway 1 and Pathway 2** may be activated simultaneously. **Pathway 2**: After being excited to the FC $^1\pi_{S2}\pi_{S4}{}^*$ state, 2,4-dtUra relaxes to the minimum of $^1\pi_{S2}\pi_{S4}{}^*$ (Min-$^1\pi_{S2}\pi_{S4}{}^*$). It then relaxes to the minimum of $^1n_{S4}\pi_{S4}{}^*$ (Min-$^1n_{S4}\pi_{S4}{}^*$) via the conical intersection CI-$^1\pi_{S2}\pi_{S4}{}^*/^1n_{S4}\pi_{S4}{}^*$. The subsequent relaxation pathway is consistent with Pathway 1, i.e., populating the $^3\pi_{S4}\pi_{S4}{}^*$ triplet minimum (Min-$^3\pi_{S4}\pi_{S4}{}^*$) via ISC. **Pathway 4**: When 2,4-dtUra is excited by 287, 267, and 250 nm light, it populates the FC $^1\pi_{S2}\pi_{S2}{}^*$ state. After relaxing to Min-$^1\pi_{S2}\pi_{S2}{}^*$, it populates the Min-$^1n_{S2}\pi_{S2}{}^*$ state via the CI-$^1\pi_{S2}\pi_{S2}{}^*/^1n_{S2}\pi_{S2}{}^*$. It then populates the Min-$^3\pi_{S2}\pi_{S2}{}^*$ state via ISC. Upon reaching Min-$^3\pi_{S2}\pi_{S2}{}^*$, it can populate the GS via ISC; alternatively, it may first reach Min-$^3\pi_{S4}\pi_{S4}{}^*$ and then return to the GS from Min-$^3\pi_{S4}\pi_{S4}{}^*$, but this pathway is excluded due to the absence of long-lived TRPES signals. Additionally, after being excited to the $^1\pi_{S2}\pi_{S2}{}^*$ state, 2,4-dtUra may return to the GS from the Min-$^1\pi_{S2}\pi_{S2}{}^*$ state via the CI–$^1\pi_{S2}\pi_{S2}{}^*$/GS, though this pathway is unlikely to occur. **Pathway 3**: When 2,4-dtUra is excited by 310 nm pump pulses, it populates the FC $^1\pi_{S4}\pi_{S2}{}^*$ state. Since the $^1\pi_{S4}\pi_{S2}{}^*$ state is degenerate with the $^1n_{S2}\pi_{S2}{}^*$ state, it can populate the $^1n_{S2}\pi_{S2}{}^*$ state from the Min-$^1\pi_{S4}\pi_{S2}{}^*$ state via ultrafast internal conversion. It then relaxes on the $^1n_{S2}\pi_{S2}{}^*$ state through the out-of-plane bending motion of sulfur atom 2, with the subsequent pathway merging into Pathway 4.

Electronic transitions primarily localized on sulfur 4 (S4) and sulfur 2 (S2) correspond to the relaxation processes of Pathway 1 and Pathway 4. Pathway 1 is characterized by slow $^1n_{S4}\pi_{S4}{}^*/^3\pi_{S4}\pi_{S4}{}^*$ ISC and trapping at the $^3\pi_{S4}\pi_{S4}{}^*$ triplet minimum

(Min-$^3\pi_{S4}\pi_{S4}$*), whereas Pathway 4 is characterized by fast $^1n_{S2}\pi_{S2}$*/$^3\pi_{S2}\pi_{S2}$* ISC, a short-lived $^3\pi_{S2}\pi_{S2}$* triplet, and out-of-plane bending of sulfur atom 2. The high probability of the $^3\pi_{S2}\pi_{S2}$* triplet reaching the ISC region (ISC-$^3\pi_{S2}\pi_{S2}$*/GS) explains its rapid depopulation—this region directly channels population to the GS, featuring out-of-plane distortion of sulfur atom 2. The lifetime of $^3\pi_{S2}\pi_{S2}$* decreases with increasing excitation energy, as excess vibrational energy in the molecule and sulfur atom 2 modes helps overcome the barrier to the crossing point. The molecule was found to exhibit behavior similar to mono-substituted thiouracils under different excitation wavelengths. In the UVB excitation region, its triplet lifetime shows significant variation with excitation wavelength, decreasing from 0.54 at 287 to 0.33 ns at 250 nm. This result clarifies the dependence of ISC and triplet lifetime on excitation wavelength in double-thionated uracil, re-emphasizing the critical regulatory role of excitation wavelength in the photophysical properties of sulfur-substituted nucleobases.

Orr-Ewing group studied 2-tUra and 4-tUra in deuterated acetonitrile using sub-picosecond transient vibrational absorption spectroscopy [38]. It was found that after 2-tUra is excited to the S_2 state, a transient IR absorption peak centered at 1643 cm^{-1} forms within 0.3 ps. This peak is assigned to either 2-tUra molecules in the S_1 state or vibrationally excited molecules in the T_1 state, with the latter assignment being more consistent with previous research interpretations. The absorption peak centered at 1643 cm^{-1} decays with a time constant of 7.2 ± 0.8 ps, accompanied by the growth of multiple absorption peaks centered at 1234, 1410, 1424, 1443, 1511, 1626, and 1660 cm^{-1}. These peaks are assigned to two different conformations of 2-tUra, corresponding to two independent energy minima on the T_1 state PES. This experimental result strongly confirms the population of 2-tUra to two independent minima on its T_1 state PES after excitation. Experimental results for 4-tUra show that it populates the T_1 state within sub-picoseconds and undergoes vibrational cooling with a time constant of 5.2 ± 0.6 ps. However, the experimental results failed to distinguish between the two T_1 state conformers of 4-tUra predicted by calculations, as their vibrational frequencies are too similar.

In summary, 2-tUra, 4-tUra, and 2,4-dtUra can all populate the triplet state via the relaxation pathway: $^1\pi\pi$* state → $^1n\pi$* state → $^3\pi\pi$* state [33–37]. This indicates that such an excited state relaxation pathway is relatively universal among thiouracils.

There are relatively less studies on thio-substituted cytosine and derivatives. Mai et al. conducted studies on 2-thiocytosine (2-tCyt) in pH 7.4 phosphate buffer, observing an ultrafast ISC process with a triplet yield approaching 100% [39]. They also proposed a relaxation mechanism for 2-tCyt: upon excitation to the S_2 ($^1\pi\pi$*) state, 2-tCyt relaxes to the S_1 state via the S_2/S_1 CI; the S_1 state PES primarily exhibits $^1n\pi$* or $^1\pi\pi$* characteristics in different regions of the configuration space; subsequently, it populates the triplet manifold (T_2 and T_1 states) from the S_1 state via ISC, a process involving the presence of a three-state near-degenerate region among the S_1, T_2, and T_1 states. If the S_1 state populates the T_2 state, internal conversion further occurs to populate the T_1 ($^3\pi\pi$*) state.

Janicki et al. found that the presence of ribose significantly affects the photochemical properties of thiobases: the excited state lifetime of 2-tCyt is five times longer than that of 2-thiocytidine (2-tCyd) [40]. Combining quantum chemical simulations, they proposed that the reason for this significant difference is that 2-tCyt primarily occupies the long-lived $^3\pi\pi_{ring}{}^*$ state, whereas 2-tCyd mainly occupies the short-lived $^3n_S\pi^*$ state.

Su group investigated the triplet decay processes of 2-tCyt and 2-tCyd under different pH conditions using femtosecond transient absorption spectroscopy and quantum chemical calculations [41]. They found that protonation can alter the electronic structure of the triplet state, inhibit photoisomerization that causes rapid decay, and weaken the influence of the sugar moiety on the triplet state, resulting in a longer triplet lifetime of protonated 2-tCyd (2-tCydH$^+$) compared to 2tCyd.

In summary, the presence of ribose significantly prolongs the triplet lifetime of 2-tCyt [40], while protonation weakens the effect of ribose on the triplet state [41].

2.2.2 Sulfur-Substituted Purine Bases

Pollum et al. measured the singlet oxygen quantum yield of 6-thio-2′-deoxyguanosine (6-tGuo) in aqueous solution as 0.24 ± 0.02 (0.29 ± 0.02 in acetonitrile), which is relatively low [42]. They also found that the proportion of singlet oxygen generated during collisions between 6tdGuo triplet molecules and triplet oxygen ($^3O_2$) is relatively low ($S_\Delta = 37\%$). Further studies revealed that the triplet state of 6-tGuo can form a charge-transfer complex with $^3O_2$; this complex does not generate singlet oxygen via energy transfer but tends to decay to the GS through nonradiative pathways, which explains the low singlet oxygen quantum yield of 6-tGuo.

Siouri et al. investigated the excited state dynamics of 6-tGuo using resonance-enhanced multiphoton ionization (REMPI) and IR-UV double-resonance spectroscopy and found that the transition rate of 6-tGuo to the dark state (likely a triplet state) is significantly higher than that of 9-enolguanine [43].

Teles-Ferreira et al. studied 6-tGuo using sub-20 fs UV transient absorption spectroscopy, and the results showed that $S_2(\pi\pi^*) \rightarrow S_1(n\pi^*) \rightarrow T_1(\pi\pi^*)$ is the main pathway for 6-tGuo to relax to the triplet state [44]. 6-tGuo undergoes ultrafast internal conversion from the S_2 ($\pi\pi^*$) state to the S_1 ($n\pi^*$) state within ~81 fs, and the S_1 ($n\pi^*$) state serves as the main entrance for triplet population, with this process completing within 522 fs.

In addition, double substitution involving thiation and methylation can enhance the photoactivity of nucleobases and introduce more reaction channels. Zhou et al. analyzed the excited state dynamics of 6-methylthioguanine (6-Me-tGua) under excitation with UVA light of different wavelengths (310–330 nm) [45] and found that after being excited to the $^1\pi\pi^*_{FC}$ state, me6-TG can populate two states ($^1\pi\pi^*_{min}$ and $^1n_N\pi^*_{min}$ states). Both the $^1\pi\pi^*$min and $^1n_N\pi^*_{min}$ states can relax to the GS via radiative relaxation (fluorescence emission). Meanwhile, particles in the $^1n_N\pi^*_{min}$ state can also populate the $^3\pi\pi^*$ state through ISC, among which the ISC from the $^1n_N\pi^*_{min}$ state to the $^3\pi\pi^*$ state is the dominant relaxation process, with a triplet quantum yield of up to 0.6 ± 0.1. When 6-Me-tGua is excited with higher energy light (<325 nm), another photochemical channel emerges:

6-Me-tGua undergoes C–S bond cleavage to generate a thiyl radical (6-Me-tGua•). This photochemical channel and the aforementioned photophysical channel participate in photosensitizer applications in the form of radicals and triplet states, respectively.

Crespo-Hernandez group found that 2,6-dithiopurine (2,6-dtPur) has a triplet quantum yield of 1.0 ± 0.1 under UVA excitation and proposed its excited state relaxation mechanism in pH 7.4 phosphate buffer: after being excited to the $^1\pi\pi^*$ state, 2,6-dtPur populates the $^3n\pi^*$ state via ultrafast ISC, then populates the $^3\pi\pi^*$ state through internal conversion (this process may be accompanied by vibrational cooling within the $^3\pi\pi^*$ state), and finally relaxes back to the GS on a microsecond timescale [46].

For sulfur-substituted purine bases, both 6-tGuo [44] and 6-Me-tGua [45] can populate the $^3\pi\pi^*$ triplet state via the relaxation pathway: $^1\pi\pi^*$ state→$^1n\pi^*$ state→$^3\pi\pi^*$ state. However, 2,6-dtPur populates the $^3\pi\pi^*$ state via the pathway $^1\pi\pi^*$ state→$^3n\pi^*$ state→$^3\pi\pi^*$ state, which is slightly different from the former two.

2.2.3 Selenium-Substituted Nucleobases

In addition to sulfur substitution, researchers have also investigated selenium-substituted nucleobases. Selenium-substituted nucleobases refer to those in which the oxygen atom in the carbonyl group of the nucleobase is replaced by a selenium atom. Such substitution can significantly red shift the absorption spectrum of the nucleobase without sacrificing its high triplet yield [47, 48].

Crespo-Hernández group compared the photochemical relaxation mechanisms of 6-selenoguanine (6-SeGua) and 6-thioguanine (6-tGua) using femtosecond broadband transient absorption spectroscopy. They found that selenium substitution significantly modulates the ISC dynamics, resulting in a triplet decay rate of 6-SeGua that is 835 times faster than that of 6-tGua [47]. They also proposed the excited state relaxation pathway of 6-SeGua: upon excitation to the S_2 ($\pi\pi^*$) state, 6-SeGua populates the S_1 ($n\pi^*$) state via internal conversion, then undergoes ISC to the T_2 ($n\pi^*$) state, followed by vibrational relaxation to populate the T_1 ($\pi\pi^*$) state, i.e., the pathway is S_2 ($\pi\pi^*$) → S_1 ($n\pi^*$) → T_2 ($n\pi^*$) → T_1 ($\pi\pi^*$).

Chen and co-workers studied 2-selenuracil (2-SeUra) and its deprotonated form [48]. When it is excited to the S_2 ($\pi\pi^*$) state, it undergoes ultrafast internal conversion to populate the S_1 ($n\pi^*$) state, then populates the T_1 ($\pi\pi^*$) state from the S_1 ($n\pi^*$) state via ISC, and finally relaxes back to the GS from the T_1 ($\pi\pi^*$) state. The S_1 ($n\pi^*$) state of the deprotonated form of 2-SeUra can also populate the T_1 ($\pi\pi^*$) state via ultrafast ISC and relax back to the GS from its triplet state. It was found that the triplet lifetime of the deprotonated form is nearly an order of magnitude longer than that of its neutral form, which is attributed to the shortened N1–C2 bond, increasing the energy barrier on the T_1/S_0 crossing pathway. This finding further reveals the regulatory effect of molecular structure changes on the photophysical properties of selenium-substituted nucleobases, providing a theoretical basis for an in-depth understanding of the photochemical behavior of selenium-substituted nucleobases (Figure 2.4).

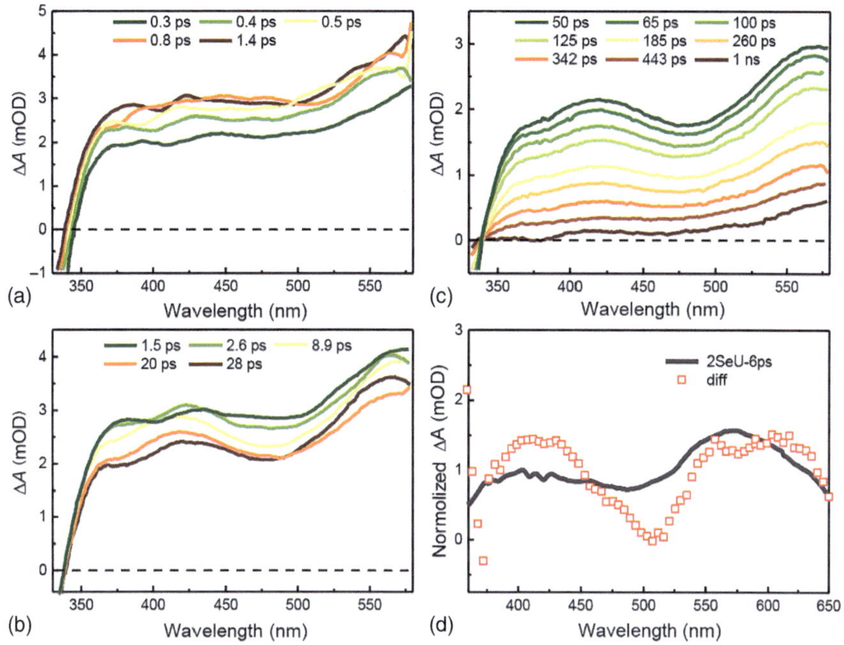

Figure 2.4 (a–c) Transient absorption spectra of 2-selenuracil (2-SeUra) under 305-nm excitation in pH = 5.4 phosphate-buffered solution. (d) Comparison between the T_1 state spectrum of 2-SeUra produced via sensitization and the transient absorption spectrum of neutral 2-SeUra at 6 ps ([48] / Royal Society of Chemistry).

2.2.4 Summary

In summary, research on the photophysical properties of thio- and selenium-substituted nucleic acid bases has yielded substantial achievements in recent years (Table 2.2). Through the application of various advanced experimental techniques, researchers have gained relatively in-depth insights into the excited state dynamics, triplet behaviors, and influencing factors of thio- and selenium-substituted nucleobases. However, current research still faces several issues that require further exploration, such as the quantitative relationship between sulfur atoms at different substitution positions and with different substitution numbers and the photophysical properties of nucleobases, as well as the photophysical behaviors of thio-substituted nucleobases in complex in vivo environments and their interaction mechanisms with biomolecules. Future studies can combine theoretical calculations and experimental techniques to explore the photophysical properties of thio-substituted nucleobases at multiple levels, including molecular level and biological systems, thereby providing a more solid theoretical foundation and technical support for their applications in fields such as phototherapy and biosensing. Meanwhile, research on other heteroatom-substituted nucleobases (e.g., selenium-substituted ones) will also offer new perspectives and directions for the development of this field, which is expected to promote further progress in the interdisciplinary field of nucleic acid chemistry and photobiology.

Table 2.2 Triplet state properties of thio- and selenium-substituted nucleobases.

References		Φ_{ISC}	τ_{ISC}	Φ_T	τ_T	Experimental conditions
[24]	4-Thiothymidine (4-tThd) 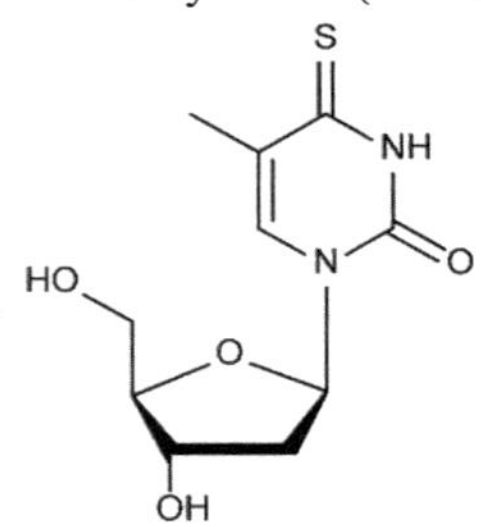	1.0 ± 0.1			4.2 ± 3.0 μs	Deaerated acetonitrile, Ar
[26]	4-tThd		0.24 ± 0.02 ps	$85 \pm 15\%$		Phosphate-buffered solutions (pH 7.4)
[28]	2-Thiothymine (2-tThy) 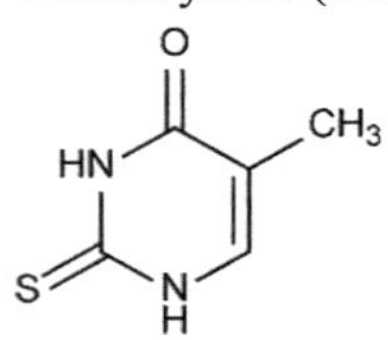			0.9 ± 0.1		Acetonitrile, Ar

	Riboside: 5-methyl-2-thiouridine (5-Me-2tUrd) 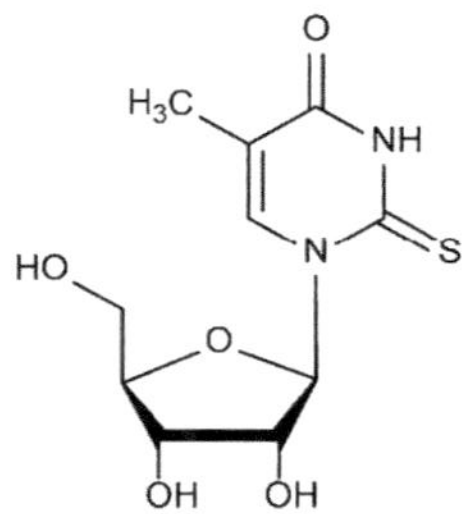 Deoxyriboside: 2-tThd 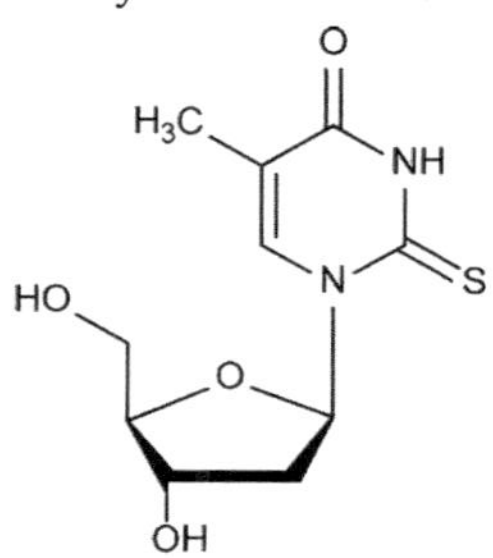		
[30]	2-tThy	620 ± 60 fs	Phosphate-buffered solution (pH 7.4)
	2-tThd	410 ± 60 fs	
	4-tThd	240 ± 20 fs	

(Continued)

Table 2.2 (Continued)

References		Φ_{ISC}	τ_{ISC}	Φ_T	τ_T	Experimental conditions
	2,4-Dithiothymine (2,4-dtThy) 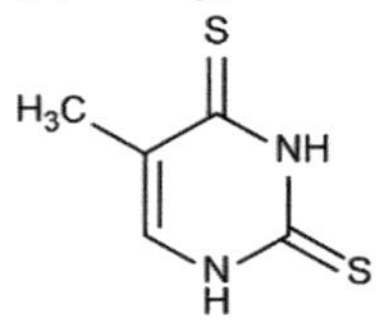		180 ± 40 fs	≳0.9		
[31]	2-tThy				1.8 ns	Gas phase
	2-Thiouracil (2-tUra) 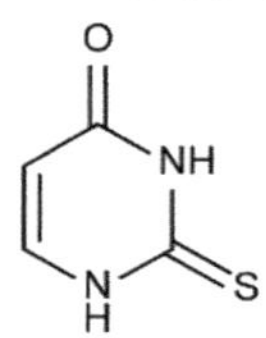				310 ps	
[29]	2-tUra	~100%	0.35 ± 0.06 ps	0.75 ± 0.20		Phosphate-buffered solution (pH 7.4)
	2-tUra		0.34 ± 0.09 ps	0.75 ± 0.20		Acetonitrile
	2-tThy		0.62 ± 0.07 ps			Phosphate-buffered solution (pH 7.4)
	2-tThy		0.32 ± 0.09 ps			Acetonitrile

[32]	2-tUra	360 ± 30 fs			Aqueous phosphate-buffered saline solution (pH 7.4)
	4-tUra 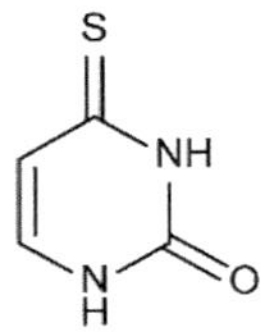	240 ± 20 fs	0.90 ± 0.15		
	2,4-Dithiouracil (2,4-dtUra) 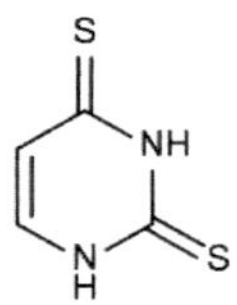	220 ± 40 fs	0.90 ± 0.15		
[33]	2-tUra	775 fs		~203 ps	Gas phase
[35]	2-tUra	207 fs ($\lambda_{ex} = 249$ nm)		48.1 ps ($\lambda_{ex} = 249$ nm)	Gas phase
		285 fs ($\lambda_{ex} = 259.9$ nm)		85.6 ps ($\lambda_{ex} = 259.9$ nm)	
		333 fs ($\lambda_{ex} = 271$ nm)		109 ps ($\lambda_{ex} = 271$ nm)	
		544 fs ($\lambda_{ex} = 283$ nm)		142 ps ($\lambda_{ex} = 283$ nm)	

(Continued)

Table 2.2 (Continued)

References		Φ_{ISC}	τ_{ISC}	Φ_T	τ_T	Experimental conditions
			775 fs ($\lambda_{ex} = 291.5$ nm)		203 ps ($\lambda_{ex} = 291.5$ nm)	
					73 ± 10 ns ($\lambda_{ex} = 268$ nm)	Acetonitrile
					69 ± 4 ns ($\lambda_{ex} = 290$ nm)	
					296 ± 23 ns ($\lambda_{ex} = 316$ nm)	
[34]	4-tUra		1794 ± 150 fs ($\lambda_{ex} = 309.02$ nm)		>1.5 ns	Gas phase
			2660 ± 250 fs ($\lambda_{ex} = 330.16$ nm)			
			3317 ± 300 fs ($\lambda_{ex} = 347.20$ nm)			
[37]	2,4-Dithiouracil (2,4-dtUra)		245 ± 20 fs ($\lambda_{ex} = 250$ nm)		0.33 ns	Gas phase
			240 ± 20 fs ($\lambda_{ex} = 267$ nm)		0.54 ns	
			350 ± 20 fs ($\lambda_{ex} = 287$ nm)		>1 ns	
			277 ± 20 fs ($\lambda_{ex} = 310$ nm)		>1.5 ns	
			420 ± 20 fs ($\lambda_{ex} = 329$ nm)		>2 ns	

		484 ± 50 fs (λ_{ex} = 344 nm) 626 ± 50 fs (λ_{ex} = 361 nm) 861 ± 50 fs (λ_{ex} = 376 nm)			
[36]	2,4-dtUra	484 ± 50 fs (λ_{ex} = 344.3 nm) 626 ± 50 fs (λ_{ex} = 361.4 nm) 861 ± 50 fs (λ_{ex} = 376 nm)		>2 ns	Gas phase
[39]	2-Thiocytosine (2-tCyt)	210 ± 50 fs	~100%		Phosphate-buffered solution (pH 7.4)
[40]	2-tCyt 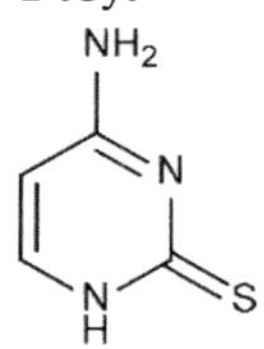 			7.1 ± 1.4 ns	Aqueous solution
	β-Anomers of 2-thiocytidine (β-2-tCyd)			1.5 ± 0.2 ns	

(Continued)

Table 2.2 (Continued)

References		Φ_{ISC}	τ_{ISC}	Φ_T	τ_T	Experimental conditions
[41]	2-tCyt				6.6 ns	Phosphate-buffered solution (pH 7.0)
	2-tCyd 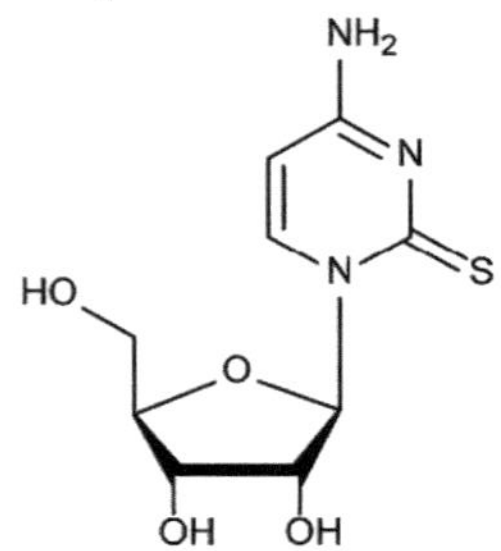				1.5 ns	
	Protonated form of 2-Thiocytosine (2-tCytH$^+$) 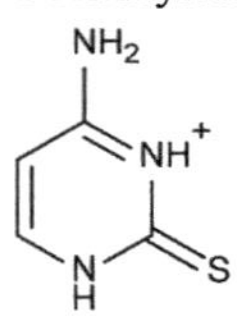				3.3 ns	Phosphate-buffered solution (pH 2.0)

	Protonated form of 2-tCydH$^+$ 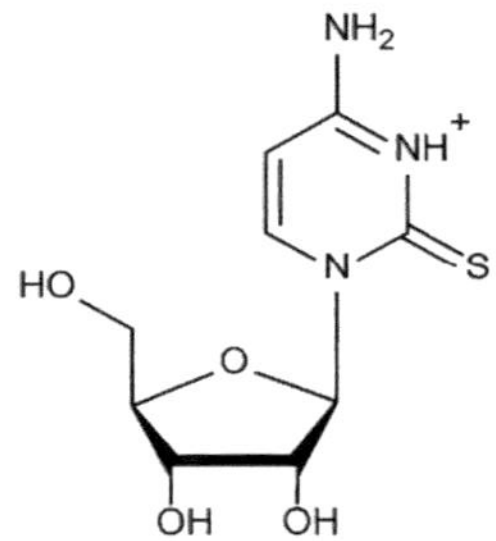			3.1 ns	
[44]	6-Thioguanine (6-tGua) 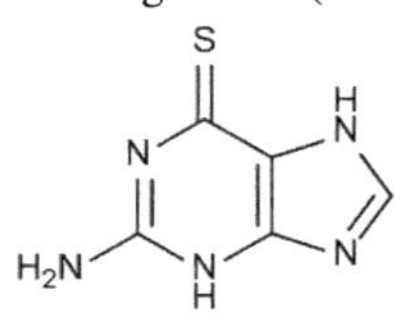	522 ± 48 fs			Phosphate-buffered solution (pH 7.4)
[45]	6-Methylthioguanine (6-Me-tGua) 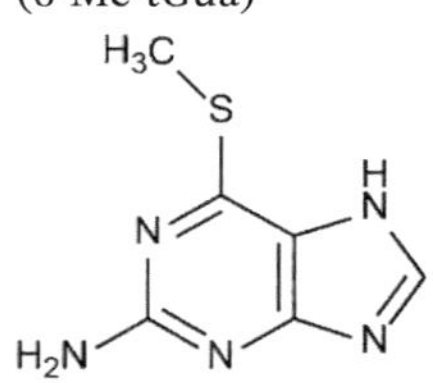	2.91 ± 0.14 ns			Phosphate-buffered solution (pH 7.0)
	6-Me-tGua			7.8 ± 0.4 μs	Deaerated water
	6-tGua		0.8 ± 0.2		Deaerated acetonitrile
	6-Me-tGua		0.6 ± 0.1		

(Continued)

Table 2.2 (Continued)

References		Φ_{ISC}	τ_{ISC}	Φ_T	τ_T	Experimental conditions
[46]	2,6-Dithiopurine (2,6-dtPur) 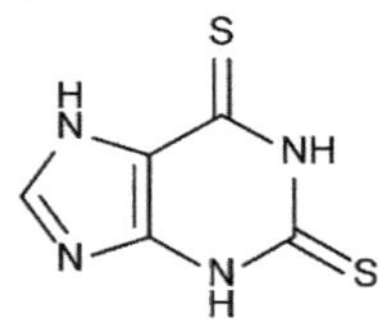			1.0 ± 0.1	0.56 ± 0.02 μs (air-saturated conditions, λ_{ex} = 325 nm) 4.0 ± 0.1 μs (nitrogen-saturated conditions, λ_{ex} = 342 nm)	Phosphate-buffered solution (pH 7.4)
[47]	6-Selenoguanine (6-SeGua) 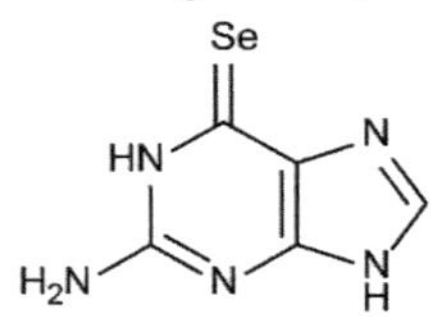		0.13 ± 0.05 ps		1.7 ± 0.1 ns (estimated by extrapolating fit past 1 ns)	Deaerated phosphate-buffered solution (pH 7.4)
	6-tGua		0.35 ± 0.04 ps			
[48]	Neutral 2-SeUra 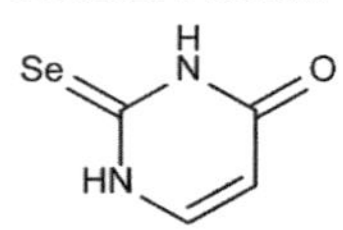			92% ± 5%	0.21 ± 0.01 ns (λ_{ex} = 305 nm)	Phosphate-buffered solution (pH 5.4)
	Deprotonated form of 2-selenuracil (de-2-SeUra II) 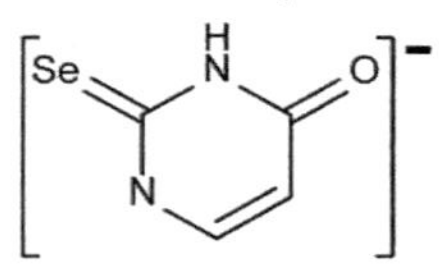				1.58 ± 0.01 ns (λ_{ex} = 340 nm)	Phosphate-buffered solution (pH 9)

2.3 Study on Triplet Excited State Dynamics of Nitrogen-Substituted Nucleobases

Nitrogen-substituted nucleobases are formed by substituting nitrogen atoms for carbon atoms at specific positions in the structure of standard nucleobases. For pyrimidine bases, the aza-substitution typically occurs at the fifth or sixth carbon atom of the pyrimidine ring, resulting in aza-pyrimidine bases such as 5-azacytosine (5-aCyt), 6-azauracil (6-aUra), and 6-azathymine (6-aThy). For purine bases, such substitution usually takes place at the 8th carbon atom of the purine ring, exemplified by 8-azaadenine (8-aAde) and 8-azaguanine (8-aGua).

The structural modification endows aza-nucleobases with more distinctive photophysical and photochemical properties, granting them significant application prospects in fields such as medicine. For instance, some aza-nucleobases have been developed into anticancer and antiviral drugs [49–51] and demonstrate efficacy in treating diverse pathologies [52, 53]; others can be combined with therapeutic approaches like chemotherapy and photodynamic therapy [54]. Studies on the excited state dynamics of aza-nucleobases not only help clarify the impact of nitrogen substitution on the excited state dynamics of standard nucleobases but also provide researchers in other fields with photophysical and photochemical information related to aza-nucleobases. Among them, the long-lived triplet states of aza-nucleobases are the key targets in excited state dynamics research, and the following is a summary of studies on the triplet states of different aza-nucleobases.

2.3.1 Nitrogen-Substituted Pyrimidine Bases

2.3.1.1 Azathymine and Its Derivatives

Suzuki research group investigated the excited state dynamics of 6-aza-2-thiothymine using nanosecond transient absorption and time-resolved phosphorescence detection (TRPD) techniques. They found that the excited singlet state of 6-aza-2-thiothymine primarily relaxes to the triplet state via ISC, with an ISC yield approaching 100% [55], indicating that ISC is the main relaxation pathway of the initially excited singlet state. Through time-dependent density functional theory (TD-DFT) calculations, they predicted that the energy gap between the two lowest singlet states, S_1 ($n\pi^*$) and S_2 ($\pi\pi^*$), of 6-aza-2-thiothymine is sufficiently small, and the two triplet states, $^3n\pi^*$ and $^3\pi\pi^*$, are close in energy and slightly lower than the S_1 ($n\pi^*$) state. The small energy gaps between the two lowest singlet states and between the two triplet states may induce vibrational coupling between states of the same spin multiplicity, causing mixing of $n\pi^*$ and $\pi\pi^*$ states and thereby enhancing spin–orbit coupling between the lowest singlet and triplet states.

Subsequently, Gobbo and Borin focused on the population mechanism of the lowest triplet state during the excited state relaxation of 6-aza-2-thiothymine using CASPT2//CASSCF calculations [56]. They found two main relaxation pathways along the minimum energy path of the S_2 $^1(\pi\pi^*)$ state: the first pathway involves excited molecules relaxing to the S_1 $^1(n\pi^*)$ state via ultrafast internal conversion at the PES crossing point between S_2 $^1(\pi\pi^*)$ and S_1 $^1(n\pi^*)$, then reaching the minimum of S_1 (S_1-min) followed by ISC to the T_1 $^3(\pi\pi^*)$ state; the second pathway involves direct ISC of molecules populated in the S_2 $^1(\pi\pi^*)$ state to the T_3 $^3(n\pi^*)$ state, followed by internal conversion of molecules in the T_3 $^3(n\pi^*)$ state to the T_1 $^3(\pi\pi^*)$ state. The calculation results by Gobbo et al. are consistent with those reported by the Suzuki group.

Chen group was the first to experimentally observe the $(n\pi^*)$ triplet state of aza-nucleobases. They explored the excited state relaxation pathways of 6-aThy and 6-azathymidine (6-aThd) using transient absorption spectroscopy and TD-DFT calculations, among other methods. They found that both compounds exhibit high triplet state generation quantum yields, which are 46% and 43%, respectively [57], and their excited state relaxation mechanisms differ. Upon excitation to the S_2 $(\pi\pi^*)$ state, 6-AThd relaxes to the S_1 $(n\pi^*)$ state via internal conversion, then to the T_1 $(\pi\pi^*)$ state via ISC. In contrast, 6-AThy has two relaxation pathways, one identical to that of 6-AThd, and the other involving direct ISC from the S_2 $(\pi\pi^*)$ state to the T_2 $(n\pi^*)$ state, followed by internal conversion to the T_1 $(\pi\pi^*)$ state.

2.3.1.2 Azauracil and Its Derivatives

Kobayashi et al. investigated the excited state properties of 6-aUra in acetonitrile and found that its ISC quantum yield and fluorescence emission can be regulated by the excitation wavelength [58]. When excited at 248 nm, 6-aUra exhibits an ISC quantum yield close to 100% with no significant fluorescence emission; in contrast, when excited at 308 nm, its ISC quantum yield decreases to approximately 93% accompanied by weak fluorescence emission. They proposed that the acceleration of the ISC process and the inhibition of ultrafast internal conversion are two possible mechanisms leading to the high ISC quantum yield.

Etinski and Marian further proposed through theoretical calculations that, in acetonitrile solution, the blue shift of the S_1 state PES of 6-aUra increases the adiabatic energy gap between S_1 and T_1, thereby accelerating ISC between S_1 and T_1 [59]. They also proposed the relaxation mechanism of 6-aUra via calculations: after being excited to the S_2 $(^1\pi\pi^*)$ state, 6-aUra undergoes ultrafast transfer to the S_1 $(^1n\pi^*)$ state, followed by ISC to the T_1 $(^3\pi\pi^*)$ state. The calculation results indicate that the triplet state is generated within approximately 30 ps in acetonitrile.

To explore the dependence of the triplet quantum yield of 6-aUra on excitation wavelength, Gobbo et al. performed theoretical calculations and found that different excitation wavelengths induce the molecule to adopt different relaxation pathways [60]. When excited at 248 nm, the molecule is excited to the S_2 $^1(\pi\pi^*)$ state, followed by two possible relaxation pathways: one is ISC to the T_2 $^3(n\pi^*)$ state, then reaching the T_1 $^3(\pi\pi^*)$ state via a CI; the other is relaxation from the S_2 $^1(\pi\pi^*)$ state to the S_1 $^1(n\pi^*)$ state through a CI, then ISC from the lowest energy point of the S_1 $^1(n\pi^*)$ state to the T_1 $^3(\pi\pi^*)$ state. When excited at 308 nm, most GS molecules are directly excited to the S_1 $^1(n\pi^*)$ state and undergo ISC to the T_1 $^3(\pi\pi^*)$ state from its lowest energy point, with only a small number of molecules returning to the GS via fluorescence emission.

Similarly, Kobayashi et al. studied 6-aUra and 6-azauridine (6-aUrd) in acetonitrile and found that 6-aUrd in acetonitrile also exhibits a triplet yield close to 100% when excited at 248 nm [61]. Moreover, compared with 6-aUra, the triplet quantum yield of 6-aUrd shows a stronger dependence on the excitation wavelength. They suggested that the presence of ribose in 6-aUrd significantly affects its relaxation dynamics on the S_1 $(n\pi^*)$ state PES.

Building on previous studies, Hua et al. investigated the excited state dynamics of 6-aUra using full-spectrum femtosecond transient absorption spectroscopy covering the deep UV to visible range [62] and reported the main relaxation pathway after excitation at 264 nm. 6-aUra is excited to the S_2 state, then decays to the S_1 state via internal conversion, followed by transition to the T_1 state from the S_1 state through picosecond-scale ISC. Hua et al. also found that the relaxation rate from S_1 to T_1 is obviously dependent on solvent polarity—the higher the solvent polarity, the faster the ISC rate. Subsequently, they proposed that 6-aUra in water, when excited by 266 nm pulses, it first populates the $^1\pi\pi^*$ state, then mainly relaxes to the $^1n\pi^*$ state, and further relaxes to the $^3\pi\pi^*$ state via ISC [63]. The $^1\pi\pi^*$ to GS (S_0) relaxation pathway is negligible.

2.3.1.3 Azacytosine

Kobayashi et al. also conducted experimental studies on another aza-pyrimidine base, 5-aCyt, and found that it has an extremely low triplet quantum yield [64]. Giussani et al. proposed through theoretical calculations that after 5-aCyt is excited to the $^1(\pi\pi^*)$ state, it can relax to the minimum energy of the $^1(\pi\pi^*)$ state ($^1(\pi\pi^*)$min) and then return to the GS via a CI-$(^1\pi\pi^*/gs)_{CI-\alpha}$; alternatively, it can decay to the minimum energy of the $^1(n_N\pi^*)$ state ($^1(n_N\pi^*)_{min}$) via the ($^1\pi\pi^*/^1n_N\pi^*$) CI, followed by relaxation to the GS through another CI-$(^1\pi\pi^*/gs)_{CI-\beta}$ [65]. They consider these two pathways as the most efficient nonradiative relaxation routes.

Borin et al. investigated the gas-phase relaxation dynamics of 5-aCyt using molecular dynamics calculations [66], revealing that it undergoes internal conversion from the $^1\pi\pi^*$ state to the $^1n_N\pi^*$ state and then relaxes back to the GS from the $^1n_N\pi^*$ state on a timescale slightly longer than 1 ps. Although the $^1n_N\pi^*$ state can populate the triplet state via ISC, the yield is only approximately 10%, indicating that ISC cannot compete with the rapid internal conversion process to the GS.

Previous studies have laid the foundation for exploring the photophysical properties of 5-aCyt, but there is a lack of femtosecond time-resolved experimental research on it. Chen group used femtosecond transient absorption spectroscopy to propose the relaxation pathways of 5-aCyt upon excitation at 266 nm. First, it can return directly to the GS from the S_2 ($\pi\pi^*$) state via internal conversion within 15 ps. Meanwhile, it can also relax to the S_1 ($n\pi^*$) state through a CI within 1.5 ps, followed by further relaxation back to the GS [67]. Subsequently, Chen research group conducted further studies on the excited state dynamics of 5-aCyt in solution using a broader range of transient absorption spectroscopy and a longer detection time window, and for the first time presented the excited state relaxation mechanisms of 5-azacytidine and 2′-deoxy-5-azacytidine in solution [68]. In this study, they found that after 5-aCyt is excited to the $\pi\pi^*$ state, it reaches the hot dark $n\pi^*$ state via ultrafast internal conversion within the instrument response time (approximately 120 fs). Then, it relaxes from the hot $n\pi^*$ state (S_1) to its minimum within a timescale of 1.3 ps, while the actual lifetime of the $n\pi^*$ state (S_1) is 17 ps. After 5-azacytidine is excited to the $\pi\pi^*$ state, its excited state population decays back to the GS through two distinct channels (the $\pi\pi^*$ state and the $n\pi^*$ state) with lifetimes of 0.86 and 5.1 ps, respectively. After 2′-deoxy-5-azacytidine is excited to the $\pi\pi^*$ state, almost all of its excited state population decays rapidly back to the GS through the $\pi\pi^*$ state. For 5-aCyt, no ISC process was detected as proposed by Borin et al. (Figure 2.5).

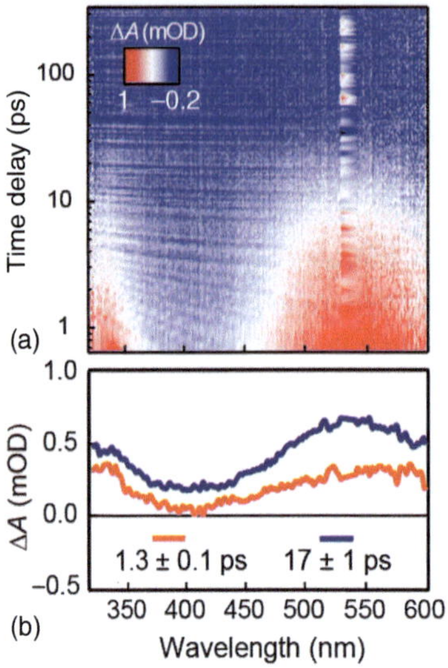

Figure 2.5 (a) Femtosecond broadband transient absorption spectra and (b) decay-associated difference spectra (DADS) of 5-azacytosine (5-aCyt) under 267 nm excitation in PBS (pH 7.4) ([68] / John Wiley & Sons.)

To summarized, experimental and theoretical studies have shown that 6-aza-2-thiothymine, 6-aThy, 6-aThd, and 6-aUra can all populate the $^3\pi\pi^*$ triplet state via the relaxation pathway: $^1\pi\pi^*$ state → $^1n\pi^*$ state → $^3\pi\pi^*$ state [56, 57, 59, 60, 63]. Among them, 6-aThy [57] and 6-aUra [60] can also populate the $^3\pi\pi^*$ triplet state through the pathway: $^1\pi\pi^*$ state → $^3n\pi^*$ state → $^3\pi\pi^*$ state. 5-aCyt has an extremely low triplet quantum yield [64, 67], although theoretical calculations suggest that it can populate the triplet state from the $^1n\pi^*$ state via ISC [66].

2.3.2 Nitrogen-Substituted Purine Bases

Kobayashi et al. measured the transient absorption spectrum using laser flash photolysis and for the first time observed the triplet–triplet absorption peak of 8-aAde at 455 nm, with a singlet oxygen quantum yield of approximately 15%[64]. However, for 8-aGua, they did not observe the presence of triplet states or the formation of sensitized singlet oxygen. They classified 8-aAde as a Type A molecule with a high ISC quantum yield, and 8-aGua as a Type B molecule with a negligible ISC quantum yield. Through theoretical calculations, they proposed that there exists a dark $^1n\pi^*$ state in the $^1\pi\pi^*$ state of Type A molecules, which plays an important role in the ISC process of aza analogs of nucleobases.

Gobbo and Borin proposed a comprehensive excited state relaxation mechanism of 8-aAde using the CASPT2//CASSCF/ANO-L calculation method [69]. They suggested that after 8-aAde is excited to the S_2 ($\pi\pi^*$) state, there are two main relaxation pathways: first, the S_2 ($\pi\pi^*$) state nonradiatively relaxes to its own lowest energy point, followed by direct ISC to the T_2 ($n\pi^*$) state, and then reaches the T_1 ($\pi\pi^*$) state through the CI between the T_2 ($n\pi^*$) and T_1 ($\pi\pi^*$) states; second, the excited molecule rapidly relaxes from the S_2 ($\pi\pi^*$) state to the S_1 ($n\pi^*$) state and, after reaching the lowest energy point of the S_1 ($n\pi^*$) state, populates the T_1 ($\pi\pi^*$) state via ISC. Both pathways can effectively generate triplet states.

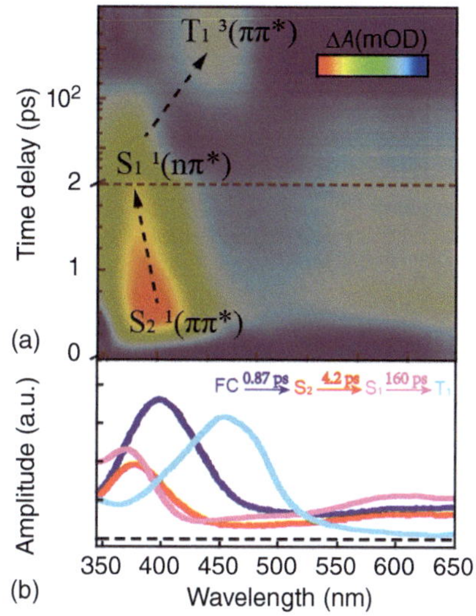

Figure 2.6 (a) Femtosecond broadband transient absorption spectra and (b) species associated spectra (SAS) of 8-azaadenine (8-aAde) under 266-nm excitation in pH = 4 buffer solutions ([70] / John Wiley & Sons).

Previous experimental studies on the excited state dynamics of 8-aAde and 8-aGua still need further improvement. Chen research group conducted detailed studies on the excited state dynamics of 8-aAde in buffer solutions of different pH values using femtosecond transient absorption spectroscopy and femtosecond fluorescence up-conversion spectroscopy [70]. The results showed that in pH = 4 buffer, the neutral form of 8-aAde, after being excited to the S_2 ($\pi\pi^*$) state by 266 nm light, first relaxes to the minimum energy point of the S_2 ($\pi\pi^*$) state and then can either relax back to the GS via internal conversion or populate the S_1 ($n\pi^*$) state. Once it relaxes to the S_1 ($n\pi^*$) state, it can reach the triplet T_1 ($\pi\pi^*$) state through ISC. In pH = 10 buffer, the deprotonated form of 8-aAde excited to the S_2 ($\pi\pi^*$) state by 266 nm radiation undergoes relaxation to the S_2 ($\pi\pi^*$) potential energy minimum, followed by quantitatively dominant GS recovery via internal conversion (Figure 2.6).

Moreover, Chen and co-workers also studied the excited state relaxation pathways of neutral 9H-8-azaguanine (neutral 9H-8-aGua) and neutral 8H-8-azaguanine (neutral 8H-8-aGua) [71]. After being excited to the $S_1(\pi\pi^*\ L_a)$ state, neutral 9H-8-AG can either relax to the $S_1(\pi\pi^*\ L_a)^{MEP}{}_{min}$ state, then relax barrierlessly from the $S_1(\pi\pi^*\ L_a)^{MEP}{}_{min}$ state to the CI with the GS and finally return to the GS, or relax to the $S_1(\pi\pi^*\ L_a)_{min}$ state, then overcome the energy barrier to relax to the CI with the GS and eventually return to the GS.

For 8-aAde, both experimental [70] and theoretical calculation [69] studies have shown that it can populate the $^3\pi\pi^*$ triplet state via the relaxation pathway: $^1\pi\pi^*$ state → $^1n\pi^*$ state → $^3\pi\pi^*$ state. In addition, experiments have found that the number of particles populating the $^3\pi\pi^*$ triplet state is related to the pH value of the solvent in which 8-aAde is located [70].

2.3.3 Summary

Overall, these studies have deepened the understanding of the excited state dynamics of aza-nucleobases and laid the foundation for their applications (Table 2.3).

Table 2.3 Triplet state properties of nitrogen-substituted nucleobases.

References		Φ_{ISC}	τ_{ISC}	Φ_T	τ_T	Experimental conditions
[55]	6-Aza-2-thiothymine (6-a-2-tThy) 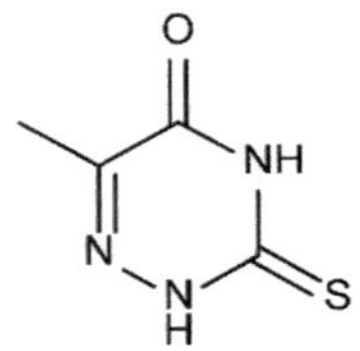	1.00 ± 0.02				Ar-saturated acetonitrile
[58]	6-Azauracil (6-aUra) 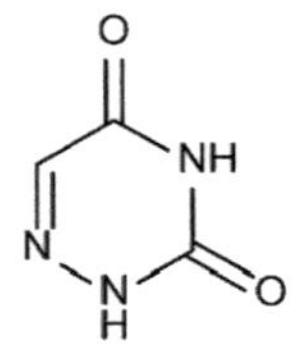	1.00 ± 0.10 ($\lambda_{ex} = 248$ nm) 0.93 ± 0.04 ($\lambda_{ex} = 308$ nm)				Ar-saturated acetonitrile
[61]	6-Azauridine (6-aUrd) 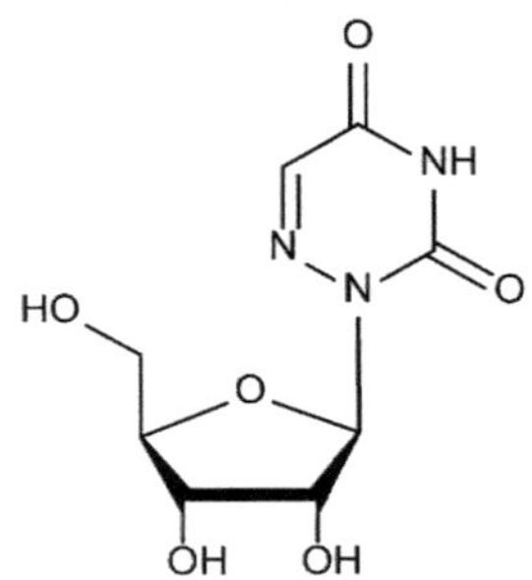	1.00 ± 0.07 ($\lambda_{ex} = 248$ nm) 0.78 ± 0.05 ($\lambda_{ex} = 308$ nm)			130 ± 10 ns	Ar-saturated acetonitrile

[62]	6-Azauracil (6-aUra)		6.9 ± 0.1 ps			N-butanol
			6.1 ± 0.1 ps			Ethanol
			5.2 ± 0.1 ps			Methanol
			5.2 ± 0.1 ps			Acetonitrile
			4.2 ± 0.1 ps			Water
[63]	6-aUra		~4.0 ps			Water
[64]	8-Azaadenine (8-aAde) 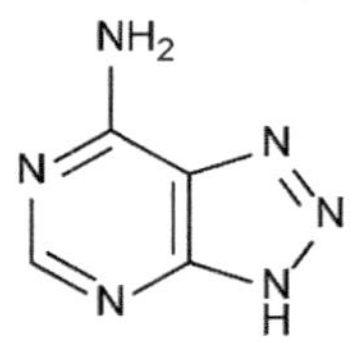				240 ± 5 ns (Triplet lifetime [1/k0])	Ar-saturated acetonitrile
[66]	5-Azacytosine (5-aCyt) 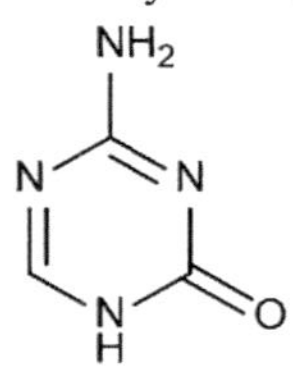	10 ± 8%	9000^{+10000}_{-5000} fs			Gas phase
[55]	6-a-2-tThy			1.00 ± 0.02		Ar-saturated acetonitrile

(Continued)

Table 2.3 (Continued)

References		Φ_{ISC}	τ_{ISC}	Φ_T	τ_T	Experimental conditions
[57]	6-Azathymine (6-aThy) 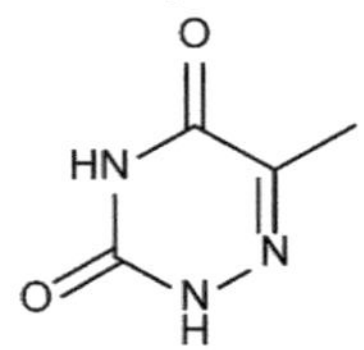				$3.2 \pm 0.4\,\mu s$	N_2-saturated acetonitrile
	6-Azathymidine (6-aThd) 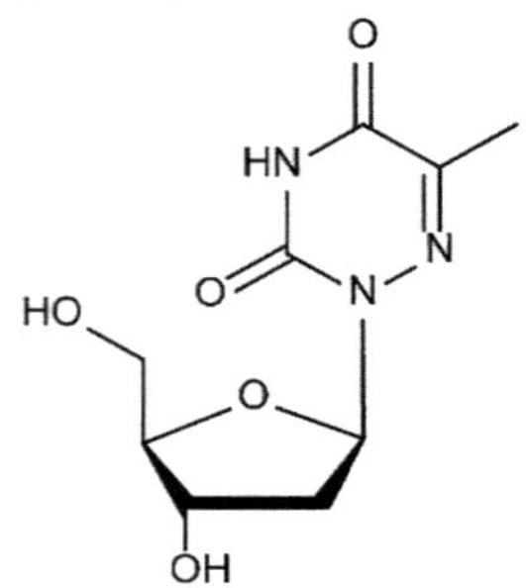				$2.7 \pm 0.5\,\mu s$	
	6-aThy			46%		Acetonitrile
	6-aThd		7.30 ps	43%		
[70]	8-Azaadenine (8-aAde)		160 ps		$6.1\,\mu s$	Buffer solutions (pH 4.0)

Among them, Chen research group has promoted the development of this field by applying advanced experimental techniques such as femtosecond transient absorption spectroscopy and femtosecond fluorescence up-conversion spectroscopy to the study of excited state dynamics of aza-nucleobases.

References

1 Lamola, A.A. and Mittal, J.P. (1966). Solution photochemistry of thymine and uracil. *Science* 154: 1560–1561. https://doi.org/10.1126/science.154.3756.1560.

2 Lamola, A.A. (1968). Excited state precursors of thymine photodimers. *Photochemistry and Photobiology* 7: 619–632. https://doi.org/10.1111/j.1751-1097.1968.tb08044.x.

3 Johns, H.E. and Whillans, D.W. (1971). Properties of the triplet states of thymine and uracil in aqueous solution. *Journal of the American Chemical Society* 93: 1358–1362. https://doi.org/10.1021/ja00735a007.

4 Szabo, A.G., Rlddell, W.D., and Yip, R.W. (1970). Detection and chemistry of the triplet state of thymine. *Canadian Journal of Chemistry* 48: 694–696. https://doi.org/10.1139/v70-114.

5 Salet, C. and Bensasson, R. (1975). Studies on thymine and uracil triplet excited state in acetonitrile and water. *Photochemistry and Photobiology* 22: 231–235. https://doi.org/10.1111/j.1751-1097.1975.tb06741.x.

6 Salet, C., Bensasson, R., and Becker, R.S. (1979). Triplet excited states of pyrimidine nucleosides and nucleotides. *Photochemistry and Photobiology* 30: 325–329. https://doi.org/10.1111/j.1751-1097.1979.tb07363.x.

7 Song, Q., Lin, W., Yao, S., and Lin, N. (1998). Comparative studies of triplet states of thymine components by acetone sensitization and direct excitation in aqueous solution at room temperature. *Journal of Photochemistry and Photobiology A: Chemistry* 114: 181–184. https://doi.org/10.1016/S1010-6030(98)00219-6.

8 Kemp, T.J., Parker, A.W., and Wardman, P. (1985). Electron-transfer quenching of triplet state thymine and uracil. *Journal of the Chemical Society, Chemical Communications* 20: 1377–1378. https://doi.org/10.1039/C39850001377.

9 Zuo, Z., Yao, S., Luo, J. et al. (1992). Laser photolysis of cytosine, cytidine and dCMP in aqueous solution. *Journal of Photochemistry and Photobiology B: Biology* 15: 215–222. https://doi.org/10.1016/1011-1344(92)85125-E.

10 Gut, I.G., Wood, P.D., and Redmond, R.W. (1996). Interaction of triplet photosensitizers with nucleotides and DNA in aqueous solution at room temperature. *Journal of the American Chemical Society* 118: 2366–2373. https://doi.org/10.1021/ja9519344.

11 Wood, P.D. and Redmond, R.W. (1996). Triplet state interactions between nucleic acid bases in solution at room temperature: intermolecular energy and electron transfer. *Journal of the American Chemical Society* 118: 4256–4263. https://doi.org/10.1021/ja954340+.

12 Marguet, S. and Markovitsi, D. (2005). Time-resolved study of thymine dimer formation. *Journal of the American Chemical Society* 127: 5780–5781. https://doi.org/10.1021/ja050648h.

13 Hare, P.M., Crespo-Hernández, C.E., and Kohler, B. (2007). Internal conversion to the electronic ground state occurs via two distinct pathways for pyrimidine bases in aqueous solution. *Proceedings of the National Academy of Sciences* 104: 435–440. https://doi.org/10.1073/pnas.0608055104.

14 Hare, P.M., Middleton, C.T., Mertel, K.I. et al. (2008). Time-resolved infrared spectroscopy of the lowest triplet state of thymine and thymidine. *Chemical Physics* 347: 383–392. https://doi.org/10.1016/j.chemphys.2007.10.035.

15 Kwok, W., Ma, C., and Phillips, D.L. (2008). A doorway state leads to photostability or triplet photodamage in thymine DNA. *Journal of the American Chemical Society* 130: 5131–5139. https://doi.org/10.1021/ja077831q.

16 Pilles, B.M., Maerz, B., Chen, J. et al. (2018). Decay pathways of thymine revisited. *The Journal of Physical Chemistry A* 122: 4819–4828. https://doi.org/10.1021/acs.jpca.8b02050.

17 Zhang, K., Wang, F., Jiang, Y. et al. (2021). New insights about the photostability of DNA/RNA bases: triplet $n\pi^*$ state leads to effective intersystem crossing in pyrimidinones. *The Journal of Physical Chemistry B* 125: 2042–2049. https://doi.org/10.1021/acs.jpcb.0c10611.

18 Banyasz, A., Douki, T., Improta, R. et al. (2012). Electronic excited states responsible for dimer formation upon UV absorption directly by thymine strands: joint experimental and theoretical study. *Journal of the American Chemical Society* 134: 14834–14845. https://doi.org/10.1021/ja304069f.

19 Wu, P., Wang, X., Pan, H., and Chen, J. (2023). Wavelength dependent excited state dynamics observed in canonical pyrimidine nucleosides. *Journal of Photochemistry and Photobiology* 18: 100211 (100211–100219). https://doi.org/10.1016/j.jpap.2023.100211.

20 Brown, I. and Johns, H. (1968). Photochemistry of uracil. Intersystem crossing and dimerization in aqueous solution. *Photochemistry and Photobiology* 8: 273–286. https://doi.org/10.1111/j.1751-1097.1968.tb05871.x.

21 Brister, M.M. and Crespo-Hernández, C.E. (2019). Excited-state dynamics in the RNA nucleotide uridine 5′-monophosphate investigated using femtosecond broadband transient absorption spectroscopy. *The Journal of Physical Chemistry Letters* 10: 2156–2161. https://doi.org/10.1021/acs.jpclett.9b00492.

22 Wang, X., Zhou, Z., Tang, Y. et al. (2018). Excited state decay pathways of 2′-deoxy-5-methylcytidine and deoxycytidine revisited in solution: a comprehensive kinetic study by femtosecond transient absorption. *The Journal of Physical Chemistry B* 122: 7027–7037. https://doi.org/10.1021/acs.jpcb.8b00927.

23 Wu, P., Wang, X., Pan, H., and Chen, J. (2022). Direct observation of excitation wavelength-dependent ultrafast intersystem crossing in cytosine nucleoside solution. *The Journal of Physical Chemistry B* 126: 7975–7980. https://doi.org/10.1021/acs.jpcb.2c05865.

24 Harada, Y., Suzuki, T., Ichimura, T., and Xu, Y. (2007). Triplet formation of 4-thiothymidine and its photosensitization to oxygen studied by time-resolved thermal lensing technique. *The Journal of Physical Chemistry B* 111: 5518–5524. https://doi.org/10.1021/jp0678094.

25 Harada, Y., Okabe, C., Kobayashi, T. et al. (2010). Ultrafast intersystem crossing of 4-thiothymidine in aqueous solution. *The Journal of Physical Chemistry Letters* 1: 480–484. https://doi.org/10.1021/jz900276x.

26 Reichardt, C. and Crespo-Hernández, C.E. (2010). Room-temperature phosphorescence of the DNA monomer analogue 4-thiothymidine in aqueous solutions after UVA excitation. *The Journal of Physical Chemistry Letters* 1: 2239–2243. https://doi.org/10.1021/jz100729w.

27 Martínez-Fernández, L., Granucci, G., Pollum, M. et al. (2017). Decoding the molecular basis for the population mechanism of the triplet phototoxic precursors in UVA light-activated pyrimidine anticancer drugs. *Chemistry–A European Journal* 23: 2619–2627. https://doi.org/10.1002/chem.201604543.

28 Taras-Goślińska, K., Burdziński, G., and Wenska, G. (2014). Relaxation of the T1 excited state of 2-thiothymine, its riboside and deoxyriboside-enhanced nonradiative decay rate induced by sugar substituent. *Journal of Photochemistry and Photobiology A: Chemistry* 275: 89–95. https://doi.org/10.1016/j.jphotochem.2013.11.003.

29 Pollum, M. and Crespo-Hernández, C.E. (2014). Communication: the dark singlet state as a doorway state in the ultrafast and efficient intersystem crossing dynamics in 2-thiothymine and 2-thiouracil. *The Journal of Chemical Physics* 140: https://doi.org/10.1063/1.4866447.

30 Pollum, M., Jockusch, S., and Crespo-Hernández, C.E. (2014). 2,4-Dithiothymine as a potent UVA chemotherapeutic agent. *Journal of the American Chemical Society* 136: 17930–17933. https://doi.org/10.1021/ja510611j.

31 Ullrich, S., Qu, Y., Mohamadzade, A., and Shrestha, S. (2022). The effect of methylation on the triplet-state dynamics of 2-thiouracil: time-resolved photoelectron spectroscopy of 2-thiothymine. *The Journal of Physical Chemistry A* 126: 8211–8217. https://doi.org/10.1021/acs.jpca.2c06051.

32 Pollum, M., Jockusch, S., and Crespo-Hernández, C.E. (2015). Increase in the photoreactivity of uracil derivatives by doubling thionation. *Physical Chemistry Chemical Physics* 17: 27851–27861. https://doi.org/10.1039/C5CP04822B.

33 Yu, H., Sanchez-Rodriguez, J.A., Pollum, M. et al. (2016). Internal conversion and intersystem crossing pathways in UV excited, isolated uracils and their implications in prebiotic chemistry. *Physical Chemistry Chemical Physics* 18: 20168–20176. https://doi.org/10.1039/C6CP01790H.

34 Mohamadzade, A., Bai, S., Barbatti, M., and Ullrich, S. (2018). Intersystem crossing dynamics in singly substituted thiouracil studied by time-resolved photoelectron spectroscopy: micro-environmental effects due to sulfur position. *Chemical Physics* 515: 572–579. https://doi.org/10.1016/j.chemphys.2018.08.011.

35 Sánchez-Rodríguez, J.A., Mohamadzade, A., Mai, S. et al. (2017). 2-Thiouracil intersystem crossing photodynamics studied by wavelength-dependent photoelectron and transient absorption spectroscopies. *Physical Chemistry Chemical Physics* 19: 19756–19766. https://doi.org/10.1039/C7CP02258A.

36 Mohamadzade, A. and Ullrich, S. (2020). Internal conversion and intersystem crossing dynamics of uracil upon double thionation: a time-resolved photoelectron spectroscopy study in the gas phase. *Physical Chemistry Chemical Physics* 22: 15608–15615. https://doi.org/10.1039/D0CP02145H.

37 Mohamadzade, A., Nenov, A., Garavelli, M. et al. (2023). Double thionated pyrimidine nucleobases: molecular tools with tunable photoproperties. *Journal of the American Chemical Society* 145: 11945–11958. https://doi.org/10.1021/jacs.2c12061.

38 Koyama, D., Milner, M.J., and Orr-Ewing, A.J. (2017). Evidence for a double well in the first triplet excited state of 2-thiouracil. *The Journal of Physical Chemistry B* 121: 9274–9280. https://doi.org/10.1021/acs.jpcb.7b06917.

39 Mai, S., Pollum, M., Martínez-Fernández, L. et al. (2016). The origin of efficient triplet state population in sulfur-substituted nucleobases. *Nature Communications* 7: 13077. https://doi.org/10.1038/ncomms13077.

40 Janicki, M.J., Kufner, C.L., Todd, Z.R. et al. (2021). Ribose alters the photochemical properties of the nucleobase in thionated nucleosides. *The Journal of Physical Chemistry Letters* 12: 6707–6713. https://doi.org/10.1021/acs.jpclett.1c01384.

41 Yan, S., Guo, W., Jie, J. et al. (2024). Protonation weakens the influence of ribose on triplet decay of 2-thiocytidine. *The Journal of Physical Chemistry Letters* 15: 11839–11846. https://doi.org/10.1021/acs.jpclett.4c02575.

42 Pollum, M., Ortiz-Rodríguez, L.A., Jockusch, S., and Crespo-Hernández, C.E. (2016). The triplet state of 6-thio-2′-deoxyguanosine: intrinsic properties and reactivity toward molecular oxygen. *Photochemistry and Photobiology* 92: 286–292. https://doi.org/10.1111/php.12563.

43 Siouri, F.M., Boldissar, S., Berenbeim, J.A., and de Vries, M.S. (2017). Excited state dynamics of 6-thioguanine. *The Journal of Physical Chemistry A* 121: 5257–5266. https://doi.org/10.1021/acs.jpca.7b03036.

44 Teles-Ferreira, D.C., Manzoni, C., Martínez-Fernández, L. et al. (2022). Ultrafast excited-state decay mechanisms of 6-thioguanine followed by Sub-20 fs UV transient absorption spectroscopy. *Molecules* 27: 1200. https://doi.org/10.3390/molecules27041200.

45 Zhou, Q., Hao, Y., Jie, J. et al. (2024). Dual functionality of 6-methylthioguanine: synergistic effects enhancing the photolability of DNA nucleobases. *JACS Au* 4: 441–453. https://doi.org/10.1021/jacsau.3c00575.

46 Ortiz-Rodríguez, L.A., Hoehn, S.J., Acquah, C. et al. (2021). Femtosecond intersystem crossing to the reactive triplet state of the 2,6-dithiopurine skin cancer photosensitizer. *Physical Chemistry Chemical Physics* 23: 25048–25055. https://doi.org/10.1039/D1CP04415J.

47 Farrell, K.M., Brister, M.M., Pittelkow, M. et al. (2018). Heavy-atom-substituted nucleobases in photodynamic applications: substitution of sulfur with selenium in 6-thioguanine induces a remarkable increase in the rate of triplet decay in 6-selenoguanine. *Journal of the American Chemical Society* 140: 11214–11218. https://doi.org/10.1021/jacs.8b07665.

48 Jin, P., Wang, X., Pan, H., and Chen, J. (2022). One order of magnitude increase of triplet state lifetime observed in deprotonated form selenium substituted uracil. *Physical Chemistry Chemical Physics* 24: 875–882. https://doi.org/10.1039/D1CP04811B.

49 Škoda, J. (1963). Mechanism of action and application of azapyrimidines. *Progress in Nucleic Acid Research and Molecular Biology* 2: 197–219. https://doi.org/10.1016/S0079-6603(08)60313-X.

50 Montgomery, J. (1959). The relation of anticancer activity to chemical structure a review. *Cancer Research* 19: 447–463.

51 Spremulli, E.N., Crabtree, G.W., Dexter, D.L. et al. (1982). Biochemical pharmacology and toxicology of 8-azaadenosine alone and in combination with 2′-deoxycoformycin (pentostatin). *Biochemical Pharmacology* 31: 2415–2421. https://doi.org/10.1016/0006-2952(82)90538-X.

52 Broughton, B.J., Chaplen, P., Knowles, P. et al. (1974). New inhibitor of reagin-mediated anaphylaxis. *Nature* 251: 650–652. https://doi.org/10.1038/251650a0.

53 El-Brollosy, N.R. (2008). Synthesis and antimicrobial evaluation of 6-azauracil non-nucleosides. *Monatshefte für Chemie-Chemical Monthly* 139: 1483–1490. https://doi.org/10.1007/s00706-008-0948-7.

54 Shnider, B.I., Frei, E. III,, Tuohy, J.H. et al. (1960). Clinical studies of 6-azauracil. *Cancer Research* 20: 28–33.

55 Kuramochi, H., Kobayashi, T., Suzuki, T., and Ichimura, T. (2010). Excited-state dynamics of 6-aza-2-thiothymine and 2-thiothymine: highly efficient intersystem crossing and singlet oxygen photosensitization. *Journal of Physical Chemistry B* 114: 8782–8789. https://doi.org/10.1021/jp102067t.

56 Gobbo, J.P. and Borin, A.C. (2013). On the population of triplet excited states of 6-aza-2-thiothymine. *The Journal of Physical Chemistry A* 117: 5589–5596. https://doi.org/10.1021/jp403508v.

57 Fu, L., Wang, Z., Liu, Y. et al. (2020). Observation of triplet nπ* state in ultrafast intersystem crossing of 6-azathymine. *Journal of Photochemistry and Photobiology A: Chemistry* 396: 112491. https://doi.org/10.1016/j.jphotochem.2020.112491.

58 Kolayashi, T., Harada, Y., Suzuki, T., and Ichimura, T. (2008). Excited state characteristics of 6-azauracil in acetonitrile: drastically different relaxation mechanism from uracil. *Journal of Physical Chemistry A* 112: 13308–13315. https://doi.org/10.1021/jp803096j.

59 Etinski, M. and Marian, C.M. (2010). Overruling the energy gap law: fast triplet formation in 6-azauracil. *Physical Chemistry Chemical Physics* 12: 15665. https://doi.org/10.1039/c0cp00106f.

60 Gobbo, J.P., Borin, A.C., and Serrano-Andrés, L. (2011). On the relaxation mechanisms of 6-azauracil. *The Journal of Physical Chemistry B* 115: 6243–6251. https://doi.org/10.1021/jp200297z.

61 Kobayashi, T., Kuramochi, H., Suzuki, T., and Ichimura, T. (2010). Triplet formation of 6-azauridine and singlet oxygen sensitization with UV light irradiation. *Physical Chemistry Chemical Physics* 12: 5140. https://doi.org/10.1039/b921568a.

62 Hua, X., Hua, L., and Liu, X. (2015). Ultrafast excited-state dynamics of 6-azauracil studied by femtosecond transient absorption spectroscopy. *The Journal of Physical Chemistry A* 119: 12985–12989. https://doi.org/10.1021/acs.jpca.5b08975.

63 Hua, X., Hua, L., and Liu, X. (2016). The methyl- and aza-substituent effects on nonradiative decay mechanisms of uracil in water: a transient absorption study in the UV region. *Physical Chemistry Chemical Physics* 18: 13904–13911. https://doi.org/10.1039/c6cp00732e.

64 Kobayashi, T., Kuramochi, H., Harada, Y. et al. (2009). Intersystem crossing to excited triplet state of aza analogues of nucleic acid bases in acetonitrile. *Journal of Physical Chemistry A* 113: 12088–12093. https://doi.org/10.1021/jp905433s.

65 Giussani, A., Merchán, M., Gobbo, J.P., and Borin, A.C. (2014). Relaxation mechanisms of 5-azacytosine. *Journal of Chemical Theory and Computation* 10: 3915–3924. https://doi.org/10.1021/ct5003175.

66 Carlos Borin, A., Mai, S., Marquetand, P., and González, L. (2017). Ab initio molecular dynamics relaxation and intersystem crossing mechanisms of 5-azacytosine. *Physical Chemistry Chemical Physics* 19: 5888–5894. https://doi.org/10.1039/c6cp07919a.

67 Zhou, Z., Zhou, X., Wang, X. et al. (2017). Ultrafast excited-state dynamics of cytosine aza-derivative and analogues. *The Journal of Physical Chemistry A* 121: 2780–2789. https://doi.org/10.1021/acs.jpca.6b12290.

68 Chen, Z., Wang, X., Jia, M. et al. (2024). Ribose and deoxyribose group alter excited-state dynamics of 5-azacytosine in solution. *Photochemistry and Photobiology* 100: 291–297. https://doi.org/10.1111/php.13696.

69 Gobbo, J.P. and Borin, A.C. (2012). On the mechanisms of triplet excited state population in 8-azaadenine. *The Journal of Physical Chemistry B* 116: 14000–14007. https://doi.org/10.1021/jp3091599.

70 Zhou, Z., Hu, Z., Zhang, X. et al. (2019). pH controlled intersystem crossing and singlet oxygen generation of 8-azaadenine in aqueous solution. *ChemPhysChem* 20: 757–765. https://doi.org/10.1002/cphc.201800969.

71 Xu, T., Hu, Z., Lv, M. et al. (2021). Hydrogen atom and water complex determine the excited state dynamics of 8-azaguanine. 544: 111118. https://doi.org/10.1016/j.chemphys.2021.111118.

3

Decoding Triplet State Dynamics in Epigenetically Modified DNA and RNA: Photophysical Mechanisms from Nucleobase Monomers to Duplex

Lina Wang and Xueli Wang

East China Normal University, State Key Laboratory of Precision Spectroscopy, 500 Dongchuan Rd., Shanghai, China

3.1 Introduction

The canonical DNA or RNA nucleobases confer exceptional photostability through ultrafast nonradiative decay pathways, with >90% of absorbed ultraviolet (UV) energy dissipated within sub-picosecond timescales via efficient nonradiative channels leading to the ground state [1–4]. This evolutionary optimization effectively suppresses competing photophysical processes, including intersystem crossing (ISC) to triplet states. However, emerging evidence demonstrates that a portion of naturally occurring nucleobases with chemical modifications, structural alterations critical for epigenetic gene regulation [5–7], fundamentally reconfigure these photodynamic paradigms. Notably, epigenetic modifications enhance photolesion formation at specific genomic loci, with 30% of UV-induced mutational hotspots in cutaneous malignancies correlating with these positions [8, 9], where triplet state intermediates are hypothesized to mediate photodamage pathways [10–12].

This chapter synthesizes experimental advances from the past decade in elucidating triplet state dynamics within epigenetically modified nucleic acids. Through hierarchical analysis spanning nucleobase monomers, dinucleotides, oligomers, and duplexes (Figure 3.1), structure–photophysical property relationships governing triplet state dynamics are delineated. Particular emphasis is placed on tracking the key step in the early stage of triplet-induced photodamage in epigenetically modified DNA and RNA systems.

3.2 Formation of Triplet State in Epigenetically Modified Nucleobase Monomer

3.2.1 Cytosine C5-Modified Nucleobases

The methylation and demethylation at the C5 position of cytosine, as illustrated in Figure 3.1, stand as the most widely studies epigenetic modifications in DNA.

Excited State Dynamics in DNA and RNA: Experimental and Computational Approaches on Triplet State, First Edition. Edited by Jinquan Chen.

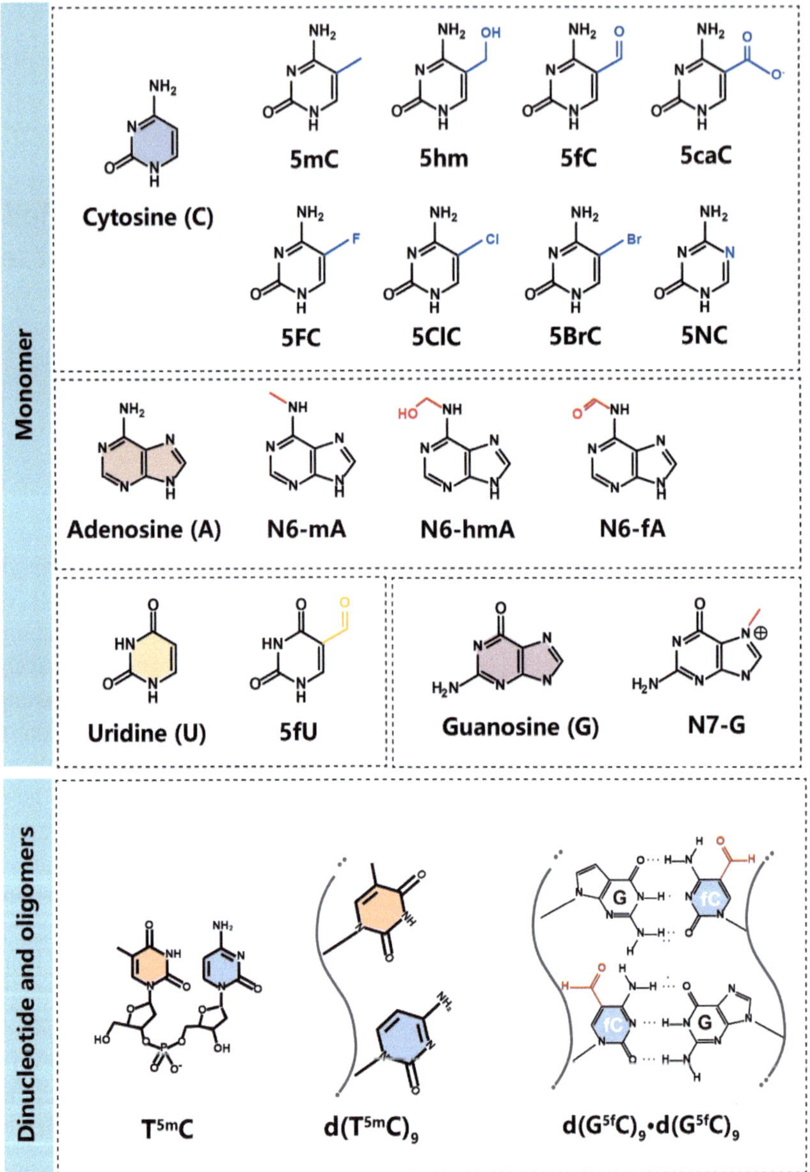

Figure 3.1 Representation of the epigenetically modified nucleic acids discussed in this review.

These modifications are essential for numerous biological processes, including embryonic development, X-chromosome inactivation, genomic imprinting, and protection against intragenomic parasites [13–15]. Furthermore, they serve as epigenetic markers for disease diagnosis [16, 17]. Due to their critical roles in genetic regulation, 5-methylcytosine (5mC) and its oxidized derivatives, namely, 5-hydroxymethylcytosine (5hmC), 5-formylcytosine (5fC), and 5-carboxycytosine (5caC), are collectively termed the "fifth to eighth" natural nucleobases [18, 19], thereby extending the classical genetic alphabet. In addition, biological systems contain various C5-substituted cytosine analogs, such as 5-fluorocytosine (5FC), 5-chlorocytosine (5ClC), and 5-bromocytosine (5BrC), which arise as metabolized DNA/RNA nucleoside byproducts and profoundly influence DNA methylation demethylation cycles by altering enzymatic recognition patterns and thermodynamic stability [20–23]. The structural adaptability of the C5 position has been strategically utilized in epigenetic drug development, with 5-azacytidine (5AC, Vidaza®) being a prime example. This clinically approved hypomethylating agent incorporates into DNA and inhibits methyltransferase activity [24, 25].

For canonical cytosine, UV excitation predominantly populates the $^1\pi\pi^*$ excited states (Figure 3.2a). A minor fraction of the excited population may transition to the $n\pi^*$ state, which acts as a critical doorway state for singlet-to-triplet ISC. However, the majority of excited state species undergo rapid nonradiative decay to the ground state (S_0) through conical intersections (CIs) linking the $^1\pi\pi^*$ states and S_0. This dominant relaxation mechanism severely constrains the achievable triplet quantum yield in these systems [12, 26–28]. Central to the ultrafast relaxation mechanism is the ethene-like CI. This CI is structurally defined by out-of-plane (oop) distortions of substituents at the C5 and C6 positions [12]. This well-characterized CI plays a pivotal role in governing the ultrafast relaxation dynamics of canonical cytosine derivatives.

Epigenetic modifications introduce chemical modification precisely at the C5 position introduces three main photophysical effects for triplet state generation (Figure 3.2b). First, such structural substitutions are proposed to modulate the accessibility of the ethene-like CI, increasing the IC lifetime because of the need to surmount a higher energy barrier to access the CI [12, 26, 29–31]. Therefore, the triplet state quantum yield can be improved by the inhibition of ultrafast IC. Second, the UV/vis absorption results show that the chemical substituents at the C5 position of epigenetically modified bases induce a discernible red shift of the lowest-lying band contributed by a $\pi \rightarrow \pi^*$ transition. This spectral shift originates from that C5 substituents exert an antibonding influence on the highest occupied molecular orbital (HOMO) through an out-of-phase orbital interaction with the C5 = C6 π-bond system [12]. This destabilization effect on the HOMO consequently reduces the energy gap between frontier molecular orbitals, thereby accounting for the observed bathochromic shift. Moreover, the red shift of $^1\pi\pi^*$ state can directly modulate the energy gap and spin–orbit coupling (SOC) between $^1\pi\pi^*$ and $^3n\pi^*$ state, facilitating direct $^1\pi\pi^* \rightarrow {}^3n\pi^*$ ISC [30, 32]. Third, heteroatom or electron-withdrawing-containing substituents at C5 create a low-lying $^1n\pi^*$ or charge-transfer state (CT state), respectively. These states can also lead to the triplet state generation by multichannel ISC [31, 32].

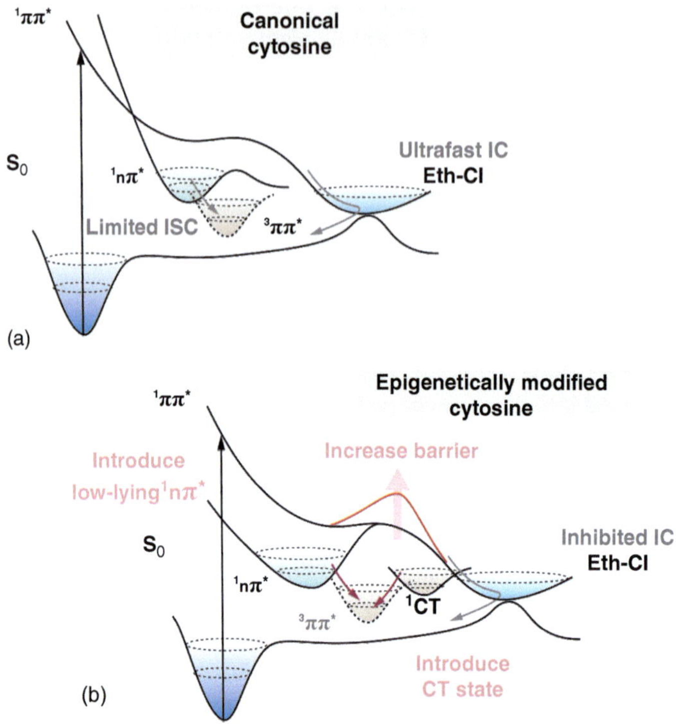

Figure 3.2 The scheme of the main excited state relaxation mechanism of canonical cytosine (a) and epigenetically modified cytosine (b).

3.2.2 Cytosine C5-Methylation

Upon 267-nm excitation, 5-methylcytidine (5mCyd) or 5-2′-deoxy-5-methylcytidine (5mdCyd) primarily populates the two bright ππ* states (S_1 and S_2). The S_2 state undergoes ultrafast internal conversion (<100 fs) to the S_1 state. A key feature of 5mdCyd is that an energy barrier (~0.1–0.3 eV) along the S_1 potential energy surface (PES). This barrier impedes access to the ethylenic twisting CI between S_1 and S_0, resulting in a fivefold longer S_1 lifetime (~7 ps) compared to dCyd (~1.4 ps) (Figure 3.3a,c) [12, 26, 28, 29]. However, methylation destabilizes nπ* states, reducing their involvement in the relaxation process and triplet state generation mediated by nπ* states [12, 29]. The primary decay pathway for 5mCyd is thus dominated by internal conversion from $S_1(\pi\pi^*)$ to S_0, with a photodegradation rate threefold slower than Cyd, enhancing its photostability (Figure 3.3b).

3.2.3 Cytosine C5-Demethylation

The excited state dynamics of 2′-deoxy-5-hydroxymethylcytidine (5hmdCyd) closely resemble those of 5mdCyd. Upon excitation at 285 nm, 5hmCyd is primarily excited to the lowest bright ππ* state. Femtosecond transient absorption (fs-TA) experiments reveal that the relaxation process involves two main steps: first, the excited state

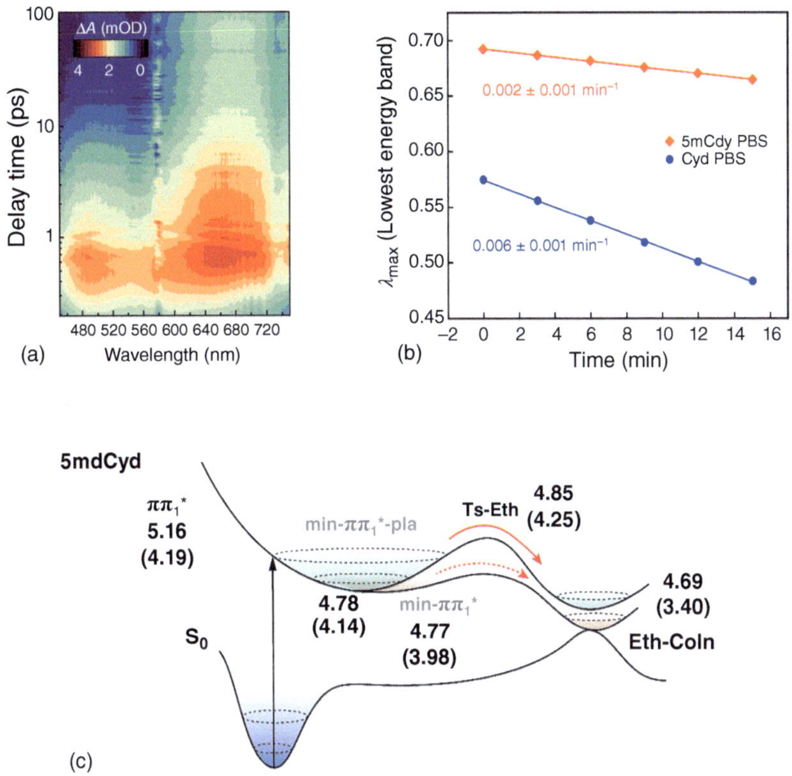

Figure 3.3 (a) The broadband fs-transient absorption spectra of 5mdCyd in buffer solution (pH = 7.4) with the excitation by 285 nm. (b) Initial rates of photodegradation for Cyd (black triangles) and 5mCyd (red circles). (c) Main deactivation pathways along the $\pi\pi^*$ for 5mdCyd predicted by TD-CAM-B3LYP/6-31G(d) calculations in water (dashed lines) and QM (MS-CASPT2(14,10)/ANO-L)/MM calculations (energies in parentheses and solid lines). Energies are relative to the S_0 (in eV).

relaxes from the Franck–Condon (FC) region to the $\pi\pi^*$ minimum with a lifetime of ~0.46 ps. Subsequently, the $\pi\pi^*$ state decays back to the ground state with a lifetime of ~2.6 ps, indicating a relatively slow internal conversion process compared with dCyd [31]. The stable excited state relaxation pathway of 5hmCyd with UV excitation also makes it less prone to triplet state generation.

The presence of the formyl group at the C5 position leads the excited state dynamics of 2′-deoxy-5-formylcytidine (5fdCyd) to be markedly different from those of 5mdCyd and 5hmCyd (Figure 3.4). The triplet quantum yield of 5fdCyd is remarkably high with a near-unity quantum yield, making it an efficient endogenous photosensitizer. Upon excitation at 285 nm, 5fdCyd is initially excited to the lowest-lying $\pi\pi^*$ state. fs-TA experiments reveal a sequential relaxation process: the $\pi\pi^*$ state decays to a dark $n_O\pi^*$ state (involving the carbonyl lone pair of the formyl group) with a lifetime of ~2.3 ps. This $n_O\pi^*$ state acts as a doorway state, facilitating efficient ISC to the triplet manifold with a lifetime of ~51 ps [31]. Moreover, time-resolved infrared (TRIR) spectroscopic data provides

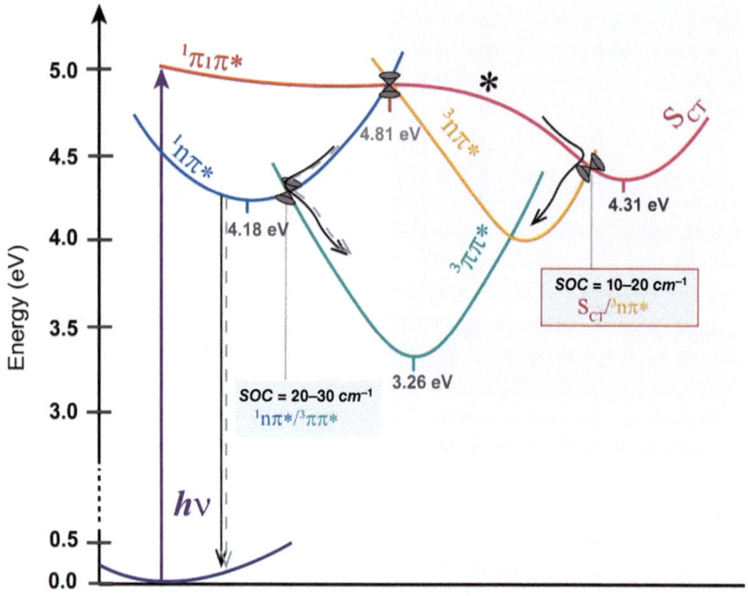

Figure 3.4 Schematic description of the main crossing and minima in 5fCyd as computed by PCM/TD-CAM-B3LYP in water.

further evidence for ultrafast ISC (<1 ps) from S_{CT}-min (vibrational marker bands: 1585 and 1670 cm^{-1}) to triplet state (1550 cm^{-1}) [32] (Figure 3.5). Additionally, the formyl group also hinders the twisting of the C5 = C6 bond, increasing the energy barrier to access the S_1/S_0 CI, thereby redirecting the relaxation pathway toward triplet formation.

Surprisingly, there is no detectable typical line of triplet state in 2′-deoxy-5-carboxylcytidine (5cadCyd) experimentally. In buffer solution at physiological pH, 5cadCyd exists in its anionic form. fs-TA experiments reveal an ultrafast relaxation process: the initially populated $\pi\pi^*$ state decays to the ground state with a lifetime of ~0.84 ps, indicating a barrierless internal conversion pathway. The kinetics in the UV region show a bleach recovery lifetime of ~3.0 ps, attributed to vibrational cooling of the hot ground state. These results are corroborated by the findings obtained from sub-30-fs-TA spectroscopy [30]. No low-lying $n\pi^*$ state that is isoenergetic with $\pi\pi^*$ state in the FC region is considered the primary reason for the lack of observable ISC.

3.2.4 Cytosine C5-Halogen

Steady state spectral results show that the halogen substation on cytidine C5 position induces a red shift of the lowest-lying absorption band due to destabilization of the $\pi\pi^*$ state and introduces weak intramolecular charge transfer (ICT) character, manifested in broad emission spectra with large Stokes shifts (>6100 cm^{-1}). TA signal reveals that excitation by UV irradiation results in a bifurcation event for 5-halogen

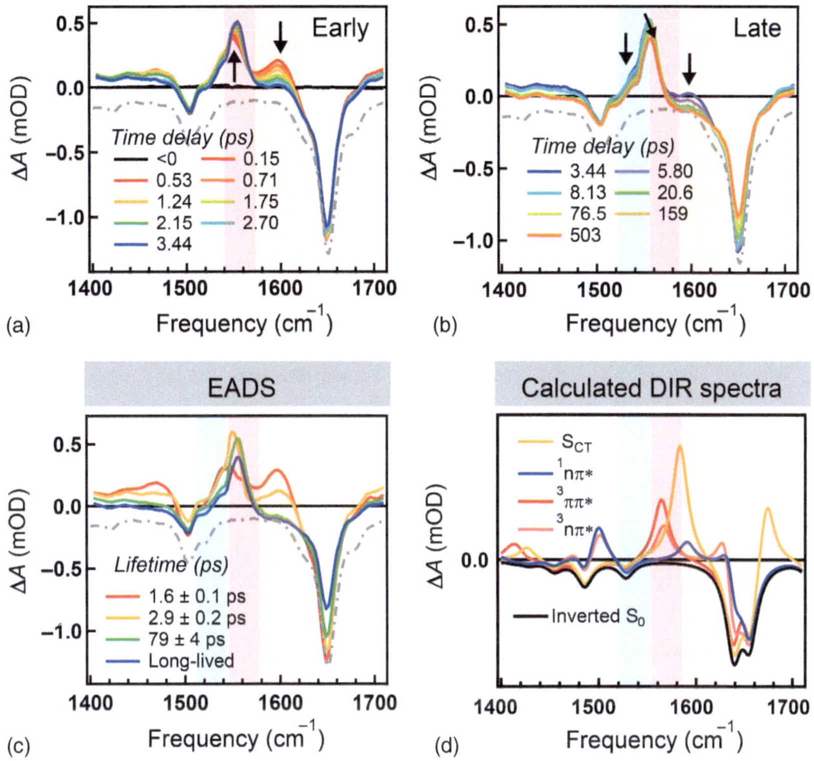

Figure 3.5 TRIR spectra at indicated delay times for deuterium-exchanged 5fdCyd in D_2O buffer (a, b) under 285-nm excitation. The inverted dot-dashed line shows the corresponding Fourier-transform infrared spectroscopy (FTIR) spectrum. Evolution-associated difference spectra (EADS) of 5fdCyd in D_2O. (c) Calculated DIR spectra of S_{CT}-min, S_1 $^1n\pi^*$-min, T_2 $^3n\pi^*$-min, and T_1 $^3\pi\pi^*$-min in D_2O (d).

cytidines that leads to ultrafast relaxation to ground state together with ICT states (Figure 3.6). These ICT states, absent in canonical cytosine, exhibit picosecond-scale lifetimes (3–31 ps) that decrease with heavier halogens, reflecting enhanced non-radiative decay [33]. Despite the heavy-atom effect, ISC to triplet states remains inefficient in halogenated derivatives due to small SOCs (SOC <5 cm^{-1} for S_1/T_1) [6].

3.2.5 Cytosine N5-Modification

The photophysical properties of 5AC and its derivatives, 5-azacytidine (5ACyd), and 2′-deoxy-5-azacytidine (5AdCyd) exhibit complex relaxation pathways that are highly sensitive to molecular substitutions [34, 35]. In 5AC, fs-TA studies reveal a two-step decay pathway: an ultrafast IC (~1.3 ps) from the initially populated bright $\pi\pi^*$ state to a lower lying dark $n\pi^*$ state, followed by a slower (17 ps) relaxation, with most of the population decay to the ground state and little to the triplet state. This $n\pi^*$ state, primarily localized on nitrogen lone pairs ($n_N\pi^*$), acts as a bottleneck, prolonging the excited state lifetime compared to canonical cytosine [34].

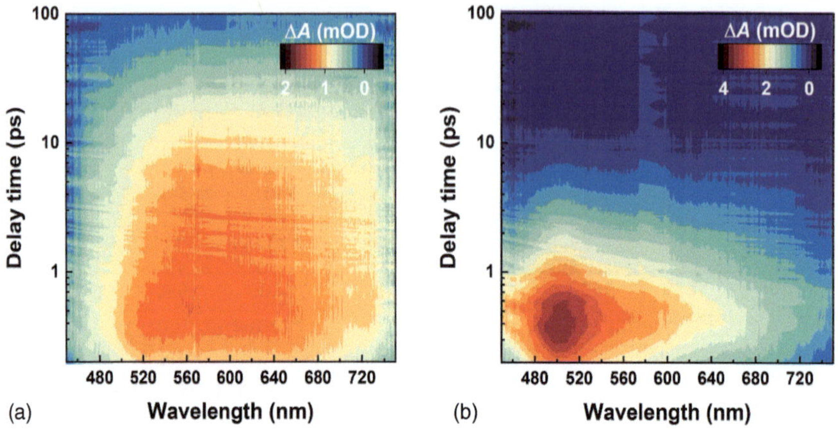

Figure 3.6 The broadband fs-transient absorption spectra of 5FdCyd (a) and 5BrdCyd (b) in buffer solution (pH = 7.4) with the excitation by 285 nm.

Substitution at the N1 position with ribose or deoxyribose groups drastically alters the relaxation dynamics [35]. In 5ACyd, the $\pi\pi^*$ state decays via two pathways: a direct return to S_0 (~0.9 ps) and a slower $n\pi^*$-mediated route (~5 ps). In 5-AdCyd, only a single ultrafast decay (~1.3 ps) is observed, suggesting suppression of the $n\pi^*$ pathway due to altered electronic coupling. These changes are attributed to sugar-induced shifts in the $n\pi^*$ state energy, reducing its accessibility and promoting faster ground state recovery. The $n_N\pi^*$ state in these molecules plays a dual role: it hinders direct $S_2(\pi\pi^*) \rightarrow S_0$ decay but also weakly facilitates ISC to the triplet manifold.

3.3 Uracil C5-Modified Nucleobases

5-Formyluracil (5fU), well-known oxidative derivative of thymine, is increasingly recognized as a significant epigenetic marker in cellular processes, particularly due to its ability to form cross-links with histones in vivo [36]. Moreover, the presence of 5fU in DNA may also influence chromatin structure and gene expression [26, 37].

From a photochemical and photophysical perspective, 5fU shows similar properties to that of 5fC. First, the extended absorption of 5fU into the UVA region means that it can be activated by a broader range of UV light, potentially increasing the risk of DNA damage and mutations [38]. Second, several studies reported that irradiation of 5fU leads to a triplet excited state with high energy (314 kJ mol^{-1}) and capable of sensitizing the formation of cyclobutane pyrimidine dimers (CPDs) in DNA via a triplet–triplet (TT) energy transfer [38, 39]. Existing flash photolysis data reveals that the triplet state of 5fU exhibits a broad TA band centered at 440–460 nm, with a lifetime of approximately 12 ms in a nitrogen-saturated solution [39]. To the best of our knowledge, there are currently no ultrafast spectroscopic data to explain the mechanism of triplet state generation in 5fU. However, quantum chemical calculations suggest that efficient ISC in 5fU is facilitated through the coupling of in-plane ring deformation [39].

3.4 Adenine N6-Modified Nucleobases

Adenine undergoes extensive epigenetic remodeling through N6-position modifications, constituting a critical regulatory layer in eukaryotic gene expression and RNA metabolism [40, 41]. The prototypical N6-methyladenine (N6-mA) modification functions as a dynamic epigenetic mark in DNA and messenger RNA (mRNA) [40, 42–44], exerting regulatory control through three primary mechanisms: (i) modulating DNA/RNA structural conformations, (ii) altering DNA/RNA–protein interaction landscapes, and (iii) influencing nucleic acid stability parameters. These molecular effects collectively govern essential biological processes, including mRNA splicing fidelity, translation efficiency modulation, and innate immune responses to viral pathogens. Furthermore, the m6A modification cycle is enzymatically regulated through oxidative demethylation catalyzed by FTO (fat mass and obesity-associated protein) and AlkB family demethylase enzymes, generating reactive intermediates such as N6-hydroxymethyladenine (N6-hmA) and N6-formyladenine (N6-fA)[44]. These modifications demonstrate developmental stage-specific patterning during cellular differentiation and show marked dysregulation in viral pathogenesis and neoplastic transformation. On the other hand, N6-mA and its metastable derivatives exhibit unique photophysical and photochemical properties that may mediate UV radiation responses. Recent studies by Xiang et al. revealed a novel UV-induced DNA damage repair mechanism involving N6-mA, where N6-mA-containing RNA rapidly accumulates at damage sites within 2 min post-UV exposure, subsequently recruiting polymerase κ (Pol κ) to facilitate repair [45].

For canonical adenine nucleobase, studies reported that upon UV excitation, it undergoes ultrafast IC to the ground state through CIs between the $\pi\pi^*$ excited state and S_0, involving ring puckering at the C2 and C6 positions [46, 47]. This intrinsic decay mechanism efficiently suppresses triplet generation, yielding ISC efficiency. Critical to this suppression is the low energy barrier at the $\pi\pi^*/S_0$ CI that enables rapid nonradiative relaxation. However, N6-epigenetic modifications fundamentally reconfigure these photophysical pathways. Analogous to the effects observed in cytosine, epigenetic modifications alter adenine's excited state dynamics via two key mechanisms. First, substituent-induced distortion of the purine ring system elevates the energy of the $\pi\pi^*$ state relative to the CI, increasing the barrier for IC-mediated ground state recovery. This energetic destabilization extends the excited state lifetime, creating the possibility for alternative relaxation channels. Second, electron-withdrawing groups (e.g. formyl in N6-mA) stabilize low-lying dark states—notably $n\pi^*$ and ICT states, which may serve as kinetic doorway states preceding ISC.

3.4.1 Adenine N6-Methylation

N6-mA and its ribose derivative N6-methyladenosine (N6-mAdo) exhibit distinct excited state relaxation dynamics upon UV excitation, as revealed by femtosecond time-resolved spectroscopy and quantum chemical calculation [31, 48]. Under 266-nm excitation, N6-mA is primarily excited to the $\pi\pi^*$ (L_a) state, which

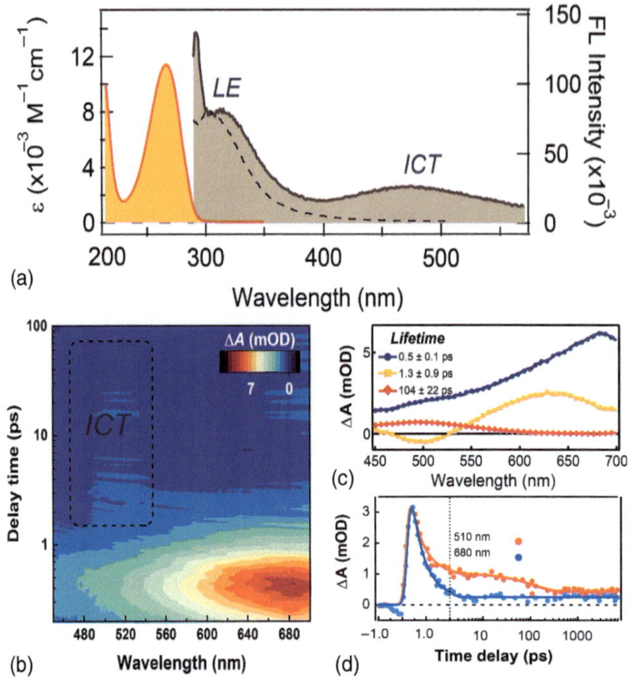

Figure 3.7 (a) Steady-state absorption and emission spectra of N6-mA. The black dashed line is the emission spectrum of adenine. (b) Broadband transient absorption spectra of N6-mA with excitation by 266 nm. (c) DADS extracted from the global fitting of the data shown in panel B. (d) Representative transient absorption kinetics.

undergoes ultrafast relaxation with a lifetime of 0.30 ± 0.06 ps, attributed to nonradiative decay facilitated by C2 puckering of the purine ring. A secondary decay component of 1.6 ± 0.2 ps is assigned to the ππ* (L_b) state. Furthermore, a longer-lived excited state species can be detected with a lifetime of 107 ± 20 ps, which should be contributed by the ICT state (Figure 3.7b–d).

The ICT state, characterized by electron donation from the methylamino group to thc purine ring, displays pronounced sensitivity to solvent hydrogen-bonding interactions. This solvent-dependent behavior is substantiated by a kinetic isotope effect (KIE ≈ 2.5) in deuterated aqueous (D_2O), where the ICT lifetime exhibits a significant enhancement to 263 ± 36 ps. The dual fluorescence emission: a short-wavelength band (313 nm) from the local excited (LE) state and a long-wavelength band (475 nm) from the delocalized excited (ICT) state, provides further evidence for the assignment (Figure 3.7a). The generation of the ICT state is likely due to a ~30 kcal mol^{-1} decrease in the ionization potential after replacing the hydrogen atom with a methyl group on the amino moiety [49]. In contrast, the ribose substitution in N6-mAdo quenches the ICT state, resulting in adenine-like excited state dynamics dominated by ultrafast internal conversion (0.56 ± 0.04 ps for ππ* (L_a) and 3.0 ± 0.4 ps for ππ* (L_b)), suggesting retained photostability in DNA/RNA contexts. Notably, since direct spectroscopic evidence for triplet formation in both N6-mA and N6-MAdo is still lacking, methylation can enhance photoionization propensity has been reported [50]. Nanosecond transient absorption (ns-TA) studies reveal that 6MeAde undergoes one-photon ionization at 266 nm with a quantum yield ($\Phi_{1hv} = 1.4 \times 10^{-3}$), generating hydrated electrons and radical cations. This is attributed to methylation-induced reduction of ionization potentials. The coexistence of long-lived ICT states and facile photoionization suggests that adenine N6-methylation may play a dual role in epigenetic regulation and photochemical DNA/RNA damage, though further studies are needed to elucidate triplet state contributions and downstream photolesion pathways.

3.4.2 Adenine N6-Demethylation

The photophysical behavior of N6-hydroxymethyladenosine (N6-hmAdo) and 6-formyladenosine (N6-fAdo) demonstrates remarkably distinct excited state relaxation dynamics compared with both adenosine and N6-mAdo. Wang et al. report a distinct relaxation pathway involving ICT state or low-lying $n\pi^*$ and subsequent triplet formation. Both N6-hmAdo and N6-fAdo demonstrate a significantly prolonged excited state lifetime and efficient triplet generation, as evidenced by fs-TA, ns-TA, and quantum chemical calculations.

Upon UV excitation (275 nm), N6-hmAdo exhibits solvent-sensitive dual fluorescence emissions: a short-wavelength band (~320 nm) attributed to a LE $\pi\pi^*$ state and a red-shifted emission (~465 nm) characteristic of an ICT state. Fs-ns TA spectroscopy identifies four decay components with the lifetimes to be 0.4 ps, 2.1 ps, 1.2 ns, and 26 ns (in air, 180 ns in N_2). The two ultrafast decays found in the TA measurement match very well with those seen in the fluorescence up-conversion (FUPC) kinetics and are assigned to the IC to ground state, associated with C2—H oop torsion and C6–NH2 oop motion, respectively. It should be noted that these two decay pathways are very similar to that of adenosine. But the substituent hinders the oop motion of the amino group, resulting in a longer lifetime for the C6-puckering channel. The longer-lived species with 1.2 ns lifetime are contributed by the ICT state, as it exhibits solvent-dependent lifetimes and amplitude. This finding is consistent with what was reported on N6-mA, wherein a long-lived emissive CT state exhibiting a lifetime on the order of hundreds of picoseconds was directly detected [31]. Oxidation of the methyl group to a hydroxymethyl moiety further reduces the ionization potential and stabilizes the CT state in hm6A, resulting in a nanosecond-scale lifetime. Crucially, the extended lifetime of ICT state enables efficient SOC-mediated ISC to triplet states. Time-dependent density functional theory (TDDFT) calculations reveal significant SOC values (5.43 and 3.51 cm^{-1}) between the ICT state (S_1) and nearby triplet states (T_3 and T_4), facilitating ISC. Structural distortions at the ICT minimum—specifically, the twisted dihedral angles of the $NHCH_2OH$ group (C5–C6–N7–C18: −78.6°)—enhance angular momentum changes during charge recombination, promoting spin-flip transitions via the spin–orbit charge-transfer ISC (SOCT-ISC) mechanism.

Upon UV excitation (267 nm), N6-fAdo displays ultrafast dynamics characterized by sequential transitions between excited states. Global analysis of the fs-TA data identifies four lifetimes (0.4, 7.3, 130 ps, and a long-lived offset >7 ns). ns-TA experiments confirm the long-lived species as a triplet state, with oxygen-sensitive lifetimes of 95 ns (air) and 348 ns (N_2). Notably, the triplet quantum yield of N6-fAdo was measured at 27%, significantly higher than adenosine (10^{-3} level at 77 K). Comparative analysis with adenosine and N6-hmAdo suggests the 0.4 ps lifetime component likely originates from oop twisting of the C2—H bond, facilitating a CI between the bright $\pi\pi^*$ state and ground state. The longer 7.3 ps component similarly corresponds to ground state repopulation. This extended lifetime may arise from C6–NH2 oop torsion, analogous to the behavior of adenosine but is further prolonged by the electronic effects of the formyl group. The 130 ps component corresponds to an intermediate dark $n\pi^*$ state (S_1) that mediates triplet formation. The calculated 9.83 cm^{-1} SOC strength of S_1min/T_2 further supports this assignment. While ISC remains unreported in natural purine bases, N6-fAdo demonstrates a pyrimidine-like ISC mechanism mediated by a low-lying $n\pi^*$ state. Remarkably, N6-fAdo exhibits a triplet quantum yield exceeding pyrimidines by an order of magnitude at room temperature and surpasses adenosine's yield (77 K) by two orders of magnitude.

3.5 Guanosine N7-Modified Nucleobases

N7-methylguanosine (m7G), characterized by methylation at the N7 position of guanine, is an important RNA modification involved in epigenetic regulation that is commonly observed in both prokaryotic and eukaryotic organisms. This modification predominantly localizes to RNA 5′ cap structures and internal regions of mRNAs, ribosomal RNAs (rRNAs), and transfer RNAs (tRNAs) and microRNAs [27, 51, 52]. Catalyzed by methyltransferases such as RNA methyltransferases (RNMT), the METTL1/WDR4 complex, and WBSCR22 (BUD23), m7G deposition regulates key processes, including RNA maturation, nucleocytoplasmic trafficking, transcriptional elongation, and translation fidelity [28, 52, 53]. These regulatory roles further modulate ribosomal biogenesis and translational efficiency through indirect mechanisms. Emerging evidence implicates METTL1/WDR4-dependent m7G-methylation as a critical driver in oncogenesis, with demonstrated pathological significance in hepatocellular carcinoma, non-small cell lung cancer, and urothelial carcinoma progression [27].

The excited state dynamics of m7G were studied by Wang and coworkers by femtosecond time-resolved TA, FUPC together with quantum chemical calculation [35]. Substitution of a methyl group at the N7 position significantly changes the p*K*a of m7G (p*K*a = 7.06) compared to guanosine, enabling the coexistence of cationic and zwitterionic forms under physiological pH conditions. To disentangle their distinct excited state relaxation mechanism, studies were carried out in buffered solutions at pH 5.0 (cationic form) and pH 9.0 (zwitterionic form). For the cationic form, TA and FUPC experiments reveal biphasic decay kinetics: a short-lived component (2.0 ps) and a longer-lived component (206 ps) (Figure 3.8a,c). The 2.0 ps lifetime corresponds to the initial La $\pi\pi^*$ state. The subsequent 206 ps component reflects a stabilized La state minimum, characterized by structural reorganization

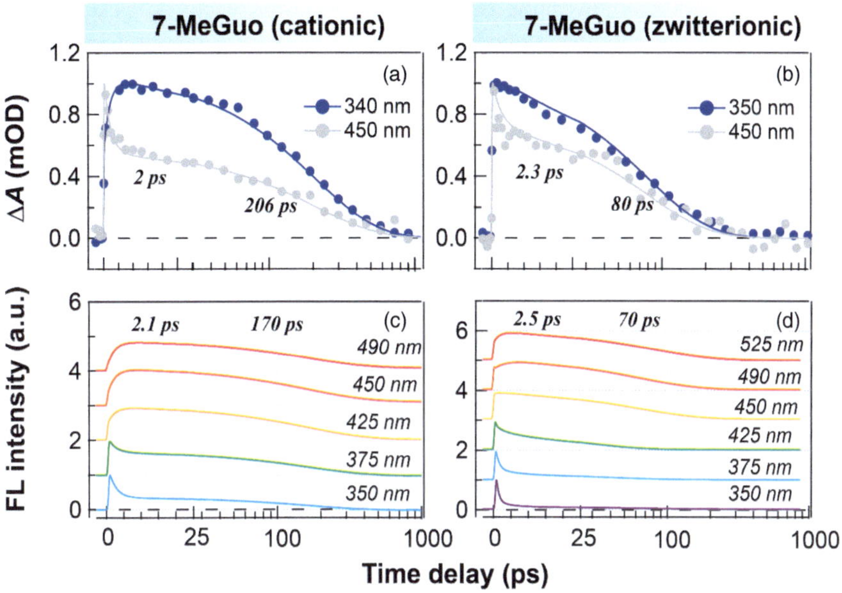

Figure 3.8 Representative kinetics of (a, c) m7G (cationic form) and (b, d) m7G (zwitterionic form) from TA data excited at 285 nm (top) and FUPC data under 295-nm excitation (bottom).

involving oop displacements of N1—H and C—H bonds—a phenomenon also observed in protonated guanosine derivatives (GMPH+) [54, 55]. In contrast, the zwitterionic form displays accelerated relaxation, with lifetimes of 2.3 and 80 ps (Figure 3.8b,d). Deprotonation at N1 destabilizes the La state minimum, as shown by red-shifted fluorescence and reduced lifetimes. Structural analyses suggest that N1 deprotonation brings the optimal structure of S_1 closer to the CI region. This proximity likely facilitates nonradiative decay, shortening the excited state lifetime.

While the primary focus of current studies centers on singlet-state dynamics, the potential involvement of triplet states warrants discussion. In m7G, the planar La $\pi\pi^*$ state and rigidified structure may suppress ISC by minimizing spin–orbit interactions, consistent with the absence of long-lived components (>1 ns) in TA data. However, N7-methylation in Guo has been shown to promote proton transfer in Watson–Crick pairs (e.g. Guo-Cyt), a process that may transiently populate CT states capable of ISC [35, 56]. Further studies employing triplet sensitizers or time-resolved spectral measurements are needed to clarify this aspect.

3.6 Formation of Triplet State in Epigenetically Modified Dinucleotides and Single-Stranded Oligomers

Compared to nucleobases monomers, the formation of dimers or single-stranded oligonucleotides introduces $\pi\pi$-stacking interactions that significantly diversify the photophysical pathways governing excited state relaxation. Upon UV excitation, such systems exhibit complex dynamics, including the emergence of delocalized excimer states, exciplex states, or CT states, which exhibit lifetime orders of magnitude longer than those of isolated bases. These extended lifetimes arise from reduced nonradiative decay rates and the stabilization of electronically coupled states. Critically, epigenetic modifications have the potential to modulate these dynamics by altering the geometries of $\pi\pi$-stacking and electronic interactions, which further influence the generation of triplet state. On the one hand, epigenetic modifications may perturb intermolecular $\pi\pi$-stacking, disrupting or enhancing the delocalization of electronic states. Such delocalized excited state can act as a gateway to triplet formation by amplifying SOC. On the other hand, delocalized electronic states may paradoxically exert photoprotective effects. Efficient charge separation in rigid, well-stacked systems could suppress ISC by an effective CT pathway.

A striking example of the nuanced effects of epigenetic modifications on photostability lies in the behavior of 5mC. While 5mC monomer exhibit enhanced photostability compared to unmodified cytosine (C), C5-methylation on cytosine can lead to poor photostability in DNA. Notably, although 5mC constitutes less than 5% of bases in the human genome, it is linked to approximately 30% of mutational hotspots in skin tumors [8, 9]. Due to the biological relevance, a number of experimental and computational studies have been devoted in the past decade to elucidate the related photodamage mechanism. The results indicate that the increased mutation rate stems from the preferential formation of thymine CPDs after deamination of T5mC CPDs, a process more efficient than with unmethylated cytosine [57, 58]. Quantum chemical analyses suggest that C5-methylation induces structural modifications in cytosine, altering conformational equilibria and potentially modulating photodamage susceptibility [12, 59]. From the experimental view, Markovitsi and coworkers reported that methylation enhances the quantum yields of cis, syn and trans, syn-CPDs in TCG trimers by 50% and 90%, respectively. However, in single-stranded TC oligomers, the predominant photoproduct remains cis, syn-CPD, with methylation exerting minimal influence on CPD yield [59].

To track the initial stage of CPD formation in 5mC-containing DNA dinucleotides and oligomers, Jin et al. investigated the excited state dynamics of T5mC dimers through time-resolved spectroscopic analysis combined with quantum chemical calculation [60]. Furthermore, the study was extended to $d(T5mC)_9$ oligomers to systematically evaluate how secondary structural configurations modulate excited state behavior. First, the TDDFT-optimized T5mC and $d(T5mC)_9$ structures were analyzed through molecular dynamics (MD) simulations. The results show that T5mC dimer exhibited high flexibility, with only 6.4% cis- and 0.6% trans-conformations maintaining π–π stacking. In contrast, $d(T5mC)_9$ oligomers showed enhanced stacking stability (>60% cis-population), while trans-conformations were sterically prohibited. Nonetheless, methylation increased stack/unstack ratios across both systems, as confirmed by comparative MD analysis. For the steady-state absorption, methylation induces a minor red shift relative to unmodified counterparts. And the emission spectra reveal a CT emission tail above 400 nm, consistent with reported

CT-state emissions in other DNA systems [59, 61]. TDDFT calculations identify bright $\pi\pi^*$ states in T5mC and TC dimers. Natural transition orbital (NTO) analyses demonstrate CT characteristics in S_1 states of both cis- and trans-configured T5mC/TC dimers, while T1 states exhibit contrasting behaviors: CT features persist in T5mC, whereas TC adopts LE characteristics.

The excited state dynamics of T5mC and d(T5mC)$_9$ exhibit distinct triplet state behaviors governed by structural and environmental factors. fs-ns TA (Figure 3.9) and TRIR spectroscopy reveal that both systems undergo triphasic relaxation with lifetimes of ~6 ps (τ_1, monomer-like $\pi\pi^*$ state), ~60–80 ps (τ_2, singlet CT state), and a long-lived component (τ_3, >10 ns). However, critical differences emerge in triplet formation pathways and yields. In the T5mC dimer, τ_3 is assigned to a triplet CT state (1488, 1522, 1571 cm^{-1} via TRIR signatures matching calculated vibrational modes of triplet CT state). The sequential conversion from singlet CT state to triplet

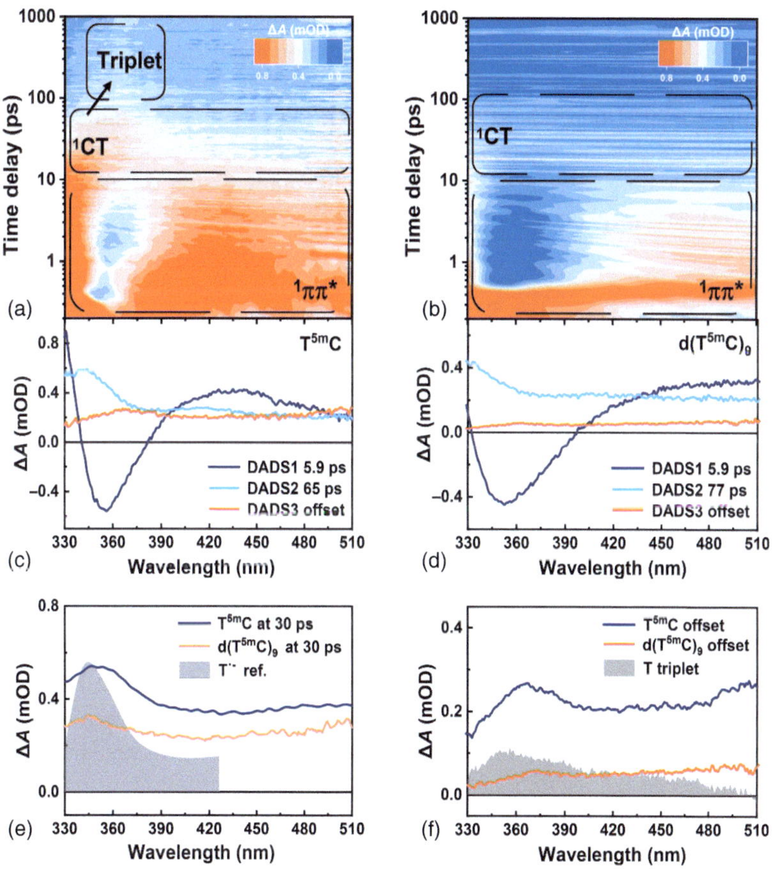

Figure 3.9 Femtosecond UV–vis transient absorption (TA) spectra and decay-associated difference spectra (DADS) obtained from a global fitting analysis of (a, b) T5mC and (c, d) d(T5mC)$_9$ in buffer solution excited at 267 nm. (e) Comparison of TA spectra delayed at 30 ps in T5mC and d(T5mC)$_9$ with the T radical anion (T) spectra (reproduced from Ref. [44]). (f) Comparison of DADS3 in T5mC and d(T5mC)$_9$ with T triplet (T1 state) measured under the same condition.

CT can be confirmed by kinetic correlations. Methylation amplifies triplet yield in T5mC compared to unmethylated TC, as evidenced by stronger long-lived TA/TRIR signals. This enhancement arises from methylation-induced stabilization of stacked nucleobase geometries, which promotes CT-state formation and lowers ISC energy barriers (<10 meV) between singlet CT and triplet CT configurations. In contrast, the single-stranded $d(T5mC)_9$ oligomer shows minimal triplet yield (~5.4% amplitude) despite similar τ_1 and τ_2 lifetimes. TRIR and MD simulations attribute this suppression to a 96-meV energy barrier between S1-MIN and the S1/T1 crossing point, hindering ISC in the structurally constrained oligomer. Further differentiation is observed in CT-state origins: TRIR spectra confirm that $d(T5mC)_9$ exclusively adopts cis-configuration singlet CT states, whereas T5mC accommodates both cis- and trans-geometries.

In summary, the findings from experimental and calculated work reconcile prior discrepancies in methylation-dependent photodamage: while flexible dimers benefit from enhanced ISC due to 5mC-induced stacking, structured oligomers suppress triplet-mediated pathways due to conformational constraints. This mechanistic insight underscores the dual role of 5mC in DNA photostability, where its mutagenic potential depends not only on chemical modification but also on secondary structure dynamics. The direct observation of CT-driven ISC pathways advances understanding of epigenetic phototoxicity, emphasizing that triplet CT states may work as central players in 5mC-related skin carcinogenesis.

3.7 Formation of Triplet State in Epigenetically Modified DNA Duplex

The introduction of epigenetic modifications into DNA duplexes, such as 5fC, can also profoundly alter the photophysical behavior by creating pathways for long-lived triplet state formation. These modifications can function as endogenous photosensitizers, increasing the susceptibility of DNA to UV-induced damage through triplet-mediated processes. The studies by Wang and coworkers provide comprehensive experimental and computational insights into the mechanisms governing triplet state generation in modified DNA duplexes [35].

Wang and coworkers conducted a detailed investigation of the excited state dynamics in the alternating duplex $d(G^{5f}C)_9 \cdot d(G^{5f}C)_9$ using fs-TA, TRIR spectroscopy together with quantum chemical calculation [35]. The experiments were performed with 285-nm excitation in D_2O buffer at room temperature, with an estimated average of 0.3 excitations per duplex to avoid multi-photon artifacts. The fs-TA spectra revealed a broad excited state absorption (ESA) band with maxima below 350 nm (peak I) and at 410 nm (peak II). Over the subsequent 20 ps, these ESA bands decay synchronously while peak II undergoes a slight blue shift. Notably, the intensity ratio of peak I to peak II decreases from 2.9 at 500 fs to 1.7 by 20 ps. The persistence of absorption signals beyond 3 ns confirms the presence of long-lived excited states. Global fitting of the TA data identified three decay components: 3.7 ± 0.1 ps (τ_1), 11.1 ± 0.2 ps (τ_2), and a long-lived component (τ_3).

The τ_1 and τ_2 lifetimes were attributed to the decay of initially populated singlet states, while τ_3 corresponded to the formation of a triplet species. ns-TA measurements under N_2-purged conditions further resolved two distinct triplet states: a localized triplet on 5fC (39 ± 6 ns) and a delocalized intrastrand triplet CT state (^{3}Intra-CT) with a lifetime of 256 ± 22 ns. The localized triplet exhibited an ESA spectrum matching that of the 5fC monomer, with a broad peak at 470 nm. In contrast, the ^{3}Intra-CT state showed a characteristic ESA profile resembling the guanine radical cation, with strong absorption below 450 nm and a shoulder at 500 nm. TRIR spectroscopy provided additional evidence for these assignments (Figure 3.10), with marker bands at 1580 cm^{-1} ($^3\pi\pi^*$-5fC) and 1705 cm^{-1} (^{3}Intra-CT) appearing during the first 20 ps and persisting beyond 3 ns. The triplet yield was quantified as 8 ± 3%, lower than in 5fC monomers due to competing singlet CT pathways that promote ground state recovery. These findings offer critical insights into how epigenetic modifications influence DNA excited state dynamics and how secondary structures modulate triplet state formation and decay. This work also establishes a foundation for investigating UV-induced photodamage mechanisms in epigenetically modified nucleic acids.

Notably, Schrodter et al. systematically investigated the photodamaging potential of 5fC triplet states using precisely engineered ternary DNA architectures [62]. These innovative constructs spatially separated the photosensitizer (5fC) from the designated damage site (TT step) through variable AT-pair spacers (n = 0–5), enabling controlled analysis of remote energy transfer mechanisms. The study employed high-performance liquid chromatography (HPLC) and polyacrylamide gel electrophoresis (PAGE) to quantitatively assess CPD formation following irradiation with 310 nm LEDs, which selectively targets 5fC's absorption shoulder while minimizing direct DNA excitation. Experimental results demonstrated that 5fC exhibited remarkable photosensitizing efficiency, generating CPD yields up to 37% through a distance-dependent mechanism involving both direct Dexter-type energy transfer (for proximal sites) and triplet energy hopping (across longer AT-pair bridges). Crucially, this work provided the first experimental evidence that naturally occurring epigenetic modifications can function as endogenous photosensitizers, establishing a novel pathway for sunlight-induced DNA damage.

3.8 Conclusion

This chapter has systematically examined how epigenetic modifications fundamentally alter the photophysical properties of nucleic acids, creating pathways for triplet state formation that challenge conventional paradigms of DNA photostability. First, through hierarchical analysis spanning nucleobase monomers to duplex DNA, we have established that those chemical modifications at critical positions (C5 of cytosine, N6 of adenine) introduce new relaxation channels that facilitate ISC to long-lived triplet states. The most striking examples emerge from formylated bases (5fC and 5FU), where carbonyl groups create low-lying nπ^* and CT states that serve as efficient gateways for triplet formation with near-unity quantum yields and

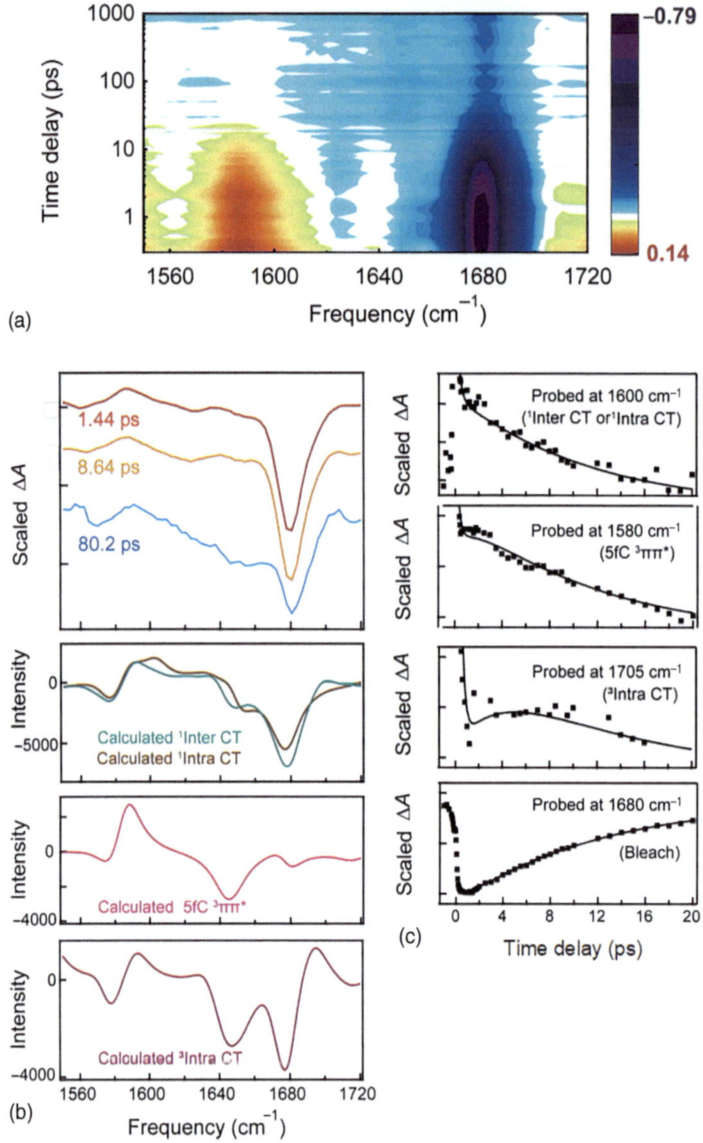

Figure 3.10 (a) Broadband TRIR spectra of d(G^{5f}C)$_9$·d(G^{5f}C)$_9$ in a D_2O-buffered solution with 285-nm excitation. (b) TRIR spectra at 1.44, 8.64, and 80.2 ps (signals offset vertically and scaled for clarity) compared with the computed difference IR spectra for ^{1}Intra-CT and ^{1}Inter-CT singlet minima and 5fC ($^3\pi\pi^*$) and ^{3}Intra-CT triplet minima. Frequencies are scaled by a factor of 0.955, and simulated band broadening used Gaussian functions of half-width at half-maximum of 10 cm^{-1}. (c) Kinetic traces probed at 1600, 1580, 1705, and 1680 cm^{-1}.

microsecond lifetimes. Second, the structural context of epigenetic modifications emerges as a critical determinant of their photochemical impact. While 5mC demonstrates enhanced photostability in monomeric form, its incorporation into DNA facilitates triplet-mediated damage through stabilization of stacked geometries that promote CT states. This paradox highlights the complex interplay between chemical modification and higher-order structure in determining DNA photostability. Similarly, the contrasting behaviors of modified bases in flexible dimers vs constrained oligomers underscore how chromatin architecture may modulate photodamage susceptibility in vivo.

References

1 Crespo-Hernández, C.E., Cohen, B., Hare, P.M., and Kohler, B. (2004). Ultrafast excited-state dynamics in nucleic acids. *Chemical Reviews* 104 (4): 1977–2019.
2 Middleton, C.T., Harpe, K.D.L., Su, C. et al. (2009). DNA excited-state dynamics: from single bases to the double helix. *Annual Review of Physical Chemistry* 60 (1): 217–239.
3 Kohler, B. (2010). Nonradiative decay mechanisms in DNA model systems. *The Journal of Physical Chemistry Letters* 1 (13): 2047–2053.
4 Improta, R. and Douki, T. (2021). *DNA photodamage: From light absorption to cellular responses and skin cancer*, 21. Royal Society of Chemistry.
5 Reik, W. (2007). Stability and flexibility of epigenetic gene regulation in mammalian development. *Nature* 447 (7143): 425–432.
6 Chang, X.-P., Fan, F.-R., Liu, K. et al. (2025). Quantum mechanics/molecular mechanics studies on mechanistic photophysics of epigenetic C5-halogenated DNA nucleosides: 2′-deoxy-5-chlorocytidine and 2′-deoxy-5-bromocytidine in aqueous solution. *Physical Chemistry Chemical Physics* 27: 7139–7150.
7 Holoch, D. and Moazed, D. (2015). RNA-mediated epigenetic regulation of gene expression. *Nature Reviews Genetics* 16 (2): 71–84.
8 Denissenko, M.F., Chen, J.X., Tang, M.S., and Pfeifer, G.P. (1997). Cytosine methylation determines hot spots of DNA damage in the human P53 gene. *Proceedings of the National Academy of Sciences of the United States of America* 94 (8): 3893–3898.
9 Pfeifer, G.P., You, Y.H., and Besaratinia, A. (2005). Mutations induced by ultraviolet light. *Mutation Research* 571 (1): 19–31.
10 Tommasi, S., Denissenko, M.F., and Pfeifer, G.P. (1997). Sunlight induces pyrimidine dimers preferentially at 5-methylcytosine bases. *Cancer Research* 57 (21): 4727.
11 Mitchell, D.L. (2000). Effects of cytosine methylation on pyrimidine dimer formation in DNA. *Photochemistry and Photobiology* 71 (2): 162–165.
12 Martinez-Fernandez, L., Banyasz, A., Esposito, L. et al. (2017). UV-induced damage to DNA: effect of cytosine methylation on pyrimidine dimerization. *Signal Transduction and Targeted Therapy* 2 (17): 021.

13 Reik, W. and Walter, J. (2001). Genomic imprinting: parental influence on the genome. *Nature Reviews Genetics* 2 (1): 21–32.

14 Li, E. (2002). Chromatin modification and epigenetic reprogramming in mammalian development. *Nature Reviews Genetics* 3 (9): 662–673.

15 Neri, F., Krepelova, A., Incarnato, D. et al. (2013). Dnmt3L antagonizes DNA methylation at bivalent promoters and favors DNA methylation at gene bodies in ESCs. *Cell* 155 (1): 121–134.

16 Shree, R. and Ha, G. (2025). Epigenetic footprints in prenatal cell-free DNA reveal risk of preeclampsia. *Nature Medicine* 31: 1081–1082.

17 Baylin, S.B. and Jones, P.A. (2011). A decade of exploring the cancer epigenome—biological and translational implications. *Nature Reviews Cancer* 11 (10): 726–734.

18 Tahiliani, M., Koh, K.P., Shen, Y. et al. (2009). Conversion of 5-methylcytosine to 5-hydroxymethylcytosine in mammalian DNA by MLL partner TET1. *Science* 324 (5929): 930–935.

19 Ito, S., Shen, L., Dai, Q. et al. (2011). Tet proteins can convert 5-methylcytosine to 5-formylcytosine and 5-carboxylcytosine. *Science* 333 (6047): 1300–1303.

20 Whiteman, M., Jenner, A., and Halliwell, B. (1997). Hypochlorous acid-induced base modifications in isolated calf thymus DNA. *Chemical Research in Toxicology* 10 (11): 1240–1246.

21 Valinluck, V., Liu, P., Kang, J.I. Jr. et al. (2005). 5-halogenated pyrimidine lesions within a CpG sequence context mimic 5-methylcytosine by enhancing the binding of the methyl-CpG-binding domain of methyl-CpG-binding protein 2 (MeCP2). *Nucleic Acids Research* 33 (9): 3057–3064.

22 Wojciechowski, M., Czapinska, H., Krwawicz, J. et al. (2024). Cytosine analogues as DNA methyltransferase substrates. *Nucleic Acids Research* 52 (15): 9267–9281.

23 Valinluck, V. and Sowers, L.C. (2007). Endogenous cytosine damage products alter the site selectivity of human DNA maintenance methyltransferase DNMT1. *Cancer Research* 67 (3): 946–950.

24 Gailhouste, L., Liew, L.C., Hatada, I. et al. (2018). Epigenetic reprogramming using 5-azacytidine promotes an anti-cancer response in pancreatic adenocarcinoma cells. *Cell Death & Disease* 9 (5): 468.

25 Krečmerová, M. and Otmar, M. (2012). 5-azacytosine compounds in medicinal chemistry: current stage and future perspectives. *Future Medicinal Chemistry* 4 (8): 991–1005.

26 Wang, X., Zhou, Z., Tang, Y. et al. (2018). Excited state decay pathways of 2′-deoxy-5-methylcytidine and deoxycytidine revisited in solution: a comprehensive kinetic study by femtosecond transient absorption. *The Journal of Physical Chemistry B* 122 (28): 7027–7037.

27 Luo, Y., Yao, Y., Wu, P. et al. (2022). The potential role of N7-methylguanosine (m7G) in cancer. *Journal of Hematology & Oncology* 15 (1): 63.

28 Haag, S., Kretschmer, J., and Bohnsack, M.T. (2015). WBSCR22/Merm1 is required for late nuclear pre-ribosomal RNA processing and mediates N7-methylation of G1639 in human 18S rRNA. *RNA* 21 (2): 180–187.

29 Hoehn, S.J., Krul, S.E., Pogharian, M.M. et al. (2023). Photochemical stability of 5-methylcytidine relative to cytidine: photophysical insight for mRNA therapeutic applications. *The Journal of Physical Chemistry Letters* 14 (48): 10856–10862.

30 Kabaciński, P., Romanelli, M., Ponkkonen, E. et al. (2021). Unified description of ultrafast excited state decay processes in epigenetic deoxycytidine derivatives. *The Journal of Physical Chemistry Letters* 12 (45): 11070–11077.

31 Wang, X., Yu, Y., Zhou, Z. et al. (2019). Ultrafast intersystem crossing in epigenetic DNA nucleoside 2′-deoxy-5-formylcytidine. *The Journal of Physical Chemistry B* 123 (27): 5782–5790.

32 Wang, X., Martínez-Fernández, L., Zhang, Y. et al. (2021). Solvent-dependent stabilization of a charge transfer state is the key to ultrafast triplet state formation in an epigenetic DNA nucleoside. *Chemistry – A European Journal* 27 (42): 10932–10940.

33 Xu, R., Hu, Z., Wang, X. et al. (2020). Intramolecular charge transfer in 5-halogen cytidines revealed by femtosecond time-resolved spectroscopy. *The Journal of Physical Chemistry B* 124 (13): 2560–2567.

34 Zhou, Z., Zhou, X., Wang, X. et al. (2017). Ultrafast excited-state dynamics of cytosine aza-derivative and analogues. *Journal of Physical Chemistry A* 121 (14): 2780–2789.

35 Chen, Z., Wang, X., Jia, M. et al. (2024). Ribose and deoxyribose group alter excited-state dynamics of 5-azacytosine in solution. *Photochemistry and Photobiology* 100 (2): 291–297.

36 Zou, G., Liu, C., Zeng, W. et al. (2020). Regulable DNA–protein interactions *in vitro* and vivo at epigenetic DNA marks. *CCS Chemistry* 2 (2): 54–63.

37 Zou, G., Zhang, K., Yang, W. et al. (2021). 5-formyluracil targeted biochemical reactions with proteins inhibit DNA replication, induce mutations and interference gene expression in living cells. *Chinese Chemical Letters* 32 (10): 3252–3256.

38 Aparici-Espert, I., Garcia-Lainez, G., Andreu, I. et al. (2018). Oxidatively generated lesions as internal photosensitizers for pyrimidine dimerization in DNA. *ACS Chemical Biology* 13 (3): 542–547.

39 Francés-Monerris, A., Lineros-Rosa, M., Miranda, M.A. et al. (2020). Photoinduced intersystem crossing in DNA oxidative lesions and epigenetic intermediates. *Chemical Communications* 56 (32): 4404–4407.

40 Wu, T.P., Wang, T., Seetin, M.G. et al. (2016). DNA methylation on N6-adenine in mammalian embryonic stem cells. *Nature* 532 (7599): 329–333.

41 Boulias, K. and Greer, E.L. (2023). Biological roles of adenine methylation in RNA. *Nature Reviews Genetics* 24 (3): 143–160.

42 Liang, Z., Geng, Y., and Gu, X. (2018). Adenine methylation: new epigenetic marker of DNA and mRNA. *Molecular Plant* 11 (10): 1219–1221.

43 Sun, Q., Huang, S., Wang, X. et al. (2015). N6-methyladenine functions as a potential epigenetic mark in eukaryotes. *Bioessays* 37 (11): 1155–1162.

44 Boulias, K. and Greer, E.L. (2022). Means, mechanisms and consequences of adenine methylation in DNA. *Nature Reviews Genetics* 23 (7): 411–428.

45 Xiang, Y., Laurent, B., Hsu, C.-H. et al. (2017). RNA m6A methylation regulates the ultraviolet-induced DNA damage response. *Nature* 543 (7646): 573–576.

46 Cohen, B., Hare, P.M., and Kohler, B. (2003). Ultrafast excited-state dynamics of adenine and monomethylated adenines in solution: implications for the non-radiative decay mechanism. *Journal of the American Chemical Society* 125 (44): 13594–13601.

47 Barbatti, M., Aquino, A.J.A., Szymczak, J.J. et al. (2010). Relaxation mechanisms of UV-photoexcited DNA and RNA nucleobases. *Proceedings of the National Academy of Sciences* 107 (50): 21453–21458.

48 Yang, Y., Shi, W., Chen, Y. et al. (2021). The direct evidence for ESPT route and ICT emission of N6-methyladenine in aqueous solution. *Journal of Luminescence* 229 (117): 698.

49 Aue, D.H., Webb, H.M., and Bowers, M.T. (1976). Quantitative proton affinities, ionization potentials, and hydrogen affinities of alkylamines. *Journal of the American Chemical Society* 98 (2): 311–317.

50 Zhou, Q., Guo, W., Hu, Z. et al. (2024). Can methylated purine bases act as photoionization hotspots? *Photochemistry and Photobiology* 100 (2): 368–379.

51 Chu, J.-M., Ye, T.-T., Ma, C.-J. et al. (2018). Existence of internal N7-methylguanosine modification in mRNA determined by differential enzyme treatment coupled with mass spectrometry analysis. *ACS Chemical Biology* 13 (12): 3243–3250.

52 Zhang, L.-S., Liu, C., Ma, H. et al. (2019). Transcriptome-wide mapping of internal N7-methylguanosine methylome in mammalian mRNA. *Molecular Cell* 74 (6): 1304–1316.e8.

53 Lin, S., Liu, Q., Lelyveld, V.S. et al. (2018). Mettl1/Wdr4-mediated m7G tRNA methylome is required for normal mRNA translation and embryonic stem cell self-renewal and differentiation. *Molecular Cell* 71 (2): 244–255.e5.

54 Karunakaran, V., Kleinermanns, K., Improta, R., and Kovalenko, S.A. (2009). Photoinduced dynamics of guanosine monophosphate in water from broad-band transient absorption spectroscopy and quantum-chemical calculations. *Journal of the American Chemical Society* 131 (16): 5839–5850.

55 Krul, S.E., Hoehn, S.J., Feierabend, K.J., and Crespo-Hernández, C.E. (2021). Excited state dynamics of 7-deazaguanosine and guanosine 5′-monophosphate. *The Journal of Chemical Physics* 154 (7): 075103.

56 Kou, Y., Koag, M.-C., and Lee, S. (2015). N7 methylation alters hydrogen-bonding patterns of guanine in duplex DNA. *Journal of the American Chemical Society* 137 (44): 14067–14070.

57 Vu, B., Cannistraro, V.J., Sun, L., and Taylor, J.S. (2006). DNA synthesis past a 5-methylc-containing cis-syn-cyclobutane pyrimidine dimer by yeast pol η is highly nonmutagenic. *Biochemistry* 45 (30): 9327–9335.

58 Song, Q., Sherrer, S.M., Suo, Z., and Taylor, J.-S. (2012). Preparation of site-specific T=mCG cis-syn cyclobutane dimer-containing template and its error-free bypass by yeast and human polymerase η. *Journal of Biological Chemistry* 287 (11): 8021–8028.

59 Esposito, L., Banyasz, A., Douki, T. et al. (2014). Effect of C5-methylation of cytosine on the photoreactivity of DNA: a joint experimental and computational study of TCG trinucleotides. *Journal of the American Chemical Society* 136 (31): 10838–10841.

60 Jin, P., Wang, J.-N., Wang, X. et al. (2023). Tracking the early stage of triplet-induced photodamage in a DNA dimer and oligomer containing 5-methylcytosine. *The Journal of Physical Chemistry B* 127 (31): 6878–6886.

61 Banyasz, A., Esposito, L., Douki, T. et al. (2016). Effect of C5-methylation of cytosine on the UV-induced reactivity of duplex DNA: conformational and electronic factors. *Journal of Physical Chemistry B* 4232–4242.

62 Schrödter, M. and Wagenknecht, H.-A. (2024). Natural epigenetic DNA modifications cause remote DNA photodamage. *Journal of the American Chemical Society* 146 (30): 20742–20749.

4

Triplet-Mediated Photochemistry of Nucleic Acid Bases

Di Song[1], *Shuyi Yan*[2], *and Hongmei Su*[2]

[1] *Beijing University of Posts and Telecommunications (BUPT), School of Physical Science and Technology, 10 Xitucheng Rd., Beijing, P. R. China*
[2] *Beijing Normal University, College of Chemistry, No. 19, Xinjiekouwai St., Beijing, P. R. China*

4.1 Introduction

Nucleic acids, which include deoxyribonucleic acid (DNA) and ribonucleic acid (RNA), are essential biological macromolecules that function as the carriers of genetic information in living organisms and play a crucial role in gene expression, storage, and transmission [1, 2]. In nature, harmful ultraviolet (UV) radiation is one of the significant environmental factors that cause damage to DNA/RNA. The absorption of UV light by nucleic acid molecules or their photochemical reactions with photosensitizers can lead to cellular mutations, carcinogenesis, and even cell death [3].

Nucleic acid bases serve as the dominant chromophores responsible for the photochemical and photophysical properties of the nucleic acids upon irradiation (Figure 4.1). UV absorption of nucleic bases initially populates singlet-excited states that relax rapidly to the ground electronic state primarily via ultrafast internal conversion (IC) due to the existence of conical intersections (CIs) [4–7]. This rapid relaxation from the excited state protects nucleic acid bases from photochemical reaction, which is likely a result of natural selection during the extended course of molecular evolution. Nevertheless, despite their inherent photostability, nucleic acid bases remain susceptible to a range of photochemical reactions when exposed to UV radiation. This can result in damage to individual monomers, oligonucleotides, and even the larger structures of DNA and RNA [8, 9].

Among the damage caused by these photochemical reactions, the most prevalent is the formation of cyclobutane pyrimidine dimers (CPDs), which occurs through photoaddition between adjacent pyrimidine bases [10–12]. In addition to CPDs, photoaddition can also lead to the formation of 6-4 photoproducts and spore photoproducts (SPs) [13–16]. The formation of 6-4 photoproducts involves reactions between the C—C and C—O double bonds of adjacent pyrimidine bases along the DNA strand [13]. SPs, on the other hand, refer to a specific type of pyrimidine dimer,

Excited State Dynamics in DNA and RNA: Experimental and Computational Approaches on Triplet State, First Edition. Edited by Jinquan Chen.

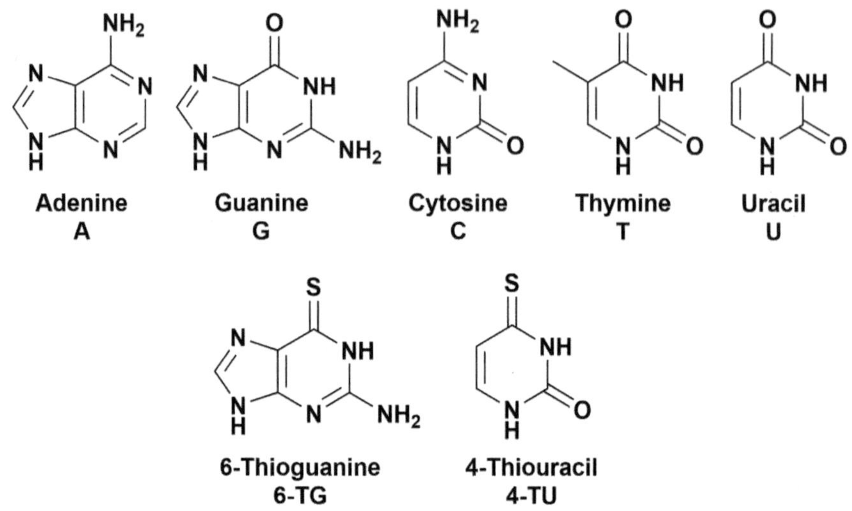

Figure 4.1 Molecular structure of five canonical nucleobases and two typical thio-bases.

5-thyminyl-5,6-dihydrothymine, which is formed by linking the allylic carbon to the C5 of a neighboring thymidine. This process is particularly common in bacterial spores, such as those found in *Bacillus* species [14–16]. Another significant form of damage to nucleic acids is caused by photo-oxidation reaction. This photochemical reaction can occur either through the direct oxidation of DNA bases by UV light or through reactions with singlet oxygen, which is produced as a result of the photoactivation of photosensitizers. Moreover, UV radiation can also promote the formation of DNA–protein crosslinks (DPCs), where proteins become covalently associated with DNA [17–19]. These photoinduced crosslinks are particularly detrimental because they can block DNA replication and transcription, physically impeding essential DNA-related processes [20].

The photochemical reactions that cause damage to nucleic acids involve intricate processes of excited states. As reported, these reactions take place not only in the singlet states but also in the triplet states. Although singlet states exhibit short lifetimes and intersystem crossing (ISC) demonstrates limited efficiency, converging evidence from both experimental and theoretical studies has established the presence of triplet states, albeit with exceptionally low quantum yields [21–23]. Given that triplet states have a longer lifetime in photoreactions, they serve as crucial precursors for the formation of various photochemical products [24, 25]. Therefore, understanding the photochemical reaction mechanisms of nucleic acid bases in the triplet state is essential for unraveling nucleic acid damage. The significance of triplet-mediated photochemical reactions is particularly pronounced in thio-bases (Figure 4.1). The introduction of the heavy sulfur atom significantly enhances the triplet quantum yield of thio-bases, approaching nearly 100% [26, 27]. This high triplet population not only broadens the range of reactants and reaction conditions available for photochemical reactions but also opens up new possibilities for these reactions. Additionally, thio-bases exhibit a red-shift characteristic in their UV absorption spectrum, enabling them to more effectively capture and utilize longer-wavelength ultraviolet A (UVA) light. This property enables them to populate triplet states in the UVA region, thereby offering broad application prospects in photobiology and photomedicine [28–31].

Over the past years, our research group has conducted in-depth explorations into the triplet-mediated photochemistry of nucleic acid bases [32–43]. By utilizing TR IR spectroscopy, in conjunction with other experimental techniques such as UV–vis transient absorption spectroscopy, we have successfully detected the photochemical reaction processes of nucleic acid bases in real time. With the assistance of quantum chemical calculations, we have revealed molecular mechanisms to account for photoaddition, photo-oxidation, and photo-crosslinking reactions. Furthermore, we have emphasized the crucial significance of these triplet-mediated photochemical reactions in the context of nucleic acid photodamage, photorepair mechanisms, and photonic applications. In this chapter, we primarily review a series of significant advancements made by our research group in recent years, aiming to provide valuable references and insights to researchers in related fields.

4.2 Photoaddition Reactions

Under UV light excitation, pyrimidine bases undergo a [2 + 2] cycloaddition reaction involving C—C double bonds, resulting in the formation of CPDs shown in Scheme 4.1, which is one of the most abundant forms of DNA photodamage [10–12]. Various CPDs, including thymine–thymine (T<>T), thymine–cytosine (T<>C), and cytosine–cytosine (C<>C), have been discovered within the photoproducts of DNA both in vivo and in vitro. Generally, the formation of CPDs can be attributed to the addition reaction occurring at the C5—C6 bond in pyrimidine bases. This intricate process may involve multiple electronic states, such as the excited $^1\pi\pi^*$, $^1n\pi^*$, and $^3\pi\pi^*$ states, as well as the electronic ground state. Notably, nonradiative effects among these states play a pivotal role in the underlying reaction mechanism.

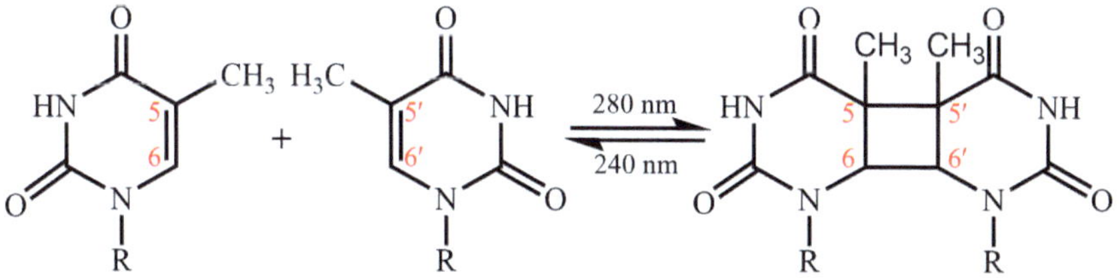

T T T<>T cyclobutane dimer

Scheme 4.1 Chemical structure of cyclobutane pyrimidine dimer (CPD) [42].

In DNA strands or oligopyrimidine nucleotides, pyrimidine bases are maintained in close proximity by the sugar phosphate backbone. In this scenario, despite its short lifetime, the singlet excited state of these bases may still serve as the primary pathway for the formation of CPDs within the strand. Kohler and coworkers [44], using femtosecond TR IR spectroscopy, observed that IR marker bands specific to the cis–syn thymine dimer are formed less than 1 picosecond (ps) after excitation of single-stranded $(dT)_{18}$ at 266 nm. The sub-picosecond formation rate of CPD strongly suggests that the reaction occurs directly from the singlet $^1(\pi,\pi)^*$ state. Theoretically, CASSCF and CASPT2 calculations [45] indicated that the photodimerization can proceed via a barrierless concerted mechanism on a singlet excited state, followed by the CPD formation on the ground state through an S_0/S_1 CI. Noticeably, the observed dimerization rate of approximately 1 ps is faster than the timescale of larger conformational changes in DNA single strands [46]. This indicates that dimerization occurs only for thymine residues that are already aligned in a reactive conformation at the moment of excitation. The above mentioned elegant ultrafast experiments and high-level ab initio calculations have shown that the singlet pathway is highly conformation-restricted, requiring two bases in proximity to be already stacked in a proper orientation for dimerization, so that the bonding process can compete with ultrafast nonradiative decay.

But what happens to the excited bases that do not adopt a proper stacking conformation for the singlet mechanism? One should know that, aside from returning to the ground state by IC, ISC via a dark doorway state $^1(n,\pi)^*$ leads to a long-lived triplet state $^3(\pi,\pi)^*$ [47, 48]. It is reported that the quantum yield of ISC for bases is low, typically not exceeding a few percent [49], yet these long-lived states are of crucial importance as they maintain photoreactivity during DNA conformational changes. When these long-lived states encounter another pyrimidine residue with a favorable orientation brought about by the frequent conformational changes of mobile single-stranded DNA, they undergo dimerization to form the CPD product. By conducting comparative femtosecond transient absorption measurements on thymidine (dT) and thymidine oligomer $(dT)_{20}$, Phillips and coworkers observed a significantly faster T_1 decay dynamics in $(dT)_{20}$, characterized by a time constant of approximately 140 ps, compared to dT, which exhibited a time constant of approximately 4 ns. This finding strongly suggests that the triplet state undergoes chemical quenching through dimerization with an adjacent thymine base, thereby serving as a primary precursor for the formation of CPDs in DNA systems [24]. Kohler and his colleagues also pointed out [50] that the ultrafast formation of CPDs is most readily explained by a singlet reaction, but this does not definitively rule out the involvement of a triplet state. They observed the transient IR spectra of the lowest triplet states of thymine and thymidine, which appeared at 1603 cm^{-1} and approximately 1700 cm^{-1}, respectively, due to carbonyl stretching vibration modes. However, due to the weak IR absorption and low product yields, the CPD photoproduct was not detected.

Given that most of the information regarding the significant participation of the triplet pathway has been obtained through monitoring triplet decay dynamics or probing products using steady-state methods, there is a need for direct experimental evidence of observing the photoproduct formation dynamics. Consequently, we focus on the direct observation of the photoproduct formation dynamics arising from [2 + 2] photocycloaddition reactions of triplet pyrimidines. By utilizing

266-nm excited pyrimidine (either uracil or thymine) with cyclopentene as model reaction systems, we have investigated the photoproduct formation dynamics of [2+2] photocycloaddition reactions of triplet pyrimidines in solution, offering mechanistic insights into this crucial DNA photodamage reaction.

Experimentally, the formation dynamics of photoproducts and the kinetics of triplet quenching are measured by combining nanosecond time-resolved Fourier transform infrared (FTIR) spectroscopy with UV–vis laser flash photolysis (LFP). In the absorption spectrum, negative peaks at 1691 and 1732 cm^{-1} were observed, which correspond to the absorption of uracil molecules. Between these two negative peaks, two positive peaks at 1712 and 1724 cm^{-1} emerged gradually over a period of 0.2–0.5 ms (Figure 4.2). Based on structural optimization and frequency analysis of the products conducted at the B3LYP/6-311+G(d,p) level, these two positive peaks were attributed to the absorption of cycloaddition products. This observation indicates a gradual formation process of the photoaddition products through a bimolecular reaction. Characteristic IR absorption bands indicative of cyclobutane photoproduct formation have also been detected, with product quantum yields determined to be approximately 0.91% for uracil and 0.41% for thymine. When compared to the measured high quenching rate constants of triplet uracil ($1.5 \times 10^9\ M^{-1} \cdot s^{-1}$) or thymine ($0.6 \times 10^9\ M^{-1} \cdot s^{-1}$) by cyclopentene, the inefficient formation of photoproducts suggests that the presence of competitive physical quenching processes on the pathway leads to these products, ultimately resulting in low product yields.

Theoretically, a nonadiabatic reaction pathway occurring through the triplet state T_1 and a diradical stepwise reaction mechanism were proposed, based on B3LYP/6-311+G(d,p) calculations. In the T_1 state, C5 atoms of the two bases first form bond, leading to a diradical intermediate P2-T-im (Figure 4.2). Once this single crosslinked biradical intermediate P2-T-im is formed on the triplet PES, the system tends to cross over to the S_0 surface via the T_1/S_0 crossing points and complete the ring-closure process thereafter, leading eventually to the cyclobutane adduct P2 S in its ground state. Compared to other adiabatic reaction pathways involving the S_1, S_0, and T_1 potential energy surfaces, the energy at the T_1/S_0 potential energy surface crossing point is only 9.8 $kcal\ mol^{-1}$ higher than that of the triplet diradical intermediate, significantly lowering the reaction barrier and making this nonadiabatic pathway energetically favorable.

In subsequent work, we further explored the reaction mechanism of the dinucleotide thymine dinucleoside monophosphate (TpT), utilizing density functional theory (DFT) (ωB97XD) and the time-dependent density functional theory (TD-DFT) (TD-ωB97XD) [36]. Our studies reveal the existence of two competitive physical quenching pathways, competing with the nonadiabatic triplet reaction pathway forming CPD. Photophysical quenching is shown to be the dominant process compared to photochemical CPD formation. Following initial excitation to the $^1\pi\pi^*$ state, the excited population is divided by the dark $^1n\pi^*$ state, predominantly relaxing to S_0 by IC and, to a lesser extent, undergoing ISC to the long-lived triplet state T_1 via the T_4 ($^3\pi\pi^*$) state, which has a negligible energy gap with the $^1n\pi^*$ state. Even in the reactive T_1 state, two newly identified physical quenching pathways

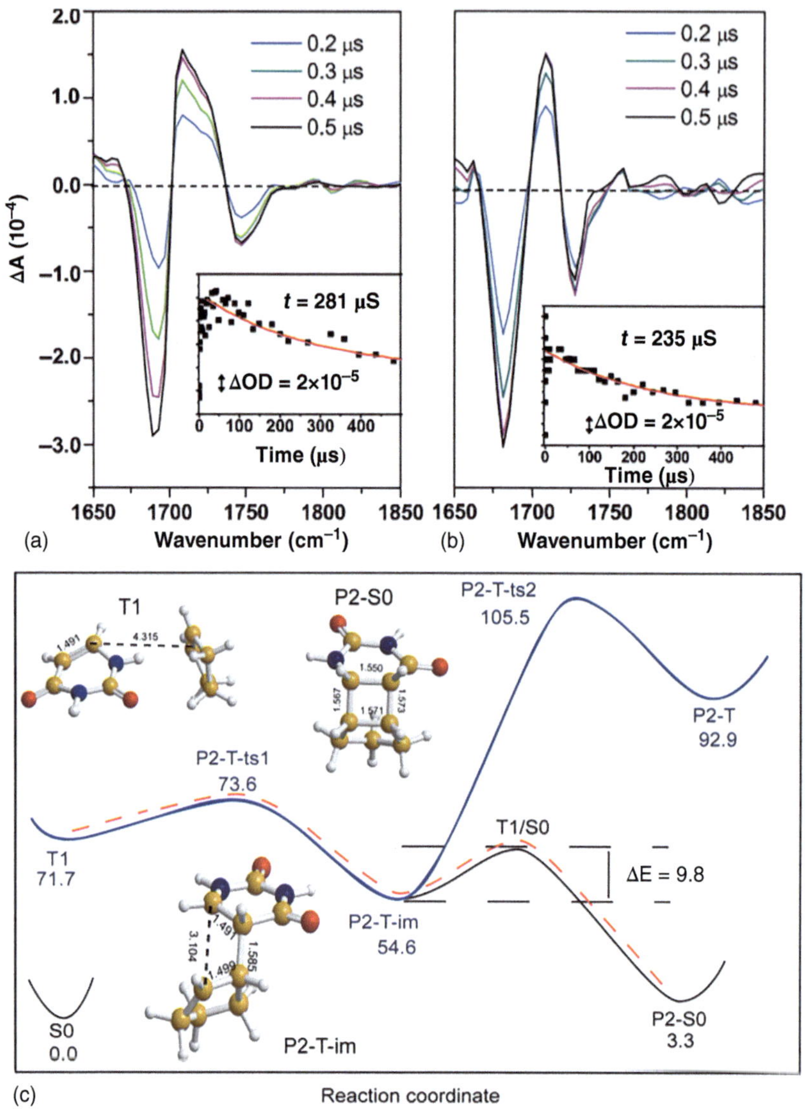

Figure 4.2 Infrared transient absorption spectra of a mixed solution of (a) uracil (0.6 mM) and (b) thymine (0.5 mM), respectively, with cyclopentene (0.10 M) in acetonitrile following laser irradiation at 266 nm. (c) Potential energy curve for the cycloaddition reaction of uracil in the T_1 state. The given energies are the differences relative to the ground state reactants, in units of kcal mol^{-1} ([32] / with permission of American Chemical Society).

lead to the conversion back to the ground state reactants via T_1/S_0 crossing points, and these pathways are also in strong competition with CPD formation. These results provide a rationale for the observed nanosecond triplet decay rates in the single-stranded $(dT)_{18}$ oligonucleotide [25], as well as the inefficiency in forming deleterious CPDs, which allows for a deeper understanding of DNA photostability.

In addition to CPD photoproducts, bacterial spores (such as *Bacillus* species) or A-form DNA produce another type of photoproducts known as SPs, when exposed to ultraviolet (UV) light. The SP is a special UV-induced pyrimidine dimer specifically formed as 5-thyminyl-5,6-dihydrothymine, characterized by the connection of an allylic carbon to the C5 atom of an adjacent thymidine (Scheme 4.2) [14–16]. It is reported that bacterial spores (e.g. *Bacillus* species) exhibit remarkable resistance to UV light compared to normal DNA, primarily due to the formation of SPs and a corresponding efficient chemical repair mechanism associated with the catalysis of a specific enzyme, spore photoproduct lyase (SPL). Upon UV irradiation of DNA, the SP (exemplified by dinucleotide TpT) rapidly forms and remains in the phase of SP-dimer for a long time, like dormancy. This dormant state effectively shields the spore DNA from further photolytic cleavage and other damaging chemical reactions. It is only when the specific enzyme (SPL) initiates the reversal process of SP-dimer formation that the spore DNA can resume its normal state and continue its life cycle.

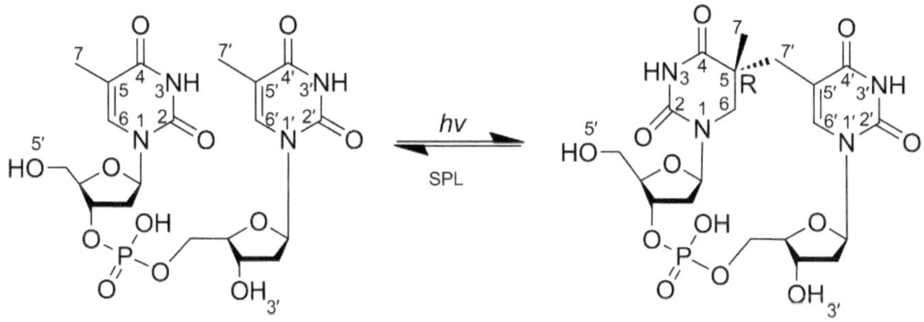

Scheme 4.2 Formation of spore photoproduct (SP) upon UV irradiation of DNA (demonstrated with the dinucleotide TpT) [33, 36].

Although the SP formation plays a vital role in elucidating DNA photolesions, the precise mechanism underlying this process remains incompletely understood. For a considerable period, there has been debate regarding whether this reaction occurs through a concerted mechanism or a stepwise mechanism involving diradicals. Vigny and coworkers proposed a concerted mechanism [51], which involves the crosslinking between the methyl group of one thymidine and the C—C double bond of another thymidine. On the other hand, Varghese and Wang suggested [52, 53] a stepwise mechanism, where the recombination of a 5-α-thyminyl radical and a 5,6-dihydrothymin-5-yl radical leads to the formation of SPs. Lin and Li further elaborated on a more detailed consecutive mechanism [54] by monitoring the hydrogen/deuterium atom transfer of two deuterium-labeled dinucleotide thymine dinucleoside monophosphate (TpT), namely [D3]TpT and [D4]TpT, after UV irradiation using ^{1}H NMR spectroscopy. They suggested that upon excitation by UV light, the C5—C6 bond of 5′-T undergoes transition into a biradical triplet excited state. Following this, a hydrogen atom from the methyl group of 3′-T migrates to the C6 atom of 5′-T, leading to the generation of two radicals: the 5-α-thyminyl radical and the 5,6-dihydrothymin-5-yl radical. These radicals subsequently recombine to produce the final product, SP.

To gain a deeper understanding of the photochemical reaction mechanism underlying SP formation, we perform theoretical calculations, selecting TpT as our model system—a starting reactant commonly used in most photoreaction experiments. TpT is chosen because it is the smallest DNA fragment capable of accommodating an SP dimer while adhering to the structural constraints of the DNA backbone. The potential energy profiles for SP formation in the possible lowest-lying triplet and singlet ground states, as well as the interactions between these two states, were explored at the B3LYP/6-311++G(d, p)//B3LYP/6-311G(d, p) level of theory. Furthermore, the reaction pathways on the singlet-excited state S_1 were investigated on the basis of TD-DFT calculations utilizing the S_0 stationary structures.

Our computational results indicate that the reaction proceeds through consecutive mechanisms via either the S_1, T_1, or S_0 electronic states, as shown in Figures 4.3 and 4.4. This process involves an initial hydrogen atom transfer from the 3′-T methyl group (specifically from the methyl's H atom) to the C_6 position of the 5′-T nucleotide, followed by the formation of a C_5—CH_2 bond that covalently links the 3′-T and 5′-T residues. However, all the adiabatic reaction pathways solely involving the S_1, T_1, or S_0 states are energetically unfavorable. In contrast, the nonadiabatic pathway that encompasses both the T_1 and S_0 states exhibits the lowest energy barrier and is thus the most energetically favorable. This nonadiabatic pathway involves an initial hydrogen atom transfer with an energy barrier of 14.2 kcal mol^{-1} (or 11.9 kcal mol^{-1} in bulk solution) along the T_1 state, leading to the formation of a triplet biradical intermediate (denoted as TI). From this intermediate, a T_1/S_0 surface crossing occurs, facilitating the transition to the S_0 state. Subsequently,

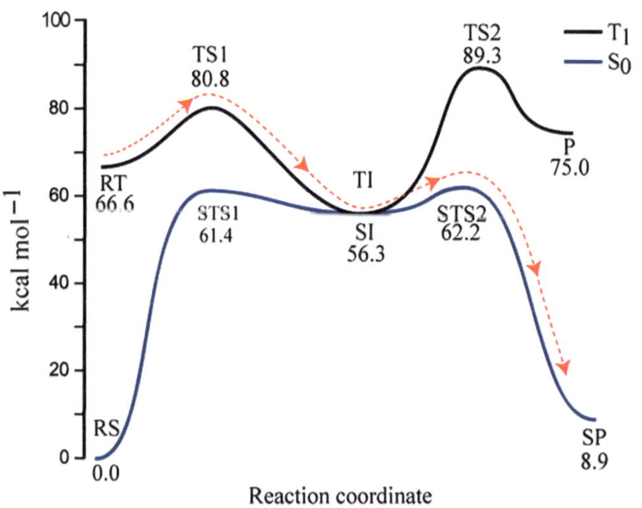

Figure 4.3 Potential energy profiles of the SP reaction pathways along the S_0 and T_1 surfaces. The relative energies are indicated in kcal mol^{-1}. The red dotted line highlights the nonadiabatic reaction pathway ([33] / with permission of American Chemical Society).

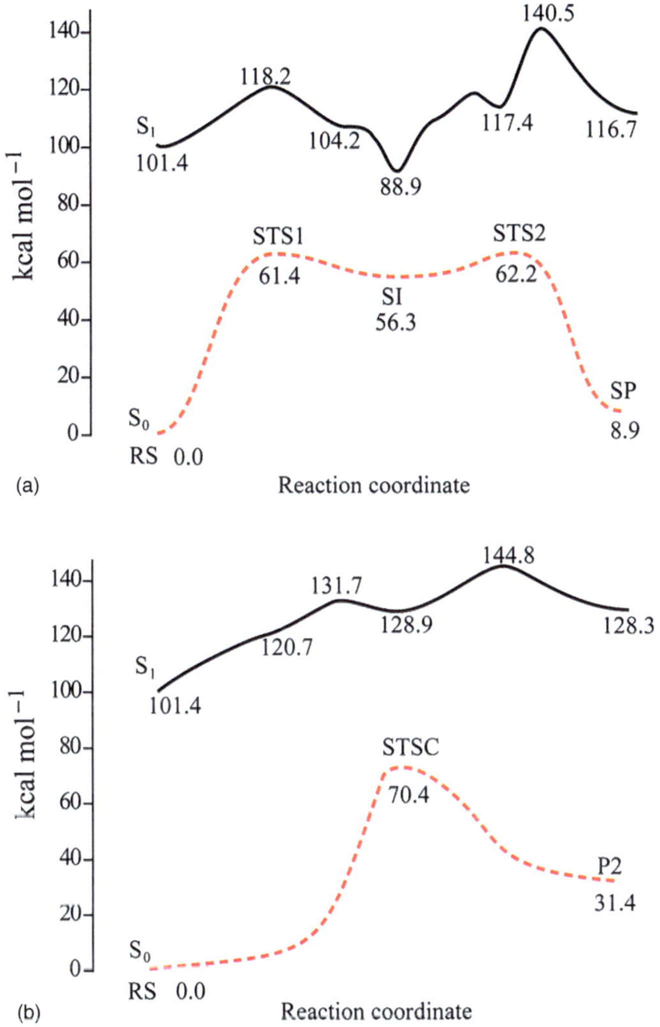

Figure 4.4 Potential energy profiles on the S_0 and S_1 surfaces: (a) consecutive reaction pathways toward SP formation and (b) concerted reaction pathways toward the C5–C6′ crosslinked product P2. The relative energies are indicated in kcal mol^{-1} ([33] / with permission of American Chemical Society).

the C_5—CH_2 bond formation proceeds readily in the S_0 state after ISC from the triplet biradical TI to the singlet biradical SI, with a relatively low energy barrier of 5.9 kcal mol^{-1} (or 3.8 kcal mol^{-1} in bulk solution). Consequently, the overall reaction rate along this nonadiabatic pathway is primarily determined by the hydrogen atom transfer step from the 3′-T methyl group to the C6 of the 5′-T residue. This finding provides a plausible explanation for the experimentally observed kinetic isotope effect of 3.5, which was observed following deuterium substitution at the 3′-T methyl group of the TpT dinucleotide. In summary, our work utilized the quantum chemical calculation method to elucidate that SP formation within the TpT dinucleotide is a stepwise mechanism involving diradicals.

In brief, the formations of CPD and SP are two forms of DNA photodamage under UV radiation. CPD, the most common photoproduct of DNA under UV exposure, arises from covalent crosslinking between adjacent pyrimidine bases, whereas SP is a crosslinking product generated by specific addition reactions between thymine bases. Despite their low proportion among the total DNA photoproducts, these two photoproducts pose a severe threat to normal cellular processes of DNA due to their high mutagenicity, affecting fundamental functions such as DNA replication, transcription, and repair. Once these photodamaged products are formed, they trigger a series of complex biological response networks, ranging from immediate acute effects (such as skin erythema and inflammatory responses) to transient immune suppression and further to long-term mutagenesis and increased risk of skin cancer. These effects not only impact individual health but may also have profound implications for the genetic stability and evolutionary potential of species.

Actually, certain living organisms have evolved mechanisms capable of repairing UV-induced DNA damage [12, 55, 56]. In recent decades, researchers have uncovered various methods for repairing CPDs. Notably, Aziz Sancar won the 2015 Nobel Prize in Chemistry for his comprehensive study of photolyase [56], which showed that some organisms can use blue light (350–500 nm) to reverse CPD damage through DNA photolyases. Zhong and coworkers [10, 57] reported that photolyase, which contains a fully reduced form of flavin adenine dinucleotide ($FADH^-$) as an active cofactor, repairs thymine dimers (T<>T) via an electron-transfer mechanism. However, photolyase is absent in humans and exhibits high costs and instability when utilized in vitro.

Interestingly, alternative solutions might be found in the so-called nanozymes. Studies showed that Fe_3O_4 nanoparticles exhibit intrinsic enzyme-mimicking activity akin to that of natural peroxidases, making them viable peroxidase mimics for oxidizing organic substrates in wastewater treatment [58]. This pioneering discovery sparked a wave of enthusiasm for utilizing nanoparticles to mimic the functions of natural enzymes [59–61]. Illustratively, gold nanoparticles (AuNPs) have demonstrated peroxidase-mimicking activity, facilitating integrin detection [59]; V_2O_5 nanowires can serve as glutathione peroxidase mimics, shielding cells from oxidative harm [60]; and MnO_2 nanoparticles mimic superoxide dismutase and catalase, effectively scavenging detrimental reactive oxygen species (ROS) [61]. These findings underscore the potential of diverse nanoparticles as natural enzyme mimics, showcasing their catalytic capabilities and promising applications in bio-nanomedicine.

Inspired by the fact that photolyase initiates CPD repair through photoinduced electron-transfer reactions, we have utilized titanium dioxide (TiO_2), a stable nanomaterial exhibiting enzyme-like properties, for repairing CPD damage. TiO_2 has been definitively proven to be a powerful and ideal photocatalyst due to its chemical stability and high reactivity. During the photocatalytic reaction, when TiO_2 absorbs energy exceeding its bandgap, interband transitions take place, generating conduction band electrons ($e^-{}_{CB}$) and valence band holes ($h^+{}_{VB}$) pairs, achieving a charge-separated state [62–66]. These charge pairs inevitably undergo processes such as trapping, recombination, and interfacial charge transfer on the surface of TiO_2 [65, 66]. In reduction reactions, TiO_2 unequivocally serves as an

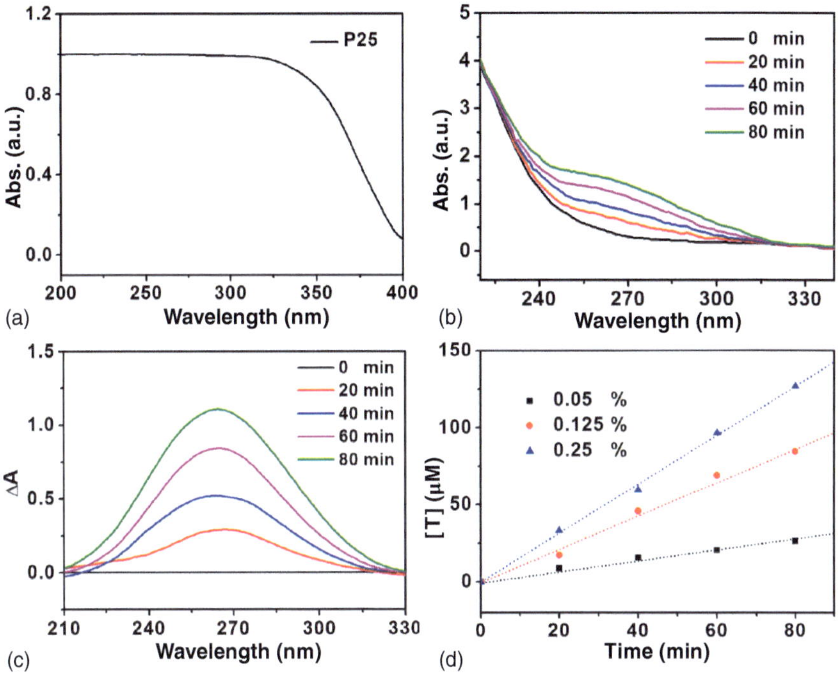

Figure 4.5 (a) Steady-state UV absorption spectrum of the P25. (b) Steady-state repair of thymine dimer by the P25 (0.25%) by illumination of the reaction mixture under UVA light. (c) Difference UV spectra from the data in panel (b). (d) Repair rates of CPD with different concentrations of P25 (0.05%, 0.125%, and 0.25%) by monitoring the repaired T bases using 266-nm absorption changes ([42] / with permission of American Chemical Society).

electron donor, with electrons in its conduction band or captured by surface defects transferring to organic substrates adsorbed on the photocatalyst surface, which is essential for the photocatalytic process.

Using steady-state UV absorption spectroscopy, we monitored the reversal process of thymine dimer T<>T (synthesized from all-thymine oligodeoxynucleotide $(dT)_8$) back to thymine monomer T under UVA radiation that exclusively excites TiO_2 particles (Figure 4.5). The obtained differential spectrum clearly demonstrates the characteristic absorption band of thymine monomer at a wavelength of 266 nm after photorepair. Further, through systematic transient UV–vis spectroscopy technique, we observed the key reactive intermediate of this photorepair process, namely, the charge-separated state (e^-_{CB}/h^+_{VB}) of P25, and tracked the reaction dynamics in real time (Figure 4.6). It is found that the decay rate of the charge-separated state on the surface of P25 was accelerated in the presence of CPD, while $(dT)_8$ had no significant effect on it. This observation aligns with the fact that CPD has a more reducible redox potential (−1.96 V vs NHE [10]) compared to the thymine base (−2.18 V vs NHE [10]), indicating that the photogenerated electrons of P25 are more inclined to transfer to CPD. TR spectroscopy provides direct evidence that the photogenerated electron of TiO_2 is transferred to T<>T, causing structural instability

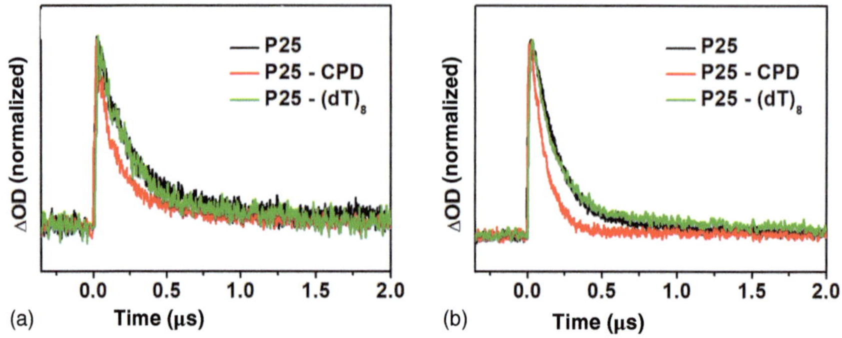

Figure 4.6 Temporal profile of the signals monitored at λ = 310 nm (a) and 400 nm (b) for the P25 (black), P25-CPD (red), and P25-$(dT)_8$ (green) in acetonitrile suspension solution recorded after 355-nm laser excitation ([42] / with permission of American Chemical Society).

and initiating the repair process. T–T$^-$ then undergoes bond cleavage to form T and T$^-$, and T$^-$ returns an electron to TiO_2, thereby completing the photorepair cycle of CPD (Scheme 4.3). Additionally, we observed that the phase composition of TiO_2 can affect its photocatalytic activity in repairing CPD. These findings provide a mechanistic foundation for developing TiO_2-based nanozymes in biological applications for repairing DNA damage.

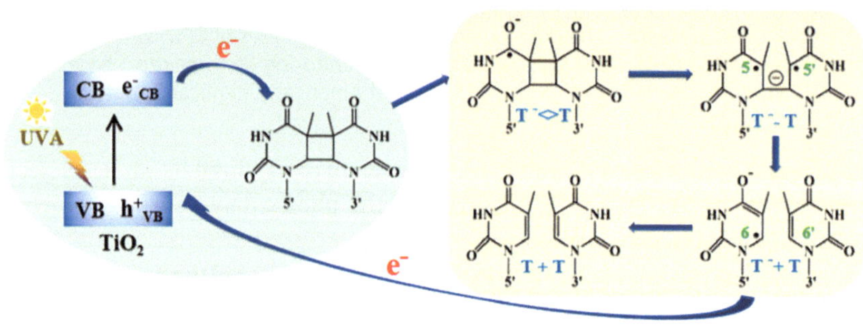

Scheme 4.3 Proposed photo-repairing mechanism of CPD by TiO_2 [42].

Our experimental data conclusively demonstrates that TiO_2 can serve as an efficient photolyase mimic for repairing CPD damage, thereby opening up vast potential applications for TiO_2 in the field of bioremediation. Beyond its exceptional photocatalytic performance, TiO_2 nanoparticles, due to their minute size and excellent biocompatibility, are easily internalized by cells without triggering any adverse immune responses in vivo [67]. Numerous studies have confirmed that TiO_2 exhibits extremely low cytotoxicity and is harmless to cell lines, making it safe for application in various fields such as sunscreens, food additives, pharmaceutical treatments, and coatings [68, 69].

Despite the limitation of TiO_2's relatively large bandgap (approximately 3.0 eV), which restricts its ability to absorb light with wavelengths shorter than 400 nm, scientists have devised a solution. By loading a variety of nanoparticles onto the TiO_2 substrate, particularly by employing single-atom catalysts (SACs) where active metal species exist stably as isolated single atoms, this challenge has been effectively addressed [70]. The introduction of SACs not only narrows the bandgap of the semiconductor catalyst but also creates impurity energy levels, thereby significantly broadening the light response range and enhancing the utilization efficiency of visible light, making solar-driven photocatalytic reactions feasible [71, 72]. Therefore, photocatalysts designed based on the TiO_2 platform will be pivotal in bioremediation, offering innovative solutions to repair DNA photodamage.

4.3 Photo-oxidation Reactions

In biological systems, there exists a class of modified bases that resemble common bases in structure but exhibit significantly different photochemical properties. Among them, thio-bases represent a unique subclass. By replacing an oxygen atom with a sulfur atom, thio-bases show distinctive photosensitive characteristics. Their absorption spectrum is significantly red-shifted, and their triplet state quantum yield is dramatically increased to nearly 100%, due to the heavy-atom effect of the sulfur atom. Such an extremely high triplet state quantum yield enables thio-bases to efficiently interact with O_2 and generate singlet oxygen through energy transfer. Singlet oxygen is highly oxidative and can selectively attack surrounding biomolecules, including lipids, proteins, and DNA. This property suggests that thio-bases could serve as potential photosensitizers for photodynamic therapy (PDT). However, singlet oxygen can also oxidize the thio-bases themselves, resulting in the accumulation of sulfonic acid oxidation products. For exogenous thio-drugs, this self-oxidation process may trigger off-target gene mutations and even induce secondary carcinogenesis. Therefore, the photo-oxidation reactions initiated by the triplet state play a significant role in the photochemistry of these thio-bases.

The thiopurines 6-mercaptopurine (6-MP), azathioprine, and 6-Thioguanine (6-TG), classified as prodrugs—substances that the body metabolically converts into active therapeutic forms—are metabolized into 6-TG and incorporated into DNA [28, 29, 73]. However, patients who have been taking these drugs for a long time and are frequently exposed to sunlight face a serious issue—a significant increase in the incidence of skin cancer. This phenomenon is not accidental but closely related to the photochemical properties of 6-TG [30, 74]. 6-TG can strongly absorb UVA light, making it an efficient photosensitizer. Under UVA irradiation, 6-TG generates ROS, particularly singlet oxygen ($^1O_2$) [30, 75, 76], which causes oxidative damage to DNA and proteins [31, 77–80]. Notably, 6-TG itself is also oxidized by $^1O_2$, which represents one of the major forms of DNA damage induced by UVA photoactivation of 6-TG [30, 31, 75]. The main products of the reaction between free 6-TG and $^1O_2$ are guanosine-6-sulfinate (G^{SO2}) and guanosine-6-sulfonate (G^{SO3}), with guanosine (G) as a minor product, as illustrated in Figure 4.7. The formation of G^{SO3} is biologically significant because it cannot form stable base pairs with any

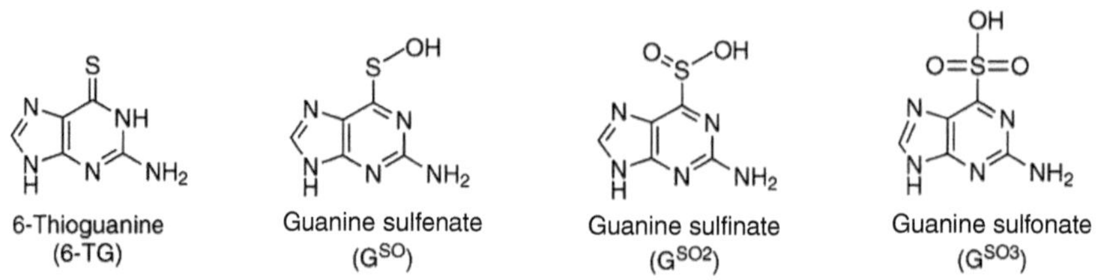

Figure 4.7 Structures of 6-TG and its oxidation products [35].

normal DNA bases. This characteristic of G^{SO3} will strongly block DNA replication and transcription, interfering with cell division and proliferation, and becoming a major factor in increasing the risk of skin cancer [29–31, 75].

Despite extensive biological research, the mechanism by which 6-TG reacts with $^1O_2$ to produce the mutagenic oxidation product G^{SO3} remains unclear. A crucial question is whether there exists an initial, unstable, partially oxidized intermediate—guanine sulfenic acid (G^{SO})—that was previously believed to undergo further oxidation to form G^{SO2} and G^{SO3}. However, this presumptive intermediate, G^{SO}, has never been observed experimentally. The failure to detect G^{SO} in experimental studies [30, 75] may suggest its instability or indicate the existence of different and unknown reaction pathways leading to the formation of G^{SO2} and G^{SO3}, necessitating further investigation. In addition, we found that the major reaction product of 6-TG with $^1O_2$, G^{SO3}, exhibits prominent fluorescence [30, 31], which prompted us to propose a scheme utilizing 6-TG as a sensitive fluorescence probe for $^1O_2$. Given the significance of $^1O_2$ in understanding its associated biological functions and PDT, the development of sensitive and specific $^1O_2$ probes is crucial [81, 82]. Direct monitoring of $^1O_2$ would yield valuable insights. To implement this novel method for detecting $^1O_2$, it is imperative to understand the reaction mechanism between 6-TG and $^1O_2$ and to determine its reaction stoichiometry.

To elucidate the reaction mechanisms, we have explored the potential energy profiles leading to G^{SO3} formation using DFT at the B3LYP/6-311+G** level, both in vacuum and with the solvent effect (H_2O) simulated using the polarizable continuum model (PCM) (Figures 4.8 and 4.9) [35]. Our calculated results reveal that the formation of G^{SO3} from the highly reactive oxidation of 6-TG by $^1O_2$ likely proceeds through a series of intermediates: G^{SOOH}, G^{SO2}, and G^{SO4}, without involving the previously anticipated intermediate, G^{SO}. The addition of $^1O_2$ to the sulfur atom of 6-TG initiates the formation of the peroxy intermediate, G^{SOOH}, which subsequently converts to G^{SO2} by intramolecular rearrangement. Further oxidation of G^{SO2} by $^1O_2$ yields G^{SO4}, which reacts with G^{SO2} or H_2O to produce the most stable oxidation product, G^{SO3}. The newly identified $G^{SOOH} \rightarrow G^{SO2} \rightarrow G^{SO4} \rightarrow G^{SO3}$ pathways involve low energy barriers (8.7 kcal mol^{-1}) and are the most energetically favorable. Conversely, the anticipated G^{SO} pathway is unlikely to occur due to the high energy barrier (35.5 kcal mol^{-1}) associated with its formation.

In addition to the main pathway of $G^{SOOH} \rightarrow G^{SO2} \rightarrow G^{SO4} \rightarrow G^{SO3}$, we also identified a side pathway with a bit higher energy barrier (14.3 kcal mol^{-1}): $G^{SOOH} \rightarrow G$. The predicted mechanisms were further confirmed through joint

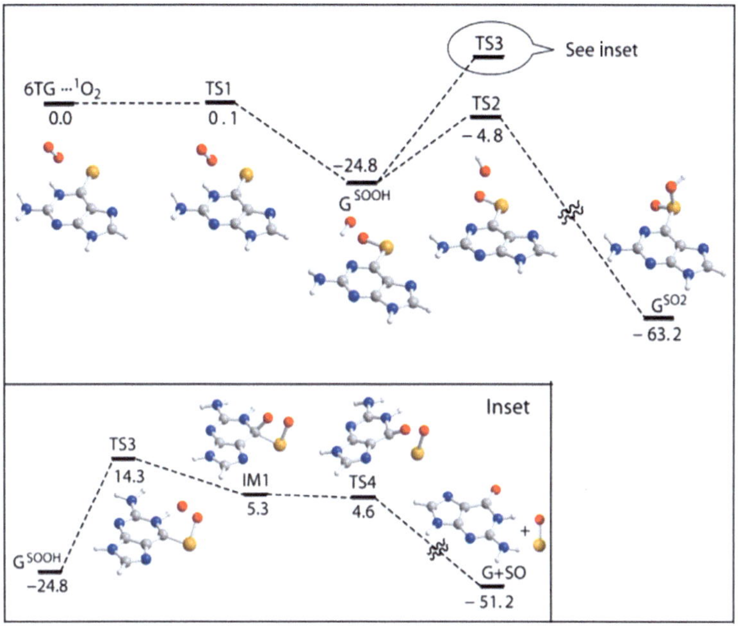

Figure 4.8 Potential energy profiles for the 6-TG reaction with $^1O_2$ toward the formation of the partially oxidized product G^{SO2} and the minor product G. The energies (in kcal mol^{-1}) are obtained at the B3LYP/6-311+G** level with the solvent effect simulated by PCM. All energies given are relative to the reactant complex of 6-TG with $^1O_2$. Carbon, oxygen, nitrogen, sulfur, and hydrogen atoms are denoted with gray, red, blue, yellow, and white balls, respectively ([35] / with permission of American Chemical Society).

fluorescence experimental measurements of reaction stoichiometry and rate constants. The measured fast rate constant of $4.9 \times 10^9\ M^{-1} \cdot s^{-1}$ supports the low-barrier $G^{SOOH} \rightarrow G^{SO2} \rightarrow G^{SO4} \rightarrow G^{SO3}$ mechanism over the high-barrier G^{SO} mechanism. The observed stoichiometry of 0.58 aligns with the predicted values for the $G^{SOOH} \rightarrow G^{SO2} \rightarrow G^{SO4} \rightarrow G^{SO3}$ pathways, further validating this mechanism. Moreover, the reaction rate constant and stoichiometry provide quantitative kinetic data for the reaction between 6-TG and $^1O_2$.

The newly elucidated reaction mechanisms explain the absence of G^{SO} in previous detection and rationalize the observed product distribution, with G^{SO2} and G^{SO3} as the major products and G as a minor product. Furthermore, the low energy barriers and fast reaction rates underscore the high reactivity of $^1O_2$ in oxidizing 6-TG. From both mechanistic and kinetic perspectives, our findings offer new chemical insights into the high phototoxicity of 6-TG in DNA associated with the increased risk of skin cancer in patients treated with thiopurine drugs upon UVA light exposure and point to usage of 6-TG as a sensitive fluorescence probe for the quantitative detection of $^1O_2$, particularly in DNA-related biological environments due to its easy integration into DNA.

Aside from 6-TG, the photo-oxidation mechanisms were also explored for thiopyrimidine bases. David S. Kliger and his coworkers studied photophysical properties

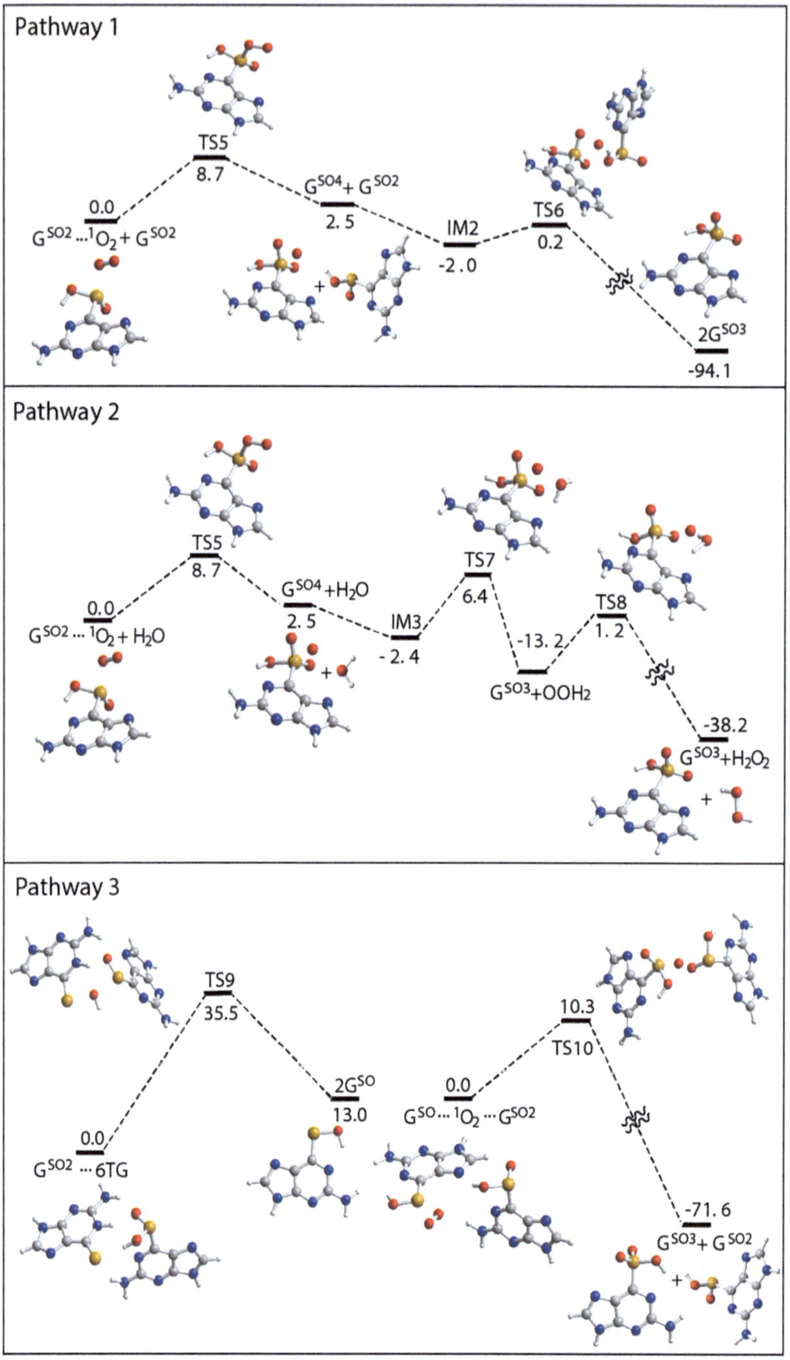

Figure 4.9 Potential energy profiles for the formation of the fully oxidized major product G^{SO3}, preceded by the intermediate of G^{SO2}. The energies (in kcal mol^{-1}) are obtained at the B3LYP/6-311+G** level with the solvent effect simulated by PCM. All energies given are relative to the reactant complex of 6-TG with $^1O_2$. Carbon, oxygen, nitrogen, sulfur, and hydrogen atoms are denoted with gray, red, blue, yellow, and white balls, respectively ([35] / with permission of American Chemical Society).

of 4-Thiouracil (4-TU) in aqueous solution using LFP [27]. Their systematic research determined that the ISC yield of 4-TU is 0.9 ± 0.1. Such a high triplet quantum yield can be explained by the multiconfigurational self-consistent field (MCSCF) and TD-DFT calculations, which reveal that the dark $n\pi^*$ state of 4-TU can serve as a doorway state leading to the efficient ISC to the lowest triplet $\pi\pi^*$ because the energy gap between the two states is minor [83].

The high triplet quantum yield enhances the photochemical reactivity of 4-TU, greatly expanding its application prospects in multiple fields such as biomedicine. With a strong UVA absorption maximum at ~330 nm, 4-TU serves as a site-specific optical probe in nucleic acid crosslinking research. During the photochemical crosslinking process, the C—S bond in the triplet state of 4-TU interacts with the C5—C6 bond in the ground state of pyrimidine bases, generating (5-4) or (6-4) pyrimidine—pyrimidone photoproducts [13]. By accurately detecting these crosslinked products and leveraging specific light excitation wavelengths, 4-TU can be used as a photoprobe to efficiently and precisely analyze the fine structures of DNA/RNA.

Notably, when irradiated with UVA in an aerobic environment, the triplet state of 4-TU, similar to 6-TG, can sensitize oxygen into singlet oxygen through energy transfer. Measurements show that the quantum yields of photosensitized generation of $^1O_2$ by 4-TU in acetonitrile and aqueous solutions are 0.5 and 0.2 [84], respectively. This sensitization process competes vigorously with the photochemical crosslinking reaction between the triplet state of 4-TU and other bases, thus influencing the efficiency of 4-TU as the biological photoprobe. Moreover, the generated singlet oxygen can further oxidize 4-TU to U^{SO3} [37, 85]. It has been found that UVA irradiation of dilute solutions of 4-TU can lead to the formation of uridine under aerobic conditions [86]. These complex photochemical behaviors of 4-TU present significant challenges for its practical applications. Therefore, elucidating its photophysical and photochemical properties is fundamentally essential for facilitating its applications.

We have conducted a comprehensive study of the photophysical and photochemical properties of 4-TU, employing a combination of TR-FTIR spectroscopy and DFT calculations. Our results demonstrate that 4-TU efficiently absorbs UVA radiation, leading to the formation of triplets with a remarkable quantum yield of 0.9, in good agreement with previous work in the literature [27]. In an anaerobic environment saturated with N_2, these reactive triplets primarily undergo crosslinking with the ground state of 4-TU, resulting in the formation of the (5-4)/(6-4) pyrimidine–pyrimidone product. However, under aerobic conditions with the presence of O_2, the triplet state of 4-TU serves as an energy donor, facilitating the production of highly reactive singlet oxygen through triplet–triplet energy transfer. The ROS readily reacts with 4-TU, resulting in the formation of two distinct photoproducts. One is uracil (U), produced with a yield of 0.2. Its characteristic C—O absorption band at $1693\,cm^{-1}$ is clearly observed in both TR-FTIR spectra and steady-state FTIR spectra, as shown in Figure 4.10. The other product is uracil-6-sulfonate (U^{SO3}), which exhibits fluorescence at approximately 390 nm.

The observed formation pathways and product distribution can be comprehensively explained through joint B3LYP/6-311+G(d,p) calculations. These calculations

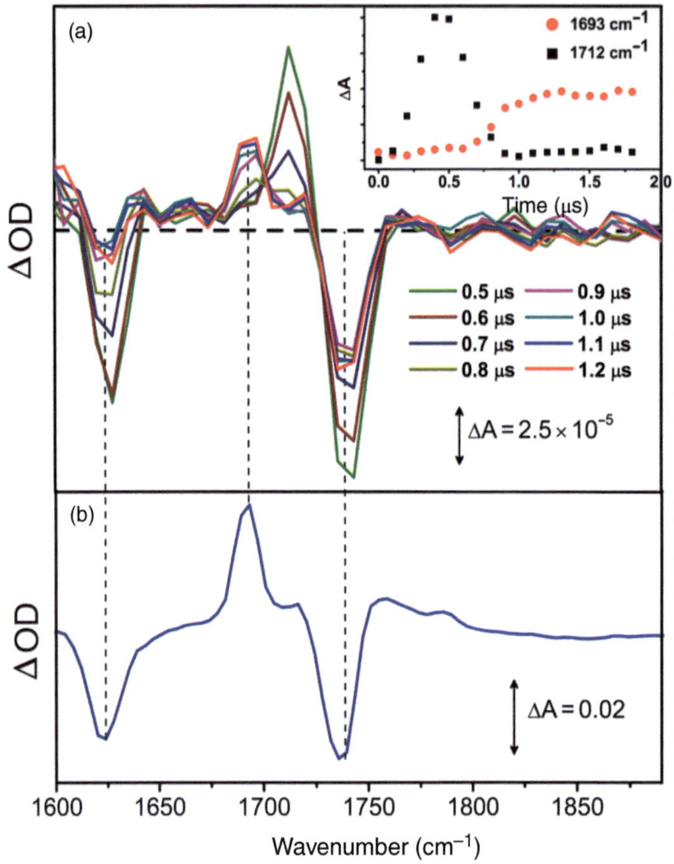

Figure 4.10 (a) Infrared transient absorption (TR-FTIR) spectra of a 1.0 mM solution of 4-TU in acetonitrile saturated with O_2 following 355-nm laser irradiation. The spectra were collected with a resolution of 16 cm^{-1}. (b) Steady-state IR difference spectrum of 4-TU (1.0 mM) in acetonitrile saturated with O_2 after 1 min of 355-nm laser irradiation. The spectrum was collected with a resolution of 8 cm^{-1} ([37] / with permission of American Chemical Society).

have revealed the mechanism behind U^{SO3} formation, which proceeds through a series of peroxy intermediates ($U^{SOOH} \rightarrow U^{SO2} \rightarrow U^{SO4} \rightarrow U^{SO3}$) (Scheme 4.4). Additionally, we have clarified the mechanism of U formation, which involves the addition of $^1O_2$ to the C—S bond of 4-TU. Notably, the energy barriers for U^{SO3} formation are lower than those for U formation. The potential energy profiles indicate that the overall reaction is energetically favorable, explaining the high reactivity of $^1O_2$ in oxidizing 4-TU.

Our findings reveal that 4-TU exhibits strong photosensitizing properties under UVA radiation, efficiently generating singlet oxygen, which holds significant implications for research areas such as photo-crosslinking, photodamage, and PDT. It shows that the ability of 4-TU to photosensitize and produce $^1O_2$ under aerobic conditions, followed by the highly reactive $^1O_2$-mediated processes, can interfere

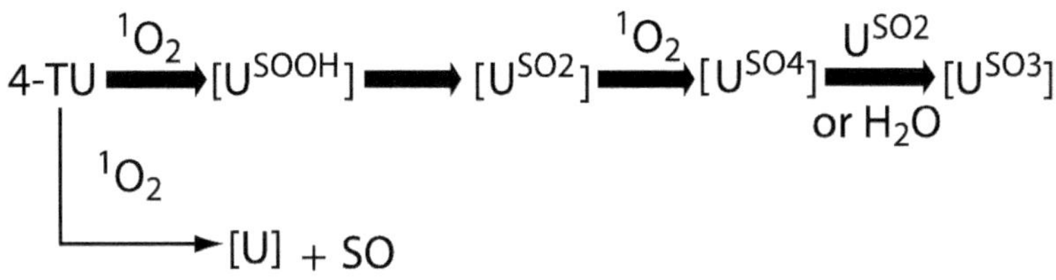

Scheme 4.4 Elucidated reaction mechanisms for 4-TU with $^1O_2$ [37].

with the photo-crosslinking reactions, affecting its performance as a site-specific photoprobe. Therefore, it is essential to carefully evaluate the role of 4-TU as a precursor for singlet oxygen generation and the subsequent oxidative reactions in relevant applications.

As our mechanistic studies reveal [35, 37], thio-bases such as 6-TG and 4-TU have emerged as significant factors causing photodamage to DNA/RNA and phototoxicity to cells, due to their high triplet quantum yields and notable photosensitizing properties. In the field of thio-base photo-applications, one of the key challenges is to effectively eliminate and significantly reduce their phototoxicity. For example, 6-TG—a widely used anticancer, anti-inflammatory, and immunosuppressive drug—can increase skin cancer incidence in patients by 50–200 times with long-term use. To address this issue, researchers have explored various strategies to alleviate or prevent the phototoxicity of 6-TG. One common approach involves using antioxidants (such as ascorbate, *N*-acetylcysteine, β-mercaptoethanol, and cysteine) to reduce the photochemical oxidation of 6-TG. However, studies have shown that, aside from ascorbic acid, most antioxidants react with 6-TG activated by UVA light and 6-TG in oligodeoxynucleotides, forming other harmful adducts through nucleophilic substitution and thereby causing additional damage. This suggests that traditional antioxidant strategies may not be sufficient to address the issue of phototoxicity [87].

In this context, we explore a new method to eliminate the phototoxicity of 6-TG by loading it onto AuNPs. AuNPs are biocompatible and possess unique physicochemical properties that make them suitable for transporting and unloading pharmaceuticals as drug carriers [88–90], especially for sulfur-containing compounds due to their strong affinity and ability to form Au—S bonds [91]. Studies have demonstrated that loading 6-TG molecules onto AuNPs can enhance the intracellular uptake of the drug via endocytosis [92]. Furthermore, the release of 6-TG from AuNPs after internalization can be triggered in response to the intracellular stimulus glutathione (GSH) [93, 94].

In our experiments, citrate-stabilized AuNPs were prepared using a conventional method [95, 96]. The narrow plasmon resonance peak observed at 520 nm in the UV–vis absorption spectrum indicates that these AuNPs are monodispersed with a diameter of approximately 15 ± 2 nm. When 6-TG is introduced, it undergoes an exchange reaction with the loosely attached citrate shell of the AuNPs. This interaction allows 6-TG to conjugate with the AuNPs via an Au—S covalent bond, ultimately leading to the aggregation of the AuNPs. The aggregated AuNPs exhibit a plasmon peak at 700 nm in the UV–vis spectrum, accompanied by a reduction in the intensity of the monodispersed AuNPs peak at 520 nm. We employed centrifugation to remove excess 6-TG, ensuring that 6-TG is fully bound to AuNPs and eliminating interference from free 6-TG on the surface chemistry of the AuNP-adsorbed 6-TG. By conducting systematic ultraviolet irradiation experiments, we have demonstrated that the photochemistry of 6-TG can be successfully modulated by AuNPs, and the phototoxicity under UVA irradiation is effectively eliminated.

Upon irradiation of the AuNP/6-TG solution with a 266 nm laser, the presence of the fluorescent product G^{SO3}, exhibiting a characteristic emission at 408 nm (Figure 4.11a), signifies the occurrence of the photo-oxidation of 6-TG. The weaker fluorescence intensity of G^{SO3} is attributed to the capability of AuNPs to quench the fluorescence of adjacent molecules [97, 98]. As the irradiation time increases, the UV–vis spectra shown in Figure 4.11b display a decrease in the intensity of the aggregation peak at 700 nm and a concomitant increase in the 520 nm monodispersion peak of AuNPs. These results indicate that the 6-TG molecules are desorbed from AuNPs after being oxidized to G^{SO3}, leading to the disaggregation of the AuNPs.

In contrast, for UVA irradiation at 355 nm, it is found that AuNPs can effectively inhibit the formation of $^1O_2$ and the subsequent oxidation of 6-TG by $^1O_2$. As illustrated in Figure 4.11c, prolonged 355 nm irradiation of the AuNP/6-TG solution does not reveal any 408 nm emission characteristic of the oxidation product G^{SO3} in its fluorescence spectrum. This indicates that, without undergoing photo-oxidation to produce G^{SO3}, 6-TG molecules remain conjugated with AuNPs, keeping the AuNPs aggregated. The negligible change in the UV–vis absorption spectrum of AuNPs (Figure 4.11d) further confirms the absence of G^{SO3} products at 355 nm. These findings suggest that AuNPs shield 6-TG from being oxidized to G^{SO3} under UVA (355 nm) irradiation, thereby preventing DNA damage caused by G^{SO3} formation. Additional control experiments using 532 nm irradiation (Figure 4.11e,f) were conducted, in which only AuNPs are excited. The absence of the fluorescent oxidation product G^{SO3} indicates that no disaggregation of the AuNPs has occurred. Hence, the excitation of AuNPs alone is not associated with the photo-oxidation of 6-TG.

With the theoretical results obtained through the MS-CASPT2//SA-CASSCF protocol, we can gain insight into the modulation effects of AuNPs on the photochemistry of 6-TG (Figure 4.11g). Under UVA (355 nm) light, the S_2 excitation corresponds to a transition from the C—S/purine-conjugated π orbital to the π^* orbital, which is closely related to the deformation of the C—S bond. Once 6-TG binds to AuNPs through the formation of an Au—S bond, the C—S double bond characteristic of 6-TG is lost and transforms into a single bond within the C—S—Au structure. Consequently, the C—S/purine-conjugated π orbital feature of 6-TG disappears. Therefore, once these molecules are bound to AuNPs, 355 nm light can no longer excite 6-TG to the S_2 state. Without being photoactivated, the photo-oxidation events of 6-TG are completely suppressed. On the other hand, UVC (266 nm) light corresponds to the S_3 excitation, which is primarily localized within the C—N (of amino group) bond. In this scenario, the formation of the Au—S bond

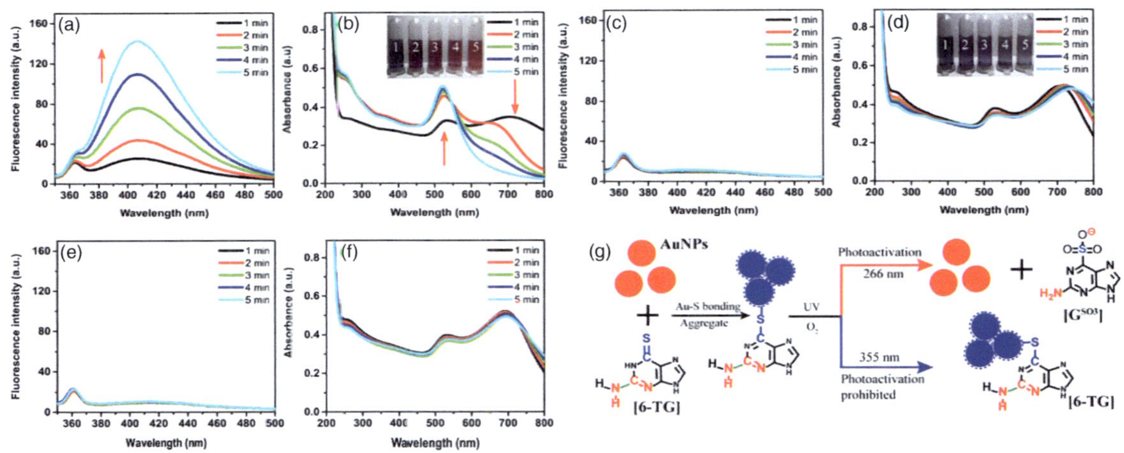

Figure 4.11 Fluorescence and UV-Vis absorption spectra of O_2-saturated AuNP (1 nM)/6-TG (2 μM) mixed solution upon irradiation of 266 nm (a, b), 355 nm (c, d), and 532 nm (e, f). The excitation wavelength for fluorescence measurements was 320 nm. Arrows indicate the intensity change of the product fluorescence or AuNP plasmon peak with irradiation time. (g) Schematic illustration of the photochemistry of 6-TG adsorbed on AuNP surface under UVC or UVA irradiation. Red and blue balls represent the monodispersed and aggregated AuNPs, respectively ([38] / with permission of American Chemical Society).

between 6-TG and AuNPs has minimal impact on the S_3 excitation. At 266 nm, 6-TG molecules can still be photoactivated to the S_3 state and undergo relaxation to the T_1 state, resulting in the formation of $^1O_2$ and subsequent oxidation. It appears that the formation of the Au—S bond between 6-TG and AuNPs only modulates the S_2 excitation and, therefore, inhibits the UVA-induced photochemistry.

Regarding the application of AuNPs to modulate the photochemistry of adsorbed molecules, we have shown that AuNPs can effectively eliminate the UVA phototoxicity side effects of 6-TG, which is a major contributor to the increased incidence of skin cancers among patients undergoing thiopurine treatment. Indeed, 6-TG belongs to an important class of drugs based on thio-bases, including 4-thiothymidine, 2-thiothymidine, 4-TU, 2-thiouracil, and 6-MP, among others. These drugs share the photochemical property of generating $^1O_2$ upon UVA absorption and photosensitization [99–101]. Therefore, the significant protective effects of AuNPs in preventing the harmful UVA phototoxicity of 6-TG and enhancing drug efficacy can, in principle, be extended to this class of thio-base drugs. Further related studies in this area would be of great interest.

4.4 Photo-Crosslinking Reactions

In addition to photoaddition and photo-oxidation reactions, DNA bases can also undergo crosslinking reactions with proteins, forming DPCs. DPCs are complex lesions where proteins are covalently bound to DNA molecules. Their large size and intricate structure can severely impede the processes of DNA replication and transcription. Compared to conventional DNA damage (such as single-strand breaks, double-strand breaks, and base damage), DPCs pose a more significant threat to genomic integrity and protein function. Therefore, elucidating the photochemical mechanisms underlying DPC formation is a key area of current research. This not only helps to understand their roles in aging, carcinogenesis, and structural biology but also provides new insights for the treatment of related diseases.

Taking 6-TG as an example, we have conducted an in-depth study of its photo-crosslinking mechanism with tryptophan (Trp), providing a necessary foundation for understanding the photo-crosslinking mechanisms between DNA and protein systems. 6-TG is an important immunosuppressive agent with anti-inflammatory and anticancer properties. Unlike canonical bases, 6-TG has a strong absorption capacity for UVA light, endowing it with unique photochemical properties. Previous studies have demonstrated that upon UVA irradiation, mixtures of free 6-TG and low-molecular-weight thiol compounds—such as cysteine, *N*-acetylcysteine, or GSH—generate addition products. These reactions do not occur directly with 6-TG, but rather involve nucleophilic substitution by an oxidized intermediate of 6-TG, guanine-6-sulfonate (G^{SO3}), which is formed through singlet oxygen ($^1O_2$)-mediated oxidation of 6-TG. However, whether 6-TG can directly crosslink with amino acids without first being oxidized to G^{SO3} remains an unresolved question.

By combining TR UV–vis absorption spectroscopy and high-performance liquid chromatography with tandem mass spectrometry (HPLC-MS/MS) measurements, we have unveiled a novel photo-crosslinking mechanism between 6-TG and Trp (Scheme 4.5). Upon UVA excitation at 355 nm, 6-TG undergoes direct covalent crosslinking with Trp via a radical mechanism involving electron transfer. The initially excited singlet state of 6-TG rapidly undergoes ISC to form a long-lived triplet state (^{3}6-TG*), as shown in Figure 4.12. This triplet state is redox-active and rapidly engages in electron transfer with Trp (with a rate constant of 10^8 $M^{-1} \cdot s^{-1}$). The electron transfer generates TrpH$^{\bullet+}$ and 6-TG$^{\bullet-}$, followed by rapid deprotonation of TrpH$^{\bullet+}$ to form Trp$^{\bullet}$. These two electron transfer products, 6-TG$^{\bullet-}$ and Trp$^{\bullet}$, subsequently undergo radical recombination, accompanied by the elimination of H_2S, ultimately forming the photoadduct 6-TG-Trp. High-resolution tandem MS (Figure 4.13) further confirms the reaction mechanism by identifying two chiral isomers of 6-TG-Trp 1 (crosslinking between 6-TG$^{\bullet-}$ and C-centered-Trp$^{\bullet}$) and one constitutional isomer of 6-TG-Trp 2 (crosslinking between 6-TG$^{\bullet-}$ and N-centered-Trp$^{\bullet}$).

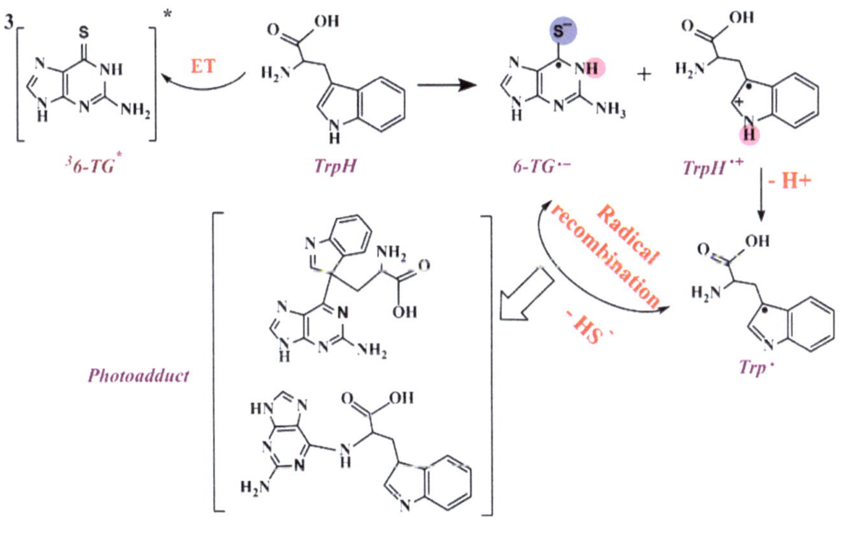

Scheme 4.5 Proposed photoreaction mechanism of 6-TG with TrpH upon 355 nm excitation at pH = 7.0 [41].

Additionally, analogous photoreactions can occur between 6-TG and peptides containing TrpH at pH 7.0, as demonstrated in the similar transient absorption spectra obtained for 6-TG with the oligopeptide WTAEGAT (W, TrpH; T, Thr; A, Ala; E, Glu; G, Gly) that contains a single TrpH upon 355-nm photoexcitation. These results indicate that the photochemical reaction of 6-TG with TrpH is largely context-independent and represents a photolesion of biological significance. Noticeably, exploration with other aromatic amino acids reveals that ^{3}6-TG* does not undergo electron transfer with either tyrosine or histidine. This is attributed to their higher oxidation potentials and insufficient driving forces (ΔG).

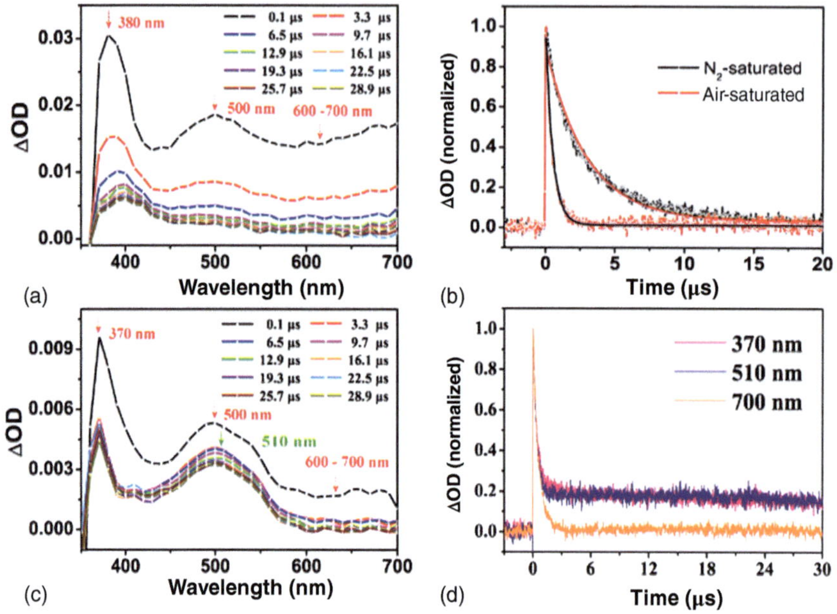

Figure 4.12 (a) Transient absorption spectra of 60 μM 6-TG in pH 7.0 phosphate buffer solution saturated with N_2 following 355-nm excitation. (b) Representative decay traces for 6-TG triplet absorption at 700 nm under air- and N_2-saturated conditions. Fitted curves are shown by red and black solid lines, respectively. (c) Transient absorption spectra of 60 μM 6-TG with 2 mM TrpH in pH 7.0 phosphate buffer solution saturated with N_2 after 355-nm excitation. (d) Normalized kinetics curves for transient absorptions at 370, 510, and 700 nm of 6-TG with TrpH after 355-nm excitation ([41] / with permission of American Chemical Society).

Thiouracil bases can form crosslinks not only with DNA but also with RNA. As a chromophore sensitive to UVA light, thiouracil bases can be incorporated into RNA molecules and selectively excited under UVA irradiation, thereby inducing efficient photo-crosslinking with adjacent amino acid residues of proteins. This crosslinking can be further mapped through appropriate sequencing procedures to reveal the direct contact sites between proteins and RNA. 4-TU, a classic photo-crosslinking probe for RNA–protein interactions, has been widely used in relevant studies due to its ability to effectively reduce the direct excitation of other biomolecules. However, the photo-crosslinking mechanism of 4-TU has not yet been fully elucidated.

Using 4-TU and tryptophan (TrpH) as a model system, we employed nanosecond TR UV–vis LFP to observe the photoinduced electron-transfer process between the triplet state of 4-TU and TrpH. The transient absorption spectra reveal characteristic peaks corresponding to the 4-TU radical anion ($4\text{-TU}^{\bullet-}$) and the TrpH radical cation ($TrpH^{\bullet+}$), as well as the neutral tryptophan radical ($Trp^{\bullet}$) formed via deprotonation of $TrpH^{\bullet+}$. These absorption features indicate that photoinduced electron transfer is the crucial initial step in the photochemical covalent crosslinking between 4-TU and TrpH. By analyzing the decay kinetics

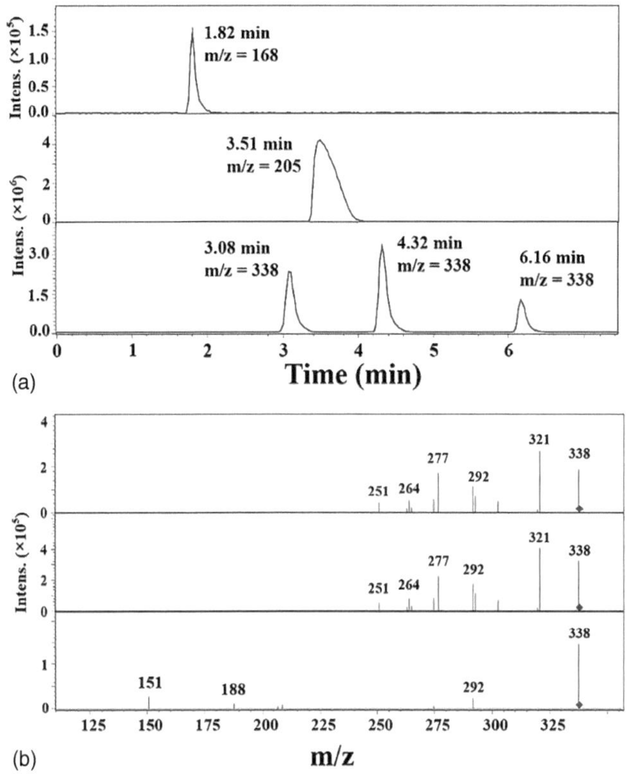

Figure 4.13 HPLC-MS isolation and characterization of the end photoproduct of 6 TG (100 μM) + TrpH (16 mM) at pH = 7.0 after 355 nm photoexcitation. The sample solution was injected into an HPLC instrument and eluted with acetonitrile/acetic acid. (a) Extracted ion chromatography (EIC) m/z 168 at 1.82 min (6-TG), m/z 205 at 3.51 min (TrpH), and m/z 338 at 3.08, 4.32, and 6.16 min (crosslinking photoadduct of 6-TG-TrpH). (b) ESI-(+)-Q-TOFMS/MS spectrum of the m/z 338 at 3.08, 4.32, and 6.16 min, respectively ([41] / with permission of American Chemical Society).

of the 4-TU triplet state, we determined a pseudo-first-order rate constant for the electron-transfer reaction of $2.88 \times 10^9\ \mathrm{L \cdot mol^{-1} \cdot s^{-1}}$, indicating that the process is diffusion-controlled. Furthermore, the electron-transfer driving force (ΔG) calculated using the Rehm–Weller empirical equation is −0.15 eV, suggesting that the reaction is thermodynamically favorable. These results show that electron transfer initiates the photochemical covalent crosslinking of 4-TU with TrpH, subsequently triggering proton transfer and radical recombination, ultimately yielding covalently crosslinked products. This finding elucidates the mechanism by which photoactivated nucleotide derivatives in the triplet state undergo photoinduced electron transfer with proteins and provides essential experimental evidence for investigating photochemical covalent crosslinking in more complex RNA–protein systems.

4.5 Conclusions

In this chapter, we have reviewed our systematic investigations of the triplet-mediated photochemistry of nucleic acid bases, emphasizing their significance in DNA and RNA photodamage, repair mechanisms, and applications in photobiology and photomedicine. By employing advanced techniques such as TR IR spectroscopy, UV–vis transient absorption spectroscopy, and quantum chemical calculations, we have elucidated the mechanisms of various photochemical reactions of nucleic acid bases under UV radiation, which involve photoaddition, photo-oxidation, and photo-crosslinking reactions. These reactions can lead to the formation of various photoproducts, such as CPDs, SPs, G^{SO3}, U^{SO3}, and DPCs, which pose a serious threat to genomic integrity and cellular function.

Additionally, we have explored the potential applications of these photochemical reactions in bioremediation and PDT. For example, our research on titanium dioxide (TiO_2) as a photocatalyst has demonstrated its potential for repairing CPD damage and mimicking natural photolyase enzymes. We have also successfully eliminated the phototoxicity of 6-TG under UVA irradiation by loading it onto AuNPs. Meanwhile, our studies on thio-bases, such as 6-TG and 4-TU, have revealed their unique photochemical properties and potential as photosensitizers and photoprobes. These findings provide new insights for developing therapeutic strategies to mitigate UV-induced DNA damage and enhance the safety and efficacy of thio-base-based drugs.

In summary, the triplet-mediated photochemistry of nucleic acid bases is a highly promising research area. It not only helps us gain a deeper understanding of the molecular mechanisms of UV-induced nucleic acid damage and repair but also offers broad prospects for the development of novel biomedical applications.

References

1 Li, M., Yin, F., Song, L. et al. (2021). Nucleic acid tests for clinical translation. *Chemical Reviews* 121 (17): 10469–10558.

2 Maurel, M.-C. and Décout, J.-L. (1999). Origins of life: molecular foundations and new approaches. *Tetrahedron* 55 (11): 3141–3182.

3 Matsumura, Y. and Ananthaswamy, H.N. (2004). Toxic effects of ultraviolet radiation on the skin. *Toxicology and Applied Pharmacology* 195 (3): 298–308.

4 Merchán, M., González-Luque, R., Climent, T. et al. (2006). Unified model for the ultrafast decay of pyrimidine nucleobases. *The Journal of Physical Chemistry B* 110 (51): 26471–26476.

5 Crespo-Hernández, C.E., Cohen, B., Hare, P.M., and Kohler, B. (2004). Ultrafast excited-state dynamics in nucleic acids. *Chemical Reviews* 104 (4): 1977–2019.

6 Perun, S., Sobolewski, A.L., and Domcke, W. (2006). Conical intersections in thymine. *The Journal of Physical Chemistry A* 110 (49): 13238–13244.

7 Matsika, S. (2004). Radiationless decay of excited states of uracil through conical intersections. *The Journal of Physical Chemistry A* 108 (37): 7584–7590.

8 Rycyna, R.E. and Alderfer, J.L. (1988). Ultraviolet irradiation of nucleic acids: formation, purification, and solution conformational analyses of the cis-syn and trans-syn photodimers of UpU. *Biochemistry* 27 (9): 3142–3151.

9 Beukers, R., Eker, A.P.M., and Lohman, P.H.M. (2008). 50 years thymine dimer. *DNA Repair* 7 (3): 530–543.

10 Liu, Z., Guo, X., Tan, C. et al. (2012). Electron tunneling pathways and role of adenine in repair of cyclobutane pyrimidine dimer by DNA photolyase. *Journal of the American Chemical Society* 134 (19): 8104–8114.

11 DiGiovanna, J.J. and Kraemer, K.H. (2012). Shining a light on xeroderma pigmentosum. *Journal of Investigative Dermatology* 132 (3): 785–796.

12 Kavakli, I.H., Ozturk, N., and Gul, S. (2019). Chapter one - DNA repair by photolyases. In: *Advances in Protein Chemistry and Structural Biology*, vol. 115 (ed. R. Donev), 1–19. Academic Press.

13 Bergstrom, D.E. and Leonard, N.J. (1972). Photoreaction of 4-thiouracil with cytosine: relation to photoreactions in *Escherichia coli* transfer ribonucleic acids. *Biochemistry* 11 (1): 1–9.

14 Slieman Tony, A., Rebeil, R., and Nicholson Wayne, L. (2000). Spore photoproduct (SP) lyase from *Bacillus subtilis* specifically binds to and cleaves SP (5-thyminyl-5,6-dihydrothymine) but not cyclobutane pyrimidine dimers in UV-irradiated DNA. *Journal of Bacteriology* 182 (22): 6412–6417.

15 Donnellan, J.E. and Setlow, R.B. (1965). Thymine photoproducts but not thymine dimers found in ultraviolet-irradiated bacterial spores. *Science* 149 (3681): 308–310.

16 Desnous, C., Guillaume, D., and Clivio, P. (2010). Spore photoproduct: a key to bacterial eternal life. *Chemical Reviews* 110 (3): 1213–1232.

17 Tretyakova, N.Y., Groehler, A.I.V., and Ji, S. (2015). DNA–protein cross-links: formation, structural identities, and biological outcomes. *Accounts of Chemical Research* 48 (6): 1631–1644.

18 Liu, S. and Wang, Y. (2015). Mass spectrometry for the assessment of the occurrence and biological consequences of DNA adducts. *Chemical Society Reviews* 44 (21): 7829–7854.

19 Shigdel, U.K., Zhang, J., and He, C. (2008). Diazirine-based DNA photo-cross-linking probes for the study of protein–DNA interactions. *Angewandte Chemie International Edition* 47 (1): 90–93.

20 Smith, K.C. and O'Leary, M.E. (1967). Photoinduced DNA-protein cross-links and bacterial killing: a correlation at low temperatures. *Science* 155 (3765): 1024–1026.

21 Wu, P., Wang, X., Pan, H., and Chen, J. (2022). Direct observation of excitation wavelength-dependent ultrafast intersystem crossing in cytosine nucleoside solution. *The Journal of Physical Chemistry B* 126 (40): 7975–7980.

22 Gut, I.G., Wood, P.D., and Redmond, R.W. (1996). Interaction of triplet photosensitizers with nucleotides and DNA in aqueous solution at room temperature. *Journal of the American Chemical Society* 118 (10): 2366–2373.

23 Wood, P.D. and Redmond, R.W. (1996). Triplet state interactions between nucleic acid bases in solution at room temperature: intermolecular energy and electron transfer. *Journal of the American Chemical Society* 118 (18): 4256–4263.

24 Kwok, W.-M., Ma, C., and Phillips, D.L. (2008). A doorway state leads to photostability or triplet photodamage in thymine DNA. *Journal of the American Chemical Society* 130 (15): 5131–5139.

25 Pilles, B.M., Bucher, D.B., Liu, L. et al. (2014). Mechanism of the decay of thymine triplets in DNA single strands. *The Journal of Physical Chemistry Letters* 5 (9): 1616–1622.

26 Reichardt, C., Guo, C., and Crespo-Hernández, C.E. (2011). Excited-state dynamics in 6-thioguanosine from the femtosecond to microsecond time scale. *The Journal of Physical Chemistry B* 115 (12): 3263–3270.

27 Milder, S.J. and Kliger, D.S. (1985). Spectroscopy and photochemistry of thiouracils: implications for the mechanism of photocrosslinking in tRNA. *Journal of the American Chemical Society* 107 (25): 7365–7373.

28 Hofbauer, G.F.L., Attard, N.R., Harwood, C.A. et al. (2012). Reversal of UVA skin photosensitivity and DNA damage in kidney transplant recipients by replacing azathioprine. *American Journal of Transplantation* 12 (1): 218–225.

29 Brem, R. and Karran, P. (2012). Multiple forms of DNA damage caused by UVA photoactivation of DNA 6-thioguanine. *Photochemistry and Photobiology* 88 (1): 5–13.

30 Zhang, X., Jeffs, G., Ren, X. et al. (2007). Novel DNA lesions generated by the interaction between therapeutic thiopurines and UVA light. *DNA Repair* 6 (3): 344–354.

31 O'Donovan, P., Perrett, C.M., Zhang, X. et al. (2005). Azathioprine and UVA light generate mutagenic oxidative DNA damage. *Science* 309 (5742): 1871–1874.

32 Yang, C., Yu, Y., Liu, K. et al. (2011). [2 + 2] photocycloaddition reaction dynamics of triplet pyrimidines. *The Journal of Physical Chemistry A* 115 (21): 5335–5345.

33 Du, Q., Zhao, H., Song, D. et al. (2012). Consecutive reaction mechanism for the formation of spore photoproduct in DNA photolesion. *The Journal of Physical Chemistry B* 116 (36): 11117–11123.

34 Du, Q., Zhao, H.M., and Su, H.M. (2013). Theoretical investigation on the absence of spore photoproduct analogue at cytosine-thymine site. *Chinese Journal of Chemical Physics* 26 (6): 661–668.

35 Zou, X., Zhao, H., Yu, Y., and Su, H. (2013). Formation of guanine-6-sulfonate from 6-thioguanine and singlet oxygen: a combined theoretical and experimental study. *Journal of the American Chemical Society* 135 (11): 4509–4515.

36 Zhao, H., Liu, K., Song, D., and Su, H. (2014). Physical quenching in competition with the formation of cyclobutane pyrimidine dimers in DNA photolesion. *The Journal of Physical Chemistry A* 118 (39): 9105–9112.

37 Zou, X., Dai, X., Liu, K. et al. (2014). Photophysical and photochemical properties of 4-thiouracil: time-resolved IR spectroscopy and DFT studies. *The Journal of Physical Chemistry B* 118 (22): 5864–5872.

38 Wang, R., Yue, L., Yu, Y. et al. (2016). Gold nanoparticles modify the photophysical and photochemical properties of 6-thioguanine: preventing DNA oxidative damage. *The Journal of Physical Chemistry C* 120 (26): 14410–14415.

39 Xia, Y., Wang, F., Wang, R.-n. et al. (2019). Reaction kinetics between thiobases and singlet oxygen studied by direct detection of the $^1O_2$ luminescence decay. *Chinese Journal of Chemical Physics* 32 (1): 93–98.

40 Qi, Q.Y., C., Xia, Y. et al. (2019). Photo-induced electron transfer between 4-thiouracil and tryptophan. *Acta Chimica Sinica* 77 (6): 515–519.

41 Dong, J., Huang, C., Guo, S. et al. (2022). Free-radical-mediated photoinduced electron transfer between 6-thioguanine and tryptophan leading to DNA-protein-like cross-link. *The Journal of Physical Chemistry B* 126 (1): 14–22.

42 Zhou, Q., Zhang, T., Jie, J. et al. (2022). TiO_2 as a nanozyme mimicking photolyase to repair DNA damage. *The Journal of Physical Chemistry Letters* 13 (47): 10929–10935.

43 Yang, C., Su, H., Sun, X., and George, M.W. (2012). Ultrafast formation of the benzoic acid triplet upon ultraviolet photolysis and its sequential photodissociation in solution. *The Journal of Chemical Physics* 136 (20): 204507.

44 Schreier, W.J., Schrader, T.E., Koller, F.O. et al. (2007). Thymine dimerization in DNA is an ultrafast photoreaction. *Science* 315 (5812): 625–629.

45 Boggio-Pasqua, M., Groenhof, G., Schäfer, L.V. et al. (2007). Ultrafast deactivation channel for thymine dimerization. *Journal of the American Chemical Society* 129 (36): 10996–10997.

46 Martínez, J.M., Elmroth, S.K., and Kloo, L. (2001). Influence of sodium ions on the dynamics and structure of single-stranded DNA oligomers: a molecular dynamics study. *Journal of the American Chemical Society* 123 (49): 12279–12289.

47 Hare, P.M., Crespo-Hernández, C.E., and Kohler, B. (2007). Internal conversion to the electronic ground state occurs via two distinct pathways for pyrimidine bases in aqueous solution. *Proceedings of the National Academy of Sciences* 104 (2): 435–440.

48 Hare, P.M., Crespo-Hernández, C.E., and Kohler, B. (2006). Solvent-dependent photophysics of 1-cyclohexyluracil: ultrafast branching in the initial bright state leads nonradiatively to the electronic ground state and a long-lived $1n\pi^*$ state. *The Journal of Physical Chemistry B* 110 (37): 18641–18650.

49 Schulman, S.G. (1992). Bioorganic photochemistry, volume 1: photochemistry and the nucleic acids. *Journal of Pharmaceutical Sciences* 81 (1): 112–112.

50 Hare, P.M., Middleton, C.T., Mertel, K.I. et al. (2008). Time-resolved infrared spectroscopy of the lowest triplet state of thymine and thymidine. *Chemical Physics* 347 (1): 383–392.

51 Costalat, R., Blais, J., Ballini, J.P. et al. (1990). Formation of cyclobutane thymine dimers photosensitized by pyridopsoralens: a triplet-triplet energy transfer mechanism. *Photochemistry and Photobiology* 51 (3): 255–262.

52 Varghese, A.J. (1970). 5-thyminyl-5,6-dihydrothymine from DNA irradiated with ultraviolet light. *Biochemical and Biophysical Research Communications* 38 (3): 484–490.

53 Wang, S.Y. (1977). Photochemistry and photobiology of nucleic acids. Volume II: biology. *The Quarterly Review of Biology* 52 (2): 192–192.

54 Lin, G. and Li, L. (2010). Elucidation of spore-photoproduct formation by isotope labeling. *Angewandte Chemie International Edition* 49 (51): 9926–9929.

55 Banaś, A.K., Zgłobicki, P., Kowalska, E. et al. (2020). All you need is light. Photorepair of UV-induced pyrimidine dimers. *Genes* 11: 1304.

56 Sancar, A. (2016). Mechanisms of DNA repair by photolyase and excision nuclease (nobel lecture). *Angewandte Chemie International Edition* 55 (30): 8502–8527.

57 Liu, Z., Tan, C., Guo, X. et al. (2011). Dynamics and mechanism of cyclobutane pyrimidine dimer repair by DNA photolyase. *Proceedings of the National Academy of Sciences* 108 (36): 14831–14836.

58 Gao, L., Zhuang, J., Nie, L. et al. (2007). Intrinsic peroxidase-like activity of ferromagnetic nanoparticles. *Nature Nanotechnology* 2 (9): 577–583.

59 Gao, L., Liu, M., Ma, G. et al. (2015). Peptide-conjugated gold nanoprobe: intrinsic nanozyme-linked immunsorbant assay of integrin expression level on cell membrane. *ACS Nano* 9 (11): 10979–10990.

60 Vernekar, A.A., Sinha, D., Srivastava, S. et al. (2014). An antioxidant nanozyme that uncovers the cytoprotective potential of vanadia nanowires. *Nature Communications* 5 (1): 5301.

61 Huang, Y., Liu, Z., Liu, C. et al. (2016). Self-assembly of multi-nanozymes to mimic an intracellular antioxidant defense system. *Angewandte Chemie International Edition* 55 (23): 6646–6650.

62 Sachs, M., Pastor, E., Kafizas, A., and Durrant, J.R. (2016). Evaluation of surface state mediated charge recombination in anatase and rutile TiO_2. *The Journal of Physical Chemistry Letters* 7 (19): 3742–3746.

63 Haisch, C., Nunes, B.N., Schneider, J. et al. (2018). Transient absorption studies on nanostructured materials and composites: toward the development of new photocatalytic systems. *Zeitschrift für Physikalische Chemie* 232 (9–11): 1469–1493.

64 Miao, T.J. and Tang, J. (2020). Characterization of charge carrier behavior in photocatalysis using transient absorption spectroscopy. *The Journal of Chemical Physics* 152 (19): 194201.

65 Ma, J., Miao, T.J., and Tang, J. (2022). Charge carrier dynamics and reaction intermediates in heterogeneous photocatalysis by time-resolved spectroscopies. *Chemical Society Reviews* 51 (14): 5777–5794.

66 Qian, R., Zong, H., Schneider, J. et al. (2019). Charge carrier trapping, recombination and transfer during TiO_2 photocatalysis: an overview. *Catalysis Today* 335: 78–90.

67 Chellappa, M., Anjaneyulu, U., Manivasagam, G., and Vijayalakshmi, U. (2015). Preparation and evaluation of the cytotoxic nature of TiO_2 nanoparticles by direct contact method. *International Journal of Nanomedicine* 10 (1): 31–41.

68 Efsa, P. O. A. A., Bampidis, V., Azimonti, G., Bastos, M.D. et al. (2021). Safety and efficacy of a feed additive consisting of titanium dioxide for all animal species (titanium dioxide manufacturers association). *EFSA Journal* 19 (6): e06630.

69 Abbas, W.A., Abdullah, I.H., Ali, B.A. et al. (2019). Recent advances in the use of TiO_2 nanotube powder in biological, environmental, and energy applications. *Nanoscale Advances* 1 (8): 2801–2816.

70 Sun, H., Tang, R., and Huang, J. (2022). Considering single-atom catalysts as photocatalysts from synthesis to application. *iScience* 25 (5): 104232.

71 Yang, K., Huang, W.-Q., Xu, L. et al. (2016). Insights into enhanced visible-light photocatalytic activity of CeO_2 doped with nonmetal impurity from the first principles. *Materials Science in Semiconductor Processing* 41: 200–208.

72 Yang, C., Zhao, Z., and Liu, Q. (2022). Regulating effect on photocatalytic water splitting performance of g-C_3N_4 via confinement of single atom Pt based on energy band engineering: a first principles investigation. *Applied Surface Science* 577: 151916.

73 Waters, T.R. and Swann, P.F. (1997). Cytotoxic mechanism of 6-thioguanine: hMutSα, the human mismatch binding heterodimer, binds to DNA containing S6-methylthioguanine. *Biochemistry* 36 (9): 2501–2506.

74 Euvrard, S., Kanitakis, J., and Claudy, A. (2003). Skin cancers after organ transplantation. *New England Journal of Medicine* 348 (17): 1681–1691.

75 Ren, X., Li, F., Jeffs, G. et al. (2010). Guanine sulphinate is a major stable product of photochemical oxidation of DNA 6-thioguanine by UVA irradiation. *Nucleic Acids Research* 38 (6): 1832–1840.

76 Zhang, Y., Zhu, X., Smith, J. et al. (2011). Direct observation and quantitative characterization of singlet oxygen in aqueous solution upon UVA excitation of 6-thioguanines. *The Journal of Physical Chemistry B* 115 (8): 1889–1894.

77 Brem, R., Daehn, I., and Karran, P. (2011). Efficient DNA interstrand crosslinking by 6-thioguanine and UVA radiation. *DNA Repair* 10 (8): 869–876.

78 Brem, R., Li, F., and Karran, P. (2009). Reactive oxygen species generated by thiopurine/UVA cause irreparable transcription-blocking DNA lesions. *Nucleic Acids Research* 37 (6): 1951–1961.

79 Gueranger, Q., Kia, A., Frith, D., and Karran, P. (2011). Crosslinking of DNA repair and replication proteins to DNA in cells treated with 6-thioguanine and UVA. *Nucleic Acids Research* 39 (12): 5057–5066.

80 Daehn, I., Brem, R., Barkauskaite, E., and Karran, P. (2011). 6-thioguanine damages mitochondrial DNA and causes mitochondrial dysfunction in human cells. *FEBS Letters* 585 (24): 3941–3946.

81 Gomes, A., Fernandes, E., and Lima, J.L.F.C. (2005). Fluorescence probes used for detection of reactive oxygen species. *Journal of Biochemical and Biophysical Methods* 65 (2): 45–80.

82 Hirakawa, K. (2009). Fluorometry of singlet oxygen generated via a photosensitized reaction using folic acid and methotrexate. *Analytical and Bioanalytical Chemistry* 393 (3): 999–1005.

83 Shukla, M.K. and Leszczynski, J. (2004). Electronic transitions of thiouracils in the gas phase and in solutions: time-dependent density functional theory (TD-DFT) study. *The Journal of Physical Chemistry A* 108 (46): 10367–10375.

84 Heihoff, K., Redmond, R.W., Braslavsky, S.E. et al. (1990). Quantum yields of triplet and $O_2(1\Delta g)$ formation of 4-thiouridine in water and acetonitrile. *Photochemistry and Photobiology* 51 (6): 635–641.
85 Hayatsu, H. and Yano, M. (1969). Permanganate oxidation of 4-thiouracil derivatives: isolation and properties of 1-substituted-2-pyrimidone-4-sulfonates. *Tetrahedron Letters* 10 (9): 755–758.
86 Favre, A., Saintomé, C., Fourrey, J.-L. et al. (1998). Thionucleobases as intrinsic photoaffinity probes of nucleic acid structure and nucleic acid-protein interactions. *Journal of Photochemistry and Photobiology B: Biology* 42 (2): 109–124.
87 Ren, X., Xu, Y.-Z., and Karran, P. (2010). Photo-oxidation of 6-thioguanine by UVA: the formation of addition products with low molecular weight thiol compounds. *Photochemistry and Photobiology* 86 (5): 1038–1045.
88 Ghosh, P., Han, G., De, M. et al. (2008). Gold nanoparticles in delivery applications. *Advanced Drug Delivery Reviews* 60 (11): 1307–1315.
89 Sun, T., Zhang, Y.S., Pang, B. et al. (2014). Engineered nanoparticles for drug delivery in cancer therapy. *Angewandte Chemie International Edition* 53 (46): 12320–12364.
90 Huo, S., Jin, S., Ma, X. et al. (2014). Ultrasmall gold nanoparticles as carriers for nucleus-based gene therapy due to size-dependent nuclear entry. *ACS Nano* 8 (6): 5852–5862.
91 Grönbeck, H., Curioni, A., and Andreoni, W. (2000). Thiols and disulfides on the Au(111) surface: the headgroup–gold interaction. *Journal of the American Chemical Society* 122 (16): 3839–3842.
92 Podsiadlo, P., Sinani, V.A., Bahng, J.H. et al. (2008). Gold nanoparticles enhance the anti-leukemia action of a 6-mercaptopurine chemotherapeutic agent. *Langmuir* 24 (2): 568–574.
93 Ock, K., Jeon, W.I., Ganbold, E.O. et al. (2012). Real-time monitoring of glutathione-triggered thiopurine anticancer drug release in live cells investigated by surface-enhanced Raman scattering. *Analytical Chemistry* 84 (5): 2172–2178.
94 Hong, R., Han, G., Fernández, J.M. et al. (2006). Glutathione-mediated delivery and release using monolayer protected nanoparticle carriers. *Journal of the American Chemical Society* 128 (4): 1078–1079.
95 Turkevich, J., Stevenson, P.C., and Hillier, J. (1951). A study of the nucleation and growth processes in the synthesis of colloidal gold. *Discussions of the Faraday Society* 11: 55–75.
96 Frens, G. (1973). Controlled nucleation for the regulation of the particle size in monodisperse gold suspensions. *Nature Physical Science* 241 (105): 20–22.
97 Chhabra, R., Sharma, J., Wang, H. et al. (2009). Distance-dependent interactions between gold nanoparticles and fluorescent molecules with DNA as tunable spacers. *Nanotechnology* 20 (48): 485201.
98 Samanta, A., Zhou, Y., Zou, S. et al. (2014). Fluorescence quenching of quantum dots by gold nanoparticles: a potential long range spectroscopic ruler. *Nano Letters* 14 (9): 5052–5057.

99 Pollum, M. and Crespo-Hernández, C.E. (2014). Communication: the dark singlet state as a doorway state in the ultrafast and efficient intersystem crossing dynamics in 2-thiothymine and 2-thiouracil. *The Journal of Chemical Physics* 140 (7): 071101.
100 Pollum, M., Jockusch, S., and Crespo-Hernández, C.E. (2014). 2,4-dithiothymine as a potent UVA chemotherapeutic agent. *Journal of the American Chemical Society* 136 (52): 17930–17933.
101 Pollum, M., Jockusch, S., and Crespo-Hernández, C.E. (2015). Increase in the photoreactivity of uracil derivatives by doubling thionation. *Physical Chemistry Chemical Physics* 17 (41): 27851–27861.

5

Drug-Sensitized Triplet States Generation in DNA

Carlos Montero-Galán, Gemma M. Rodríguez-Muñiz, and Virginie Lhiaubet-Vallet

Instituto Universitario Mixto de Tecnología Química, Universitat Politècnica de València-CSIC (UPV-CSIC), Avda de los Naranjos, s/n, Valencia, Spain

5.1 Introduction

Photosensitization by drugs is a present-day issue because modern lifestyles often associate exposure to sunlight with the presence of chemical substances in sun-exposed skin areas.

Clinically, photosensitization is defined as the appearance of abnormal cutaneous reactions triggered by irradiation doses that are considered harmless [1–4]. The photosensitizing drug is, in general, not toxic in itself, but irradiation (usually in the ultraviolet A [UVA] range) triggers its reactivity toward biological components. This adverse reaction occurs predominantly as phototoxicity that, unlike photoallergy, is not an immune-mediated response and can be reversed by withdrawal of the therapeutic agent [1–4]. More importantly, few photosensitizing drugs have also been reported to be photogenotoxic, phototumorigenic, and/or photocarcinogenic in vivo. This is the case with fluoroquinolones (FQs; Chart 5.1), which are widely used, broad-spectrum antibacterial agents. Lomefloxacin (LFX) and fleroxacin (FLX) appeared as potent photocarcinogens in mice, causing the development of malignant neoplasms like squamous cell carcinoma (SCC), whereas other FQs, such as ofloxacin (OFX) or ciprofloxacin (CPX) and the related compound nalidixic acid, also increased the development of skin tumors, but to a lesser extent [5–7]. Indeed, it has been found that FQs can be more potent phototumorigenic agents than 8-methoxypsoralen (8-MOP) [8]. Although the focus of this chapter is on those drugs that can induce undesirable photosensitization reactions, it should be mentioned that in some special cases, a benefit may be obtained, such as with psoralen-mediated therapy (PUVA) or photodynamic therapy (PDT) [9].

Photogenotoxicity is related to the formation of DNA lesions. In this context, a large variety of DNA modifications have been described, including pyrimidine (Pyr) dimers, covalent adducts, and oxidatively generated lesions [10, 11]. The type as

Excited State Dynamics in DNA and RNA: Experimental and Computational Approaches on Triplet State, First Edition. Edited by Jinquan Chen.

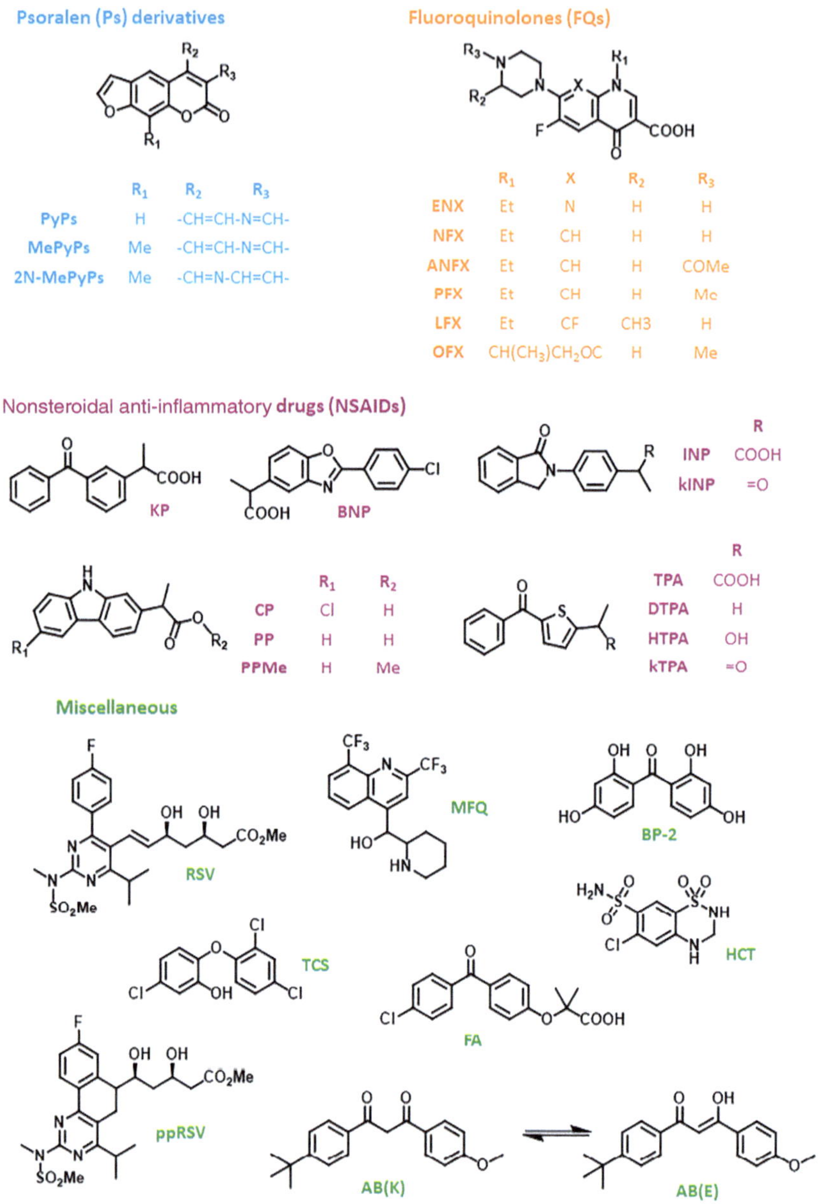

Chart 5.1 Structure of the photosensitizing compounds.

well as the relative yields of the photodamages can be regarded as a fingerprint of the photosensitization mechanism. The first step of the photosensitization process is the absorption of light by the drug (i.e. photosensitizer, Phs), usually in the UVA and/or visible range where DNA has no absorption (Figure 5.1). A short-lived singlet excited state is populated and decays by different pathways, such as fluorescence, intramolecular reactions, or nonradiative pathways, i.e. internal conversion

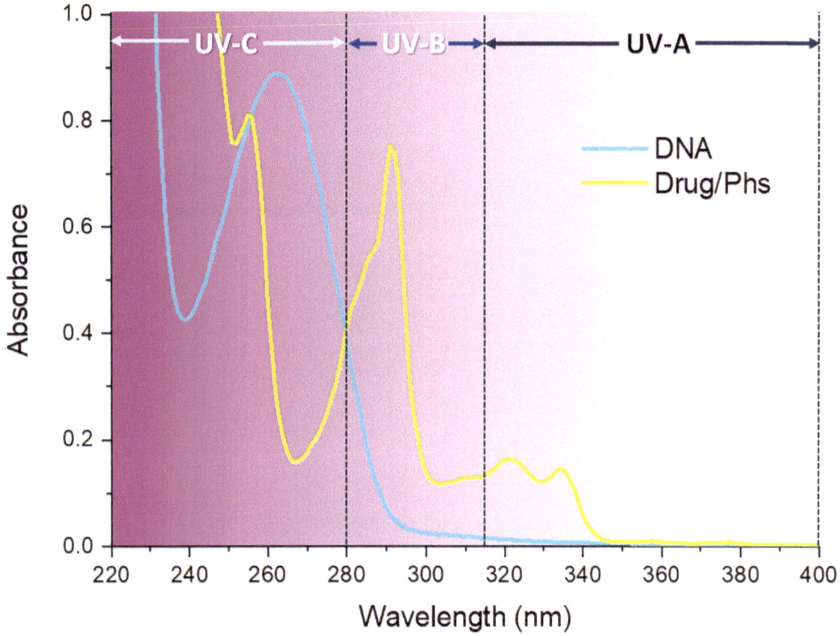

Figure 5.1 DNA and photosensitizing drug absorption spectra.

and intersystem crossing (ISC). This latter process gives rise to the triplet excited state (T_1) population, which is longer lived than the singlet manifold and thus best suited for intermolecular reactions limited by diffusion. In fact, the triplet manifold has been recognized as a biologically active species due to its central role in the photosensitization of biomolecules [3].

Photosensitization reactions are bimolecular reactions that can be classified as oxygen-dependent or oxygen-independent processes depending on whether molecular oxygen (O_2) is required throughout the reaction [12]. The first step consists of the interaction between the drug excited state (usually T_1) and the biological substrate or dissolved molecular oxygen. Oxygen-dependent reactions, also defined as photosensitized oxidations, include the well-known type I and type II processes. In the former, an electron transfer or hydrogen abstraction takes place between the excited drug and the substrate in its ground state and leads to a radical that subsequently reacts with O_2. During type II processes, singlet oxygen ($^1O_2$) is produced from an energy transfer from the triplet excited state of Phs (^{3}Phs*) to O_2. This reactive oxygen species (ROS) then attacks the biological components. Concerning the oxygen-independent reactions, a first reaction corresponds to the formation of a covalent link between the sensitizer and the substrate. However, the most reported is a triplet–triplet energy transfer (TTET, Figure 5.2) from the excited drug to the Pyr DNA bases and subsequent [2 + 2] photocycloaddition to generate the cyclobutane pyrimidine dimers (CPDs). This is the key process responsible for the T_1 formation of nucleobases (^{3}N*), and more specifically of the thymine triplet manifold (^{3}Thy*), by drugs [13, 14].

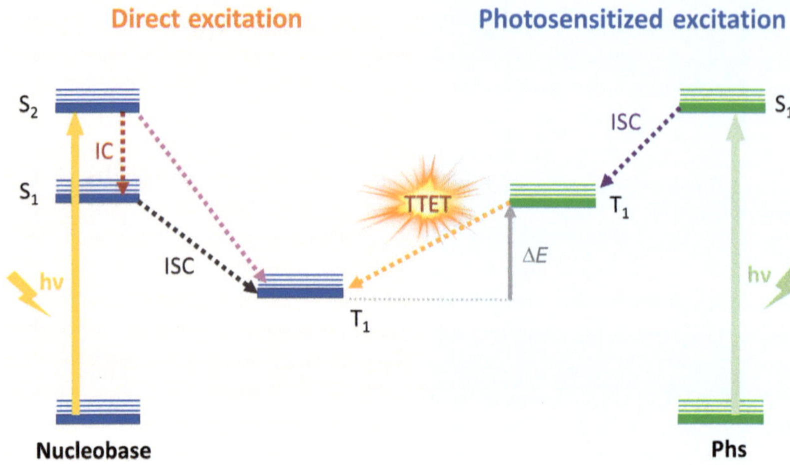

Figure 5.2 Schematic representation of triplet–triplet energy transfer (TTET) mechanism between a Phs/Drug and a nucleobase (N).

In summary, an in-depth knowledge of the photoreactivity of drugs is necessary to understand their photobiological properties and to anticipate the occurrence of photosensitivity side effects. In this chapter, the capacity of drugs to generate the triplet excited state of DNA nucleobases is addressed through a step-by-step approach that includes (i) a short description of TTET, which corresponds to the main process in $^3N^*$ photosensitization; (ii) the triplet energy of the nucleobases, which is a key parameter in TTET; (iii) the photoproducts resulting from $^3N^*$; (iv) the methodologies used to detect and quantify these photolesions; and (v) the drugs responsible for $^3N^*$ generation.

5.2 The Triplet-Triplet Energy Transfer Process

The triplet excited state of nucleobases ($^3N^*$) is populated in a very low extent under direct excitation [13], but it can be efficiently generated through an indirect process where the energy is transferred from a donor, i.e. the Phs/Drug, in its triplet excited state to the triplet of an acceptor, corresponding here to $^3N^*$ (Figure 5.2). This bimolecular mechanism is known as TTET and is generally governed by the rules of Dexter energy transfer. The first step corresponds to the absorption of a photon by the Phs that undergoes ISC to populate its triplet excited state, $^3Phs^*$ [13]. Then, the TTET to the nucleobase (N) can take place and lead to the generation of $^3N^*$. However, TTET is not possible for all the Phs/N combinations. For efficient TTET, the Phs must (i) be selectively excited (generally in the UVA); (ii) be (photo)chemically inert to avoid secondary reactions and loss of its photosensitizing potential; (iii) have an efficient ISC quantum yield ($\Phi_{ISC} \approx 1$) and a long-lived triplet excited state, with a lifetime τ_T, to improve the probability of energy transfer to the acceptor; (iv) be close enough to the Pyr as it occurs through a Dexter mechanism; and, finally,

(v) have a more energetic triplet than the $^3N^*$ to ensure the exergonicity of the process.

In this respect, the efficiency of TTET is determined by the energy transfer rate constant (k_{TTET}), which depends on the energy gap (ΔE in kJ mol^{-1}) between the triplet excited states of the donor and the acceptor (Figure 5.2). The Sandros' equation establishes the relationship between these two factors [15]:

$$k_{TTET} = k_D \frac{1}{\left(e^{-\Delta E / RT} + 1\right)}$$

where k_D is the diffusion-controlled rate constant for liquid solutions (in M^{-1} s^{-1}), T is the temperature (in K), and R is the molar gas constant ($R = 8.314$ J mol^{-1} K^{-1}).

This indirect population of the $^3N^*$, and more specifically of the Pyr triplet excited state ($^3Pyr^*$), is at the origin of key DNA photoreactivity. Therefore, ΔE plays a critical role in the photogenotoxic potential of a Phs/Drug, which justifies the importance of characterizing both $^3N^*$ and $^3Phs^*/^3Drug^*$. It is worth noting that the relative energies of $^3Phs^*$ and $^3N^*$ are crucial but are not the only parameters for an efficient TTET. In fact, when DNA is the acceptor, the binding constant and the sequence of nucleobases are also important criteria to be considered [16–18].

5.3 The Triplet Energy of Nucleobases

The relative energies of $^3N^*$ and $^3Phs^*$ are of paramount importance in assessing the impact of drug photosensitization on $^3N^*$ formation. The E_T of isolated nucleotide was first determined in the 1960s by phosphorescence experiments. Guéron and coworkers reported the emission of nucleotides in ethylene glycol:buffer glass at 77 K and provided triplet excited state energy (E_T) values from the blue edge of the emission spectra [19]. In this respect, cytidine 5′-monophosphate (CMP) and thymidine 5′-monophosphate (TMP) possess the highest and lowest E_T among the nucleobases, respectively (Figure 5.3). Interestingly, photosensitization by acetone or acetophenone was also used to populate TMP by TTET and led to the same emission spectra as direct excitation [20]. Further studies on mixtures of nucleobases, dinucleotides (dApT and TpdA), or polynucleotides (polydAT) [14, 19–24] have been among the first reports addressing the energy transfer between nucleobases, a process

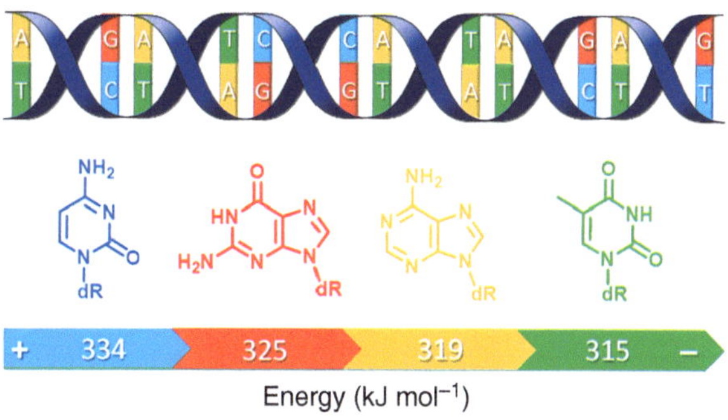

Figure 5.3 Structure and triplet excited state energy of the different nucleobases.

that is still under study. It was shown that, regardless of the excited chromophore (adenine, Thy, guanine or cytosine [Cyt]), when nucleobases are combined, the energy migrates to the one with the lower E_T (i.e. Thy) as a result of TTET [14, 19–24].

In line with this process, the emission of native DNA manifests a remarkable feature [20, 22]. Despite the higher phosphorescence quantum yield of purine (Pur), the DNA spectrum does not contain the Pur-structured band; the emission quantum yield is not as high as expected, being lower than that of adenosine 5′-monophosphate (AMP) and guanosine 5′-monophosphate (GMP); and the emission is similar to that of TMP residues. These outcomes, together with the order established for E_T(N) (Figure 5.3), led to the conclusion that the DNA structure allows an efficient energy transfer between the nucleobases, leading to population and emission from Thy, the $^3N^*$ with the lowest E_T. In agreement, a more pronounced emission has been observed for DNA with a higher adenine and Thy composition [20, 25].

Phosphorescence has brought important data and conclusions on the nucleobases and DNA triplet excited states. However, these studies performed at 77 K in a glass matrix may not exactly reflect the processes taking place at room temperature. In this context, transient absorption has provided valuable information on the energy and lifetime of $^3N^*$ in solution as well as on their interaction with a large variety of Phs, including drugs [10, 11, 14]. However, direct detection by laser flash photolysis (LFP) of the $^3N^*$ is not straightforward due to the formation in low yield of these transient intermediates, which is the result of different factors such as their low Φ_{ISC}, low molar absorption coefficient of the triplet–triplet transitions, and short lifetime [13, 14, 26]. To overcome these low signal amplitude limitations, high laser pulse energy and high concentration of nucleobase have been used, but they are not the best experimental options because they can cause nonlinear effect in the system being studied [27–29].

Redmond and colleagues adopted a simple and elegant protocol based on TTET to monitor indirectly the $^3N^*$ [26, 30]. They took advantage of the properties of well-known arylketones that exhibit high Φ_{ISC} (i.e. close to 1) and, thus, efficiently form triplet excited states of high E_T. Acetone, xanthone, acetophenone, benzophenone (BP), and indan-1-one were selected on these bases, because their E_T is higher than E_T(N) obtained by phosphorescence studies. In a first approach, the E_T of isolated nucleotides was estimated at room temperature using Sandros' equation (see Section 5.2). The biomolecular quenching rate constants for the different combinations of Phs and nucleobases provided $E_T(^3N^*)$ values in good agreement with those previously determined at 77 K [30]. Again, TMP corresponds to the nucleobase with the lowest E_T. In a second step, these results were combined with the detection of cyclobutane thymine dimer (Thy<>Thy), as a signature of ^{3}Thy* formation, in DNA (see Section 5.4), which allows classifying the Phs depending on their potential for photosensitizing Thy<>Thy formation [26]. The main conclusion is not only that the ^{3}Thy* in DNA is below that of the isolated nucleobase but also that the thermal activation energy available at room temperature permits, to some extent, an "uphill TTET" when E_T(Phs) is slightly lower than E_T(Thy). In this context, and based on the ability of Phs to generate Thy<>Thy, the E_T of ^{3}Thy* in DNA has gradually decreased from 300 kJ mol^{-1} (for methoxyacetophenones) to 290 kJ mol^{-1} (for BP) or even to a value of 267 kJ mol^{-1} (for FQs) [14]. This lower limit has been set through the use of a series of FQs sharing a similar skeleton: enoxacin (ENX), pefloxacin (PFX), norlfoxacin (NFX), OFX, rufloxacin (RFX), and its N(4′)-acetyl derivative (ANFX) (see structures in Chart 5.1) [16, 31, 32]. The experiments combining Thy<>Thy formation on supercoiled circular DNA (pBR322) and LFP have made it possible to refine the E_T values of Thy in DNA, which fall within the range defined by the E_T of NFX and ANFX (265–269 kJ mol^{-1}). It is important to keep in mind that (i) the obtained value is, indeed, the required E_T value for a compound to act as a potential DNA Phs through Thy<>Thy formation; (ii) the interaction of the drug with DNA has to be considered; and (iii) this is a mean value that depends on the type of DNA.

As a matter of fact, another aspect to consider is the impact of the sequence on the E_T(Thy) in DNA; more specifically, the effect of the flanking bases of the bipyrimidine site on CPD formation. Analysis of CPD yields in sequences NYYN (where N is any Pur or Pyr, and Y, a Pyr) revealed that formation rates of Pur flanking Thy<>Thy are lower than any of the other tetrad, and that the presence of a Pyr at the 5′ position, YYYN, gives the highest yields [33, 34]. An electron transfer process from Pur to the Thy has been proposed to explain the stronger inhibition observed for guanine than for adenine [33, 34]. Therefore, the sequence plays a crucial role in the E_T value of Thy in DNA, with -TTT- tracks acting as a sink of energy.

Unfortunately, such studies cannot be performed for the other three nucleobases. On the one hand, although Cyt<>Cyt dimers exist, their generation by TTET is less favored due to the high E_T of Cyt; and even in the case where ^{3}Cyt* is generated, an intramolecular TTET to ^{3}Thy* should take place efficiently. On the other hand, no typical reaction from the Pur triplet excited state has been reported up to now.

5.4 Photoproducts Arising from the Nucleobase Triplet Excited State

Pyr are the only nucleobases that exhibit a triplet excited state reactivity leading principally to the formation of the well-known CPDs. Consequently, CPD detection is used as a signature for ^{3}Pyr* generation during photosensitization processes.

As mentioned earlier, nucleobases absorb UV radiation, especially in the UVC and UVB ranges, and reach their reactive singlet or triplet excited state from which different photochemical reactions can take place. The main types of DNA lesions formed under direct irradiation involve two successive Pyr bases (CC, TT, TC, and CT) and correspond to CPDs, pyrimidine 6–4 pyrimidone photoproducts (6–4PPs) together with their Dewar isomers (Dew). In addition, under particular conditions, a fourth type of dipyrimidine lesion, called spore photoproduct (SP), can also be detected [14].

Among these lesions, CPDs are by far the most abundant and studied Pyr lesions. They arise from the singlet (^{1}Pyr*) or triplet excited state (^{3}Pyr*) of the nucleobase depending on the irradiation conditions. Under direct excitation, the low population of ^{3}Pyr* favors CPD formation from ^{1}Pyr*. Indeed, it has been estimated that only 0.05% of the photons absorbed by Thy gives rise to Thy<>Thy through ^{3}Thy* [35]. However, by contrast with 6–4PPs, CPDs are also obtained from an indirect process involving the excitation of a Phs that, after population of its triplet excited state, ^{3}Phs*, transfers its energy to ^{3}Pyr* (TTET process) to finally lead to CPDs. In the case of photosensitization, the yield of CPD formation is independent of Φ_{ISC} of Pyr. However, it is conditioned by the properties of the Phs (E_T and Φ_{ISC}) and the nature of the Pyr involved in the dimer. As expected, TTET to Thy is thermodynamically more favorable than to Cyt, because of Thy lowest E_T (see Sections 5.2 and 5.3) and, for that reason, homodimers formed from two Thy, Thy<>Thy, are more abundant than Cyt<>Cyt, which are hardly detectable [14, 36]. However, in the case of DNA with a high percentage of Cyt, Douki et al. reported a significant formation of Cyt<>Cyt for Phs with E_T lower than that of the isolated ^{3}Cyt*. This suggests that, as described for Thy, E_T(Cyt) is stabilized in DNA and/or that the distribution of CPD cannot be explained only considering the excitation of individual Pyr, and arises from a delocalized triplet excited state involving the Phs and at least two Pyr [36].

From a chemical point of view, CPDs are the result of a formal [2 + 2] cycloaddition giving rise to a stable four-membered cyclobutane ring structure between the C5 = C6 double bonds of two adjacent Pyr. They are referred to as homodimers when the nucleobases are identical, Thy<>Thy or Cyt<>Cyt, or heterodimers for different Pyr (Thy<>Cyt or Cyt<>Thy). Their formation involving a 1,4-biradical centered at C5 and C5′ was proposed first on the basis of computational chemistry [37, 38] and validated only a few years ago experimentally with infrared detection of the biradical hallmark [35, 39].

Several regioisomers and diastereoisomers can be obtained in solution [40]. The geometrical arrangement where the two bases are on the same side or on opposite sides with respect to the cyclobutane ring defines the *cis* or *trans* stereoisomers, respectively. Then, the configuration is called *syn* when the C5 atom of one Pyr is linked to the C5 of the other one, and consequently, the two C6 atoms are also bound to each other. By contrast, the configuration is named *anti* when C5 atoms are linked to C6 in an antiparallel orientation of the two bases. In the case of free bases, four diastereomeric homodimers can be identified. However, in nucleosides and nucleotides, the presence of different chiral centers results in a higher level of complexity. For homodimers, such as Thy<>Thy, two enantiomers exist for the *trans-syn* and *cis-anti*, while *cis-syn* and *trans-anti* diastereoisomers are *meso* forms. Thus, six diastereoisomers are isolable in the case of nucleosides. For heterodimers, a larger variety of stereoisomers is produced. *Cis-syn* and *trans-syn* CPDs have also been identified in dinucleoside monophosphate derivatives containing the different combinations of Thy and Cyt. In DNA, the *cis-syn* is the main isomer due to the structural restriction produced by the phosphodiester backbone and to π-stacking between adjacent nucleobases.

All the CPDs are not stable in solution, as those containing Cyt suffer a deamination process in the dark. In this reaction, a water molecule adds to the C4 carbon bearing the exocyclic amino group of Cyt; subsequently, a molecule of NH_3 is eliminated to finally yield an uracil-derived CPD [41]. This efficient reaction is at the origin of the high mutagenicity of Cyt<>Cyt dimer because their production ends in the generation of a uracil CPD, which, after repair, is replaced by a -TT- track. This mutation is tagged CC→TT. In addition, CPDs are photo-unstable as they can be reverted to the original bases upon exposure to UV-C radiation [33].

Although it has been reported to a lesser extent, another type of lesion may be formed from the ^{3}Pyr*. The 5-thyminyl-5,6-dihydrothymine adduct, or SP, corresponds to the most abundant UV lesion generated in bacterial spores. Although different isomers are conceivable for this Thy homodimer, the natural SP results from the addition of the methyl group located on the 3′-end to the C5 carbon located in the 5′-end, giving rise to a new chiral center with *R* absolute configuration [42]. The formation of the SP is related to three factors: (i) the low hydration level in the spore core, (ii) the presence of dipicolinic acid (DPA) in the spore core, and (iii) the conversion of DNA from B-like to A-like conformation after binding of α/β type small, acid-soluble protein. The photochemical process involved in SP formation is not fully established, but different studies point to the possibility of ^{3}Thy* photosensitization in dry films by compounds such as BP or pyridopsoralens (PyPs; Chart 5.1) [18, 43]. In addition, a recent spectroscopic study demonstrated the ability of endogenous DPA (with E_T = 328 kJ mol^{-1}) and one of its photoproducts (with E_T = 286 kJ mol^{-1}) to populate ^{3}Thy* [44].

The 6-4PP formation has been revisited a few years ago [45, 46]. These photoproducts, which represent the second major class of bipyrimidine photolesions, are formed in a two-step mechanism. First, a Paternò–Büchi photoreaction takes place with the [2 + 2] cycloaddition between the C5 = C6 double bond of the 5′-Pyr and the carbonyl or imine (for Thy or Cyt, respectively) at the C4 position of the 3′-base, which results in the formation of an unstable oxetane or azetidine intermediate. In a second step, this unstable ring cleaves to generate 6-4PPs (Figure 5.4) [47]. In contrast with what was stated previously for CPDs and SPs, there is no definitive experimental evidence for the formation of 6-4PPs through ^{3}Pyr* photosensitization. It is generally accepted that initial oxetane (or azetidine) arises from a localized ^{1}Pyr* or a charge transfer state involving two stacked Pyr [13, 48]. These 6-4PPs are unstable upon UVA or UVB irradiation and undergo a 4π sigmatropic rearrangement to give the Dewar valence isomers (Dew, Figure 5.4) [47].

However, the photoreactivity of a Pyr-derivative (*t*-butyluracil) upper nπ* triplet excited state has opened the debate (Figure 5.5) [45]. After population of the highly energetic BP upper triplet (T_n with E_T at ca. 403 kJ mol^{-1}) by multiphoton excitation at 355 nm, the T_2 (nπ*) of a Pyr-like compound is reached by an intramolecular TTET and triggers a Norrish–Yang photocyclization to finally generate a pyrimidone derivative (Figure 5.5, left) [45]. As Paternò–Büchi reactions are typical nπ* processes, this work questions the generation of 6-4PPs from T_2 (nπ*) of Pyr and opens the way to a potential photosensitization of this lesion by compounds with high E_T.

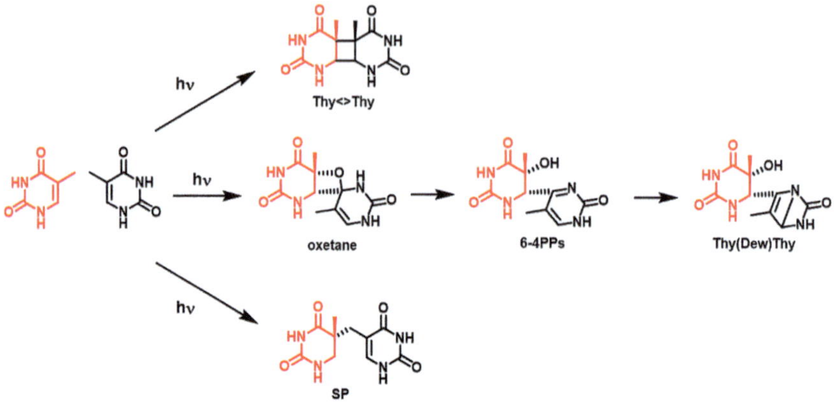

Figure 5.4 Structure of bipyrimidine lesions.

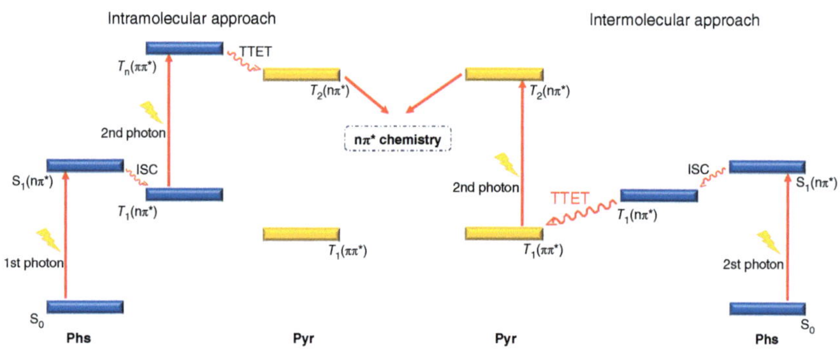

Figure 5.5 Intramolecular (left) and intermolecular (right) biphotonic approaches to investigate the reactivity of the Pyr nπ* upper triplet state.

A second biphotonic approach was described (Figure 5.5, right) [46]. This time two consecutive excitation steps take place: a first photon generates the triplet excited state of 2-methoxyacetophenone, which transfers intermolecularly its energy to the lowest ^{3}Pyr*. This latter absorbs a second photon and is promoted to a more energetic triplet level T_2 (nπ*), leading to the abovementioned Norrish–Yang reaction. This approach was also used to probe the reactivity of isolated uracil; its hydrated photoproduct was characterized as the result of its T_2 (nπ*) population [46].

It is noteworthy that some drugs give rise to triplet-mediated reactions that do not result in the generation of ^{3}Pyr* but lead to the formation of intramolecular products.

This is typically exemplified by ketoprofen (KP; Chart 5.1), a nonsteroidal anti-inflammatory drug (NSAID) containing the BP chromophore [49, 50]. The KP triplet excited state (^{3}KP*) interacts strongly with free thymidine (Thd), with a bimolecular quenching rate constant of ca. 10^8 M^{-1} s^{-1} [51, 52]. In this case,

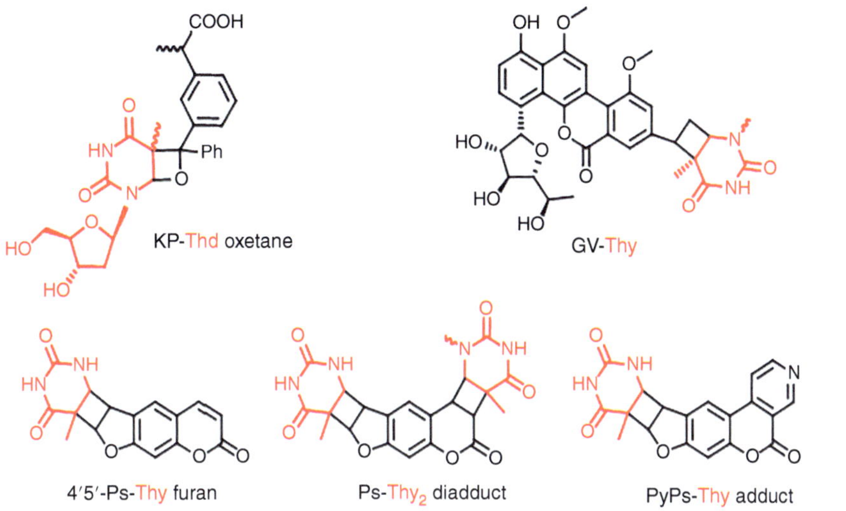

Chart 5.2 Structure of nucleobase–drug photoadducts.

the main process is not TTET, but the formation of oxetanes (KP-Thd oxetane, Chart 5.2). This is due to the E_T(KP) of ca. 290 kJ mol^{-1} (Table 5.1), which is somewhat lower than E_T(Thy). However, depending on the relative concentration between this photosensitizing drug and the nucleobase, some Thy<>Thy can be detected for KP:Thd of 1:3 ratio, probably due to an "uphill TTET" triggered by thermal activation energy available at room temperature (see Section 5.3) or due to the formation of a triplet triplex between ^{3}KP* and two Thy [81]. Interestingly, KP (or BP)-derived oxetanes have never been characterized in the presence of DNA, by contrast with the established KP (or BP) photosensitization of CPD (Table 5.1) [17, 56].

Other examples producing drug–nucleobase or drug–DNA adducts belong to the family of furanocoumarins. These compounds are widely used as Phs in photochemical therapies against different skin disorders such as psoriasis and vitiligo [82–84]. The most representative agent of this family is psolaren (Ps; Chart 5.1). It efficiently intercalates between the DNA nucleobases, and, upon UVA excitation, a [2 + 2] cycloaddition takes place between the 4′ and 5′ positions of Ps furan moiety and the C5 = C6 double bond of Thy (4′5′Ps-Thy furan, Chart 5.2). Further excitation of this monoadduct leads to another cycloaddition, causing a cross-link between the two complementary DNA strands (Ps-Thy$_2$ diadduct, Chart 5.2). Spectroscopic evidence showed that the mechanism involves the participation of the lowest triplet excited state of Ps, $T_1(\pi\pi^*)$, which is efficiently populated from $S_1(\pi\pi^*)$ [85, 86]. The formation of these adducts is, indeed, responsible for the phototherapeutic effect of Ps and its derivatives, which induces apoptosis and leads to relief of the symptoms. PyPs (Chart 5.1) are other Ps derivatives that exhibit high DNA affinity. As described for Ps, 4′,5′-monoadducts are generated in the presence of Thy (PyPs-Thy, Chart 5.2), but these photoproducts do not

Table 5.1 Photophysical properties of the photosensitizing drug triplet excited state (^{3}Drug*): triplet excited state energy E_T in kJ mol^{-1} determined by phosphorescence (P) or laser flash photolysis (LFP), T−T transition maximum in nm, bimolecular rate constant for ^{3}Drug* quenching by Thd (k_{TTET}), triplet excited state lifetime in μs. Method and substrate (Thy, Thd, DNA, animals) used for CPD determination.

Phs	E_T	λ_{TT}	k_{TTET}	τ_T	Method	Substrate
Psoralens						
MePyPs	288.5(P) [53]				HPLC HPLC PAGE	Thy [53] Thd [18] Isolated DNA [18, 54]
PyPs	290.4(P) [53]				HPLC HPLC	Thy [53] Thd [18]
2N-MePyPs	280.8(P) [53]				HPLC HPLC	Thy [53] Thd [18]
NSAIDs						
KP	290(P) [55]				Agarose gel + T4 HPLC MS/MS	Isolated DNA [49, 56, 57] Isolated DNA [36]
TPA	242(P) [58]				Agarose gel + T4	Isolated DNA [56]
DTPA	238(P) [58]		10^7 (Thd) [59]			
HTPA	242(P) [58]					
kTPA	221(P) [58]					
CP					Comet + T4	Cellular DNA [60]
PPMe	284 [61]	420 [61]	1×10^8 (Thd) [61]	5.8 [61]	NMR ^{1}H	Thd [61]
kINP	290(P) [62]	440 [62] 440 [62]	7.8×10^6 (Thd) [62]	35 [62]	PAGE + T4	Oligonucleotide [62, 63]

FQs						
ENX	273(LFP) [31]	520 [31]		0.8 [31]	HPLC MS/MS	Isolated DNA [64]
	291(P) [16],[a)]				HPLC MS/MS	Cellular DNA [64]
	301(P) [16],[b)]				Comet assay	Cellular DNA [64]
LFX	340-308(P) [34],[c)]	420 [65]			HPLC MS/MS	Isolated DNA [64]
					HPLC MS/MS	Cellular DNA [64]
					Comet assay	Cellular DNA [34]
					Immunoassay	Mice [66]
					Immunoassay	Cellular DNA [67]
					HPLC MS/MS	Isolated DNA [36]
NFX	273(LFP) [32]	620 [31]		3.5 [31, 32]	HPLC MS/MS	Isolated DNA [64]
	269(LFP) [31]				HPLC MS/MS	Cellular DNA [64]
	284(P) [16],[a)]				Agarose gel+T4	Isolated DNA [31, 32]
	299(P) [16],[b)]				HPLC MS/MS	Isolated DNA [36]
ANFX	265(LFP) [31, 32] 282(P) [16],[a)]	610 [31]		7.1 [31]	Agarose gel+T4	Isolated DNA [16]
PFX	269(LFP) [31]	620 [31]		2.2 [31]	Agarose gel+T4	Isolated DNA [31]
					Comet + immunoassay	Cellular DNA [68]

(Continued)

Table 5.1 (Continued)

Phs	E_T	λ_{TT}	k_{TTET}	τ_T	Method	Substrate
OFX	262 [31] 269(P) [16],[a)] 279(P) [16],[b)]	610 [31]		4.3 [31]	Agarose gel	Isolated DNA [69] Isolated DNA [16]
Miscellaneous						
FA	284.6(P) [55]			0.33 [55]	PAGE	Oligonucleotide [17]
ppRSV	280.1 [70]	400 [70]	3×10^7 (Thd) [70]	20 [70]		
HCT					HPLC MS/MS Immunoassay	Isolated DNA [71] Mice [71]
MFQ	237(P) [72],[d)]	430 [73]			Immunoassay	Cellular DNA [74]
TCS		320 [75]			Comet + immunoassay	Cellular DNA [76]
BP-2	293.8(P) [77]	420 [77]			Immunoassay	Cellular DNA [78]
AB(K)	292(P) [79]	380 [80]	8×10^8 (Thd) [80]	0.85 [80]		

a) In H_2O. [16]
b) In PB in the presence of DNA (10 mM) [16]
c) Range evaluated in ref. [34]
d) Determined from the maximum of emission at 500 nm. [72]

further give rise to interstrand cross-link [18, 53]. Interestingly, for various PyPs, the formation of PyPs-Thy photoadducts is accompanied by Thy<>Thy generation via TTET (see Section 5.5, Table 5.1), representing this latter lesion about one-fifth of the overall yield of PyPs-Thy [53]. In addition, SP was also characterized after irradiation of a dry film containing PyPs and Thd [18].

Gilvocarcin V (GV) is a natural antibiotic that also exhibits phototriggered anti-tumor activities through photoadduct formation. Thanks to its planar structure, GV intercalates efficiently into DNA. Upon irradiation, the complex suffers an ultrafast ISC (in less than 3 ps) that populates the GV local triplet excited state and further mediates a [2 + 2] cycloaddition between the vinyl moiety and a Thy residue, giving rise to a four-membered ring photoadduct (GV-Thy, Chart 5.2) [87].

5.5 Methodologies: Spectroscopic vs Analytical Techniques for Detection of Drug Photosensitization of $^3N^*$

5.5.1 Direct Detection: Spectroscopic Techniques

Two main techniques are generally used to detect the formation of triplet excited states, monitoring their emission (i.e. phosphorescence) or the transient absorption of their triplet–triplet transitions (T–T transition) by LFP.

Phosphorescence is a phenomenon of delayed luminescence associated with the radiative decay from the triplet excited state to the ground state. Its measurement at room temperature is challenging due to competition between emission and efficient nonradiative pathways. To circumvent this drawback, emission is monitored at low temperature, ca. 77 K, in a glass matrix of an appropriate solvent (ethanol, methylcyclohexane, etc.). Although phosphorescence gives access to the triplet excited state detection of the different nucleobases (Section 5.3), this is not a technique of choice to investigate the population of the $^3N^*$ generated by photosensitization. The main limitation is due to the solid nature of the media and, thus, to the lack of diffusion between the energy donor (drug) and the acceptor (N).

As described in Section 5.3, LFP is, in that sense, more appropriate to study the TTET process, and it was key not only to determine the E_T of the free nucleobases [26, 30] but also to provide an approximate value of E_T(Thy) in DNA [16, 31, 32]. This is a time-resolved technique that monitors the evolution of transient species absorption after excitation by a laser pulse. Time resolution of LFP is typically from hundreds of nanoseconds to hundreds of microseconds. However, the detection of the ^{3}Thy* formed through TTET is not so evident due, in part, to the low molar absorption coefficient of the T–T transition and to its short lifetime [13, 14]. As a matter of fact, direct detection of ^{3}Thy* formation by drug photosensitization has not been reported yet. Nevertheless, the formation of Thy<>Thy has been widely used as a marker to demonstrate indirectly the generation of this excited state by drug photosensitization.

5.5.2 Indirect Detection: Analytical Techniques

Different methodologies have been exploited to investigate CPD formation for the isolated nucleobases, or when Pyr is part of isolated or cellular DNA. The most frequently used methods are mentioned in the subsequent sections.

5.5.2.1 CPD Detection in Nucleosides and/or Nucleotides

The use of these simple DNA-building blocks allows a straightforward detection of the CPD formation. After irradiation, the nucleobase/drug mixture is analyzed by high-performance liquid chromatography (HPLC) coupled to a UV, mass (MS), or tandem mass (MS/MS) detector. Final assignment of CPDs is generally made by comparison with a well-characterized standard.

5.5.2.2 CPD Detection in Isolated DNA or Oligonucleotides

5.5.2.2.1 Electrophoresis Electrophoresis is well-established in molecular biology to routinely analyze and purify DNA. This technique allows the separation of DNA molecules on the basis of their size or conformation. The general principle is based on the migration, triggered by the application of an electric field, of the negatively charged DNA toward the positive electrode. Separation is achieved through the small pores formed in the agarose or polyacrylamide gel matrix. The rate of migration varies according to the length or topology of the DNA; longer or less compact DNA molecules migrate slower because they experience a greater resistance [88].

Agarose Gel Electrophoresis This is a very sensitive technique that is based on the different topologies of intact and damaged plasmid DNA. The native supercoiled circular form (referred to as Form I) is converted to the nicked relaxed form (Form II) with the generation of only one single-strand break (SSB), a subsequent cleavage in close proximity to this SSB (i.e. double-strand break, DSB) induces the formation of linear DNA (Form III). When subjected to electrophoresis through a gel matrix, smaller and/or more compact molecules migrate more readily. The supercoiled Form I moves to the positive electrode faster than Form II, with the linear Form III lying in between. Therefore, these three forms are easily separated by agarose gel electrophoresis and can be detected and quantified by staining with classical intercalating fluorescent dyes using densitometry. In addition to direct DNA strand breaks (SBs), specific base damages can be revealed after incubation with DNA repair enzymes such as *Escherichia coli* formamidopyrimidine DNA *N*-glycosylase for oxidized Pur or *E. coli* endonuclease III for oxidized Pyr. CPD is detected using T4 endonuclease V (T4 endo V) [89, 90]. Endonuclease V, encoded by the *den*V gene of bacteriophage T4, initiates DNA repair at the site of UV light-induced Pyr dimers via the combined action of a Pyr dimer-specific DNA glycosylase and an apyrimidinic-apurinic (AP) endonuclease. After binding a Pyr dimer, T4 endo V catalyzes the cleavage of the glycosylic bond of the 5′-Pyr within the dimer, leading to its removal (Figure 5.6). The phosphodiester backbone is finally cleaved via a β-elimination mechanism [89].

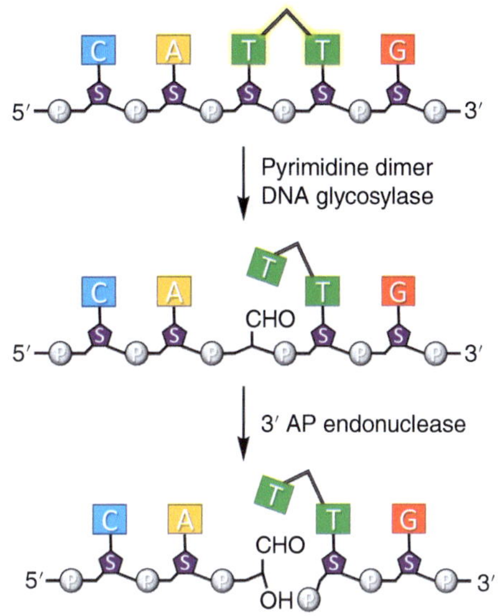

Figure 5.6 Mechanism of CPD cleavage by T4 endo V.

Polyacrylamide Gel Electrophoresis Denaturing polyacrylamide gel electrophoresis (PAGE) allows the separation of linear single-stranded DNA molecules on the basis of their molecular weight. This method can be used to analyze or purify short synthetic DNA oligonucleotides or DNA sequences. The DNA fragments to be analyzed share a common extremity, radiolabeled with ^{32}P- or tagged with a fluorophore, but vary in length. The resolution provided by the gel is high enough to separate two DNA molecules differing in size by one nucleotide as two distinct bands. The oligonucleotide sequence is then determined by standard markers able to cleave selectively a Pur or Pyr nucleobase, for example, a marker for T-sites. PAGE was originally developed for DNA sequencing applications, but due to the resolution it provides, its use has been extended to other assays, such as for establishing the formation of DNA (photo)lesions. Compared to agarose gel electrophoresis, PAGE affords a precise determination of the oligonucleotide size. Therefore, in the case of DNA photosensitization by drugs, the cleavage pattern revealed without treatment (i.e. SSB) compared to that obtained after T4 endo V treatment informs on the formation of Thy<>Thy at different TT tracks.

5.5.2.2.2 Chromatographic Analysis Analytical detection of photosensitized DNA damages can also be performed by HPLC coupled with various types of detection (UV, MS, MS/MS). To identify the type of lesion, DNA is first digested using a cocktail of enzymes, including calf spleen phosphodiesterase, nuclease P1, bovine mucosa phosphodiesterase I, and alkaline phosphatase. The extracted nucleosides are then analyzed by HPLC [64].

5.5.2.3 CPD Detection in Cellular DNA

5.5.2.3.1 Electrophoresis: Comet Assay Comet assay, also known as single-cell gel electrophoresis (SCGE) assay, is a sensitive technique for measuring DNA damage in individual eukaryotic cells [91–93]. It has been widely used for in vivo and in vitro studies to evaluate the (photo)genotoxic potential of radiations (ionizing and UV) or chemical (drugs, cosmetics, contaminants, etc.). This technique is based on the size-dependent migration of the DNA molecule when lysed cells are embedded in a low-melting agarose-microgel and submitted to an electric field. Under these conditions, the fragmented DNA moves away from the DNA nucleus to form the tail. Staining with an intercalated dye leads finally to a comet-like shape with a circular head made of intact DNA and a tail corresponding to damaged DNA. The quantitative determination of SB is obtained from the ratio between the two parts. The comet assay is an adaptable technique that, after modification of the protocol, allows detection and quantification of a wide variety of lesions inducing SSB/DSB or alkali labile sites. In addition, when combined with the use of specific antibodies or DNA repair enzymes, specific modification of Pur and/or Pyr bases can be evidenced. In this context, Thy<>Thy is revealed after incubation with T4 endo V [89, 90]. During this treatment, Thy<>Thy is converted into SSB and, thus, detected as an extra tail by comparison with nontreated samples. Immunofluorescent detection using a specific antibody can also be performed. In this case, the DNA tail moment is measured using conventional dye to determine SB, and simultaneously, Thy<>Thy formation is revealed by an indirect immunofluorescence detection using a specific monoclonal antibody [94].

In addition, the comet assay can also be used to assess cellular repair ability by investigating the removal of the comet tails after incubating the damaged cells in the dark at 37 °C for different period of time and then running the electrophoresis process.

5.5.2.3.2 Chromatographic Analysis Analysis of the damages can be performed in a similar way as described in Section 5.5.2.2.2, after extraction of DNA from the cell [95].

5.6 Photosensitization of Nucleobase Triplet by Drugs (or Cosmetics)

5.6.1 Pyridopsoralen Derivatives

PyPs (Chart 5.1) are Ps derivatives, in which the pyrone moiety is fused to a pyridine ring at the C3 = C4 double bond. Ps and its derivatives are UVA Phs used in the treatment of various skin diseases. Their antiproliferative effects are related to their ability to form monoadducts and diadducts after photocycloaddition reactions with Pyr, and especially with Thy (see Section 5.4 and Chart 5.2).

In fact, these compounds are the first reported drugs capable of photosensitizing the formation of $^3N^*$. UVA irradiation of thin film containing different PyPs in the presence of Thy (or Thd) showed the formation of Thy<>Thy (or Thd<>Thd) [18, 53]. A singlet excited state-mediated mechanism can be excluded because the S_1(PyPs) is lower than S_1(Thy), with values of ca. 303.9 and 403.3 kJ mol^{-1}, respectively. By contrast, a good correlation is found between E_T of PyPs derivatives

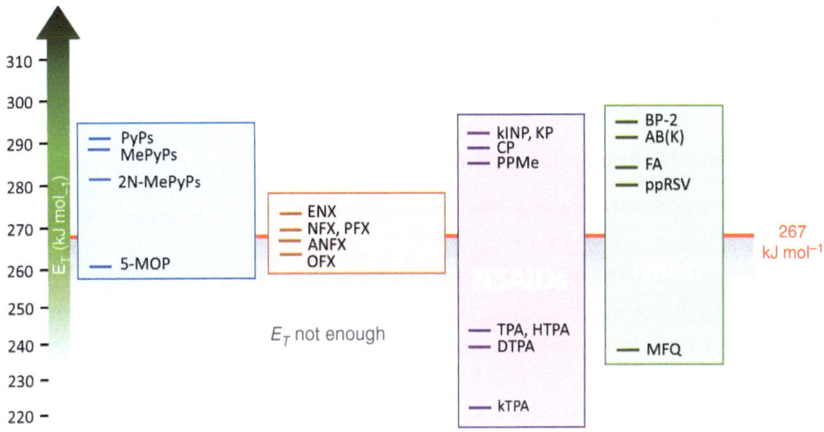

Figure 5.7 Energy diagram representing the E_T of drugs compared to the established mean value of E_T(Thy) in DNA.

(Table 5.1, Figure 5.7) and the related Thy<>Thy or Thd<>Thd formation yields, with PyPs being the most efficient Phs followed by MePyPs. A yield two orders of magnitude lower than PyPs was obtained for 2N-MePyPs, which has the lowest E_T of ca. 280.8 kJ mol^{-1}. These data together with the fact that donors with lower E_T like 3-carbethoxyPs, 5-methoxyPs (5-MOP), and 8-methoxyPs (8-MOP) do not generate Thy<>Thy in significant amounts agree with the formation of CPDs from the triplet excited state of PyPs (^{3}PyPs*) with energy higher than 288 kJ mol^{-1} [18, 53]. Further evidence for the feasibility of this process was provided by the lack of CPD formation when PsPy, MePsPy, or 2N-MePyPs was irradiated in the presence of 2′-deoxycytidine (Cyd) [18]. This can be explained by the higher E_T of Cyd relative to that of Thd (see Section 5.3 and Figure 5.3).

Interestingly, PyPs are among the few compounds that photosensitize the formation of SP [18]. The 5*R* and 5*S* diastereoisomers of SP were isolated after UVA irradiation of thin films containing the three PyPs mentioned earlier (see Section 5.4) in the presence of Thd. Again, their yields were shown to be dependent on the E_T(Phs), decreasing in the following order: PyPs > MePyPs > 2N-MePyPs. This observation supports a TTET process in SP formation.

Similar results were reported for DNA with the detection of the *cis-syn* isomer of Thy<>Thy along with the monoadduct formation, the former accounting for about one-fifth of the total yield of the furan-side PyPs adduct to Thy (Chart 5.2) in the case of MePsPy [18]. In addition, the intercalation mode of PyPs derivatives induces a sequence-dependent Thy<>Thy formation. These lesions were predominantly formed at 5′-TTTTA-3′ and 5′-TTAAT-3′ sites, which are also the strongest sites for photoaddition [18]. Further studies compare the Phs properties of MePyPs and its derivative (DH-MePsPy) designed to block the formation of monoadducts by saturating the 4′-5′-furan double bond [54]. Selective Thy<>Thy formation was evidenced at the Thy located essentially at the 5′ position of 5′-TA-3′ step. The yields, in turn, vary depending on the flanking bases. This selectivity establishes that for DH-MePyPs (and more generally PyPs), TTET does not occur during a simple collision between the Phs and DNA, but through the formation of a noncovalent complex.

5.6.2 Nonsteroidal Anti-inflammatory Drugs

The generation of photosensitized CPDs by these drugs was first hypothesized in the early 1990s. The decreased mobility of DNA Form I when irradiated in the presence of benoxaprofen (BNP), naproxen, tiaprofenic acid (TPA), or KP (Chart 5.1) was attributed to the unwinding effect induced by the [2 + 2] photocycloaddition of the adjacent Pyr [96]. Subsequent studies using T4 endo V confirmed the formation of CPDs by KP and TPA but ruled out the role of naproxen and indomethacin [17, 56, 57, 59, 97]. It is noteworthy that the two former drugs share a BP-like structure. This chromophore has been recognized as paradigmatic because it is capable of generating a wide variety of DNA damages derived from type I or type II processes, TTET, Paternò–Büchi reaction, etc. [98]. Photosensitization of a ^{32}P-labeled oligonucleotide by KP corroborates the formation of Thy<>Thy, mainly at TTT tracks [17]. However, other types of CPDs are also photoinduced by KP. HPLC-MS analysis of isolated calf thymus, *Clostridium perfringens*, or *Micrococcus luteus* DNA identified dimer formation at CT, TC, and CC sites. Their relative yields can be associated with the percentage of Cyt in the DNA sequence, with Cyt-containing CPDs being found in greater amounts in *M. luteus* than in *C. perfringens* and calf thymus DNA [36]. Carprofen (CP; Chart 5.1) photosensitizes CPDs in human keratinocytes (HaCaT cells) as revealed by the comet assay [60]. Photochemical studies performed with the methyl ester form of its dehalogenated photoproduct (PPMe, Chart 5.1) report that this drug, containing a carbazole chromophore, exhibits a double-edged sword photobehavior [61]. On the one hand, its triplet excited state is high enough (E_T of ca. 284 kJ mol^{-1}, Table 5.1 and Figure 5.7) to transfer energy to ^{3}Thy*. But, on the other hand, in its singlet excited state, PPMe is able to photoreduce BP-Thy oxetanes (as a model of 6-4PP intermediate), inducing their photocycloreversion [61]. Therefore, this compound has both DNA damaging and repair properties.

It is noteworthy that the arylpropionic NSAIDs are known to be efficiently degraded by photodecarboxylation [50]; as a consequence, the photoproducts may be important players in their photobiological reactivity. A case in point is indoprofen (INP). Photosensitized formation of Thy<>Thy was detected by gel sequencing experiments, but the lack of triplet excited state (^{3}INP*) signal in LFP questioned the direct participation of the drug in the TTET process [99, 100]. In addition, no Thy<>Thy was observed for the methyl ester derivative of INP, in which the decarboxylation process is blocked. By contrast, these lesions were produced if pre-irradiated INP solution was used [100]. Combined photobiological and photophysical studies of the different INP photoproducts (ethyl, hydroxyethyl and acetyl) revealed that the photoactive one is the acetyl derivative (kINP, Chart 5.1); this has been mainly attributed to the presence of an acetyl group on the phenyl moiety of the *N*-phenylphtalimidine, which increases the generally low ISC of this chromophore [62, 63].

Similar studies were run for TPA (Chart 5.1); in that case, both the parent drug and its three major photoproducts gave rise to Thy<>Thy formation [97]. This is in line with the bimolecular quenching rate constant, determined by LFP, for the deactivation of the triplet excited state of the ethyl photoproduct (DTPA, Chart 5.1) by Thd, $k_{TTET} = 10^7\ M^{-1}\ s^{-1}$ [59]. However, the lack of accuracy in the reported E_T values avoids to draw a conclusion on their ability to perform TTET [97].

These findings demonstrated the importance of photoproducts in the photosensitization process by drugs. Taken together with the high quantum yield of KP photodecarboxylation in neutral media, the question is raised as to the photosensitizing potential of the parent drug or, instead, of the photoproducts generated during irradiation. To the best of our knowledge, the study of the properties of KP photoproducts has not been addressed yet.

5.6.3 Fluoroquinolones

FQs are well-established antibacterial drugs with a broad-spectrum activity against Gram-positive, Gram-negative, and some atypical microbes. Their pharmacological potential is related to the inhibition of the bacterial type II topoisomerases, the DNA gyrase, and topoisomerase IV [101]. Some FQs show adverse side effects, such as skin phototoxicity, photomutagenicity, and photocarcinogenicity [5–7]. Structurally, FQs contain at least one fluorine atom at the C6 position of the bicyclic skeleton, and in some cases, a second halogen atom is present at the C8 position, such as in LFX. Their structures are shown in Chart 5.1.

Traynor et al. first described photosensitization of CPDs by LFX using the comet assay combined with T4 endo V digestion (Table 5.1) [34]. They correlated the skin tumor formation in mice with the LFX-mediated production of Thy<>Thy. The authors reasoned that although the E_T of LFX, with a value between 340 and 308 kJ mol^{-1}, is high enough to photosensitize Thy<>Thy (Figure 5.7), the low Φ_{ISC} of about 0.07 represents a limitation. This can be interpreted, with the background acquired years later, by the high photolability of LFX, which loses the F atom at the C8 position [102], and the involvement of photoproducts in the photosensitization process. Some years later, studies on *xeroderma pigmentosum* (XP) group gene-deficient mice XPA(−/−), which are defective in nucleotide excision, responsible for repairing CPDs in the organism, demonstrated the ability of LFX to induce CPDs in vivo [66]. These results, in addition, reveal the slower rate of CPD repair for XPA(−/−) mice compared to wild animals and illustrate the case of the patients with XP, who have an extremely high risk of developing skin cancer [66].

Further studies showed FQs potential to generate ^{3}Thy* by detecting Thy<>Thy not only in isolated or cellular DNA but also in mice using a different set of techniques, including HPLC-MS, electrophoresis, and immunoassays [16, 31, 32, 64, 66–69]. As a result, the list of FQs capable of inducing Thy<>Thy was extended to include ENX, NFX, PFX, and OFX (when high drug concentrations are used). The most efficient is ENX, which is in agreement with the respective E_T(FQ) values (Table 5.1) [16, 31, 32]. Interestingly, HPLC-MS experiments showed that

Thy<>Thy was the major damage in isolated DNA, as its formation rate for LFX and NFX is twice that of the main oxidatively generated lesions, 8-oxo-2′-deoxyguanine. However, this ratio was significantly lower for ENX [64]. By contrast, in cellular DNA, ENX was the most efficient FQ to photosensitize Thy<>Thy formation [64].

The structural similarity of FQs has been used to determine a mean value of E_T(Thy) in DNA (see Section 5.3, Figure 5.7) [16, 31, 32]. In these studies, an important point was raised regarding the consequences of drug interaction with DNA. As a matter of fact, close proximity between the energy donor and the acceptor is key for TTET, which is a Dexter mechanism; and the properties of ^{3}Drug*, such as E_T or Φ_{ISC}, can be modified when complexed to DNA compared to those in bulk solution (Table 5.1) [16]. Low temperature phosphorescence emission of FQs in the presence of DNA is blue shifted, and therefore, the E_T are 10–15 kJ mol^{-1} higher than in water. Moreover, ultrafast transient absorption spectroscopy run on the DNA-FQ complex concluded that TTET between ^{3}FQ* and Thy is efficient, taking place within the picosecond timescale [16].

As reported for NSAIDs, LFX and NFX also induce the formation of Cyt-derived CPDs, with formation yields related to the percentage of Cyt in the DNA under consideration [36]. To interpret the formation of Cyt<>Cyt, Douki et al. propose an additional mechanism for TTET involving the ^{3}Phs* and two adjacent Pyr [36].

5.6.4 Other Drugs and Cosmetic Agents

Miscellaneous drugs, in addition to cosmetic ingredients, are able to photoinduce ^{3}Thy* formation. These agents are listed here.

Another BP derivative, fenofibric acid (FA; Chart 5.1), which is the active metabolite of the hyperlipidemia agent fenofibrate, induces Thy<>Thy at TTT tracks of short oligonucleotides as observed by PAGE combined with T4 endo V treatment [17]. It is less efficient than the BP-like NSAID KP, as expected from its lower E_T of 284.6 kJ mol^{-1} (Table 5.1) [55].

Photophysical characterization of a photoproduct of rosuvastatin (ppRSV; Chart 5.1), another drug used to lower cholesterol levels in blood, revealed that it exhibits the requirements to photosensitize Thy<>Thy through TTET: its triplet excited state lying at 280 kJ mol^{-1} is quenched by Thd with a bimolecular rate constant of 3×10^7 M^{-1} s^{-1} (Table 5.1), k_{TTET} somewhat higher than that obtained for INP photoproduct [70]. Nevertheless, ^{3}Thd* signal was not detected due to the experimental restrictions as mentioned in Section 5.3.

Mefloquine (MFQ) is an antimalarial drug that contains a quinoline chromophore. Although it has a low-lying triplet excited state with a maximum for phosphorescence emission at 500 nm [72], CPD formation has been detected by immunofluorescence assay in keratinocytes upon UVB and sunlight irradiation in the presence of this antimalarial agent [74].

In vitro studies on isolated DNA revealed the potential of the diuretic agent hydrochlorothiazide (HCT) to photosensitize Thy<>Thy under UVA irradiation. Dimer formation was detected in vivo in the epidermis and the dermis of wild or XPA(−/−) mice. After 48 h, the CPDs were better repaired in the dermis of wild mice than in the XPA(−/−) type, which is consistent with the involvement of the nucleotide excision repair pathway to restore the original DNA sequence [71].

Triclosan (TCS) is an antimicrobial preservative used in personal care products, such as cosmetics, medical devices, and household items. This compound, detected in human breast milk, blood, and urine, causes skin irritation. Comet assay performed on human keratinocytes (HaCaT cells) evidences CPD formation under UVB or sunlight irradiation [76].

Some UV filters, containing the photoactive arylketone moiety, were also reported for their photosensitizing properties. Avobenzone (AB), or 4-*tert*-butyl-4′-methoxydibenzoylmethane, is one of the few UVA filters authorized in the market. It has a strong absorption with maximum at ca. 340–350 nm. This filter can exist in solution in different tautomeric forms (AB(E) and AB(K) in Chart 5.1). It exhibits a keto-enol equilibrium favoring the chelated enol form, AB(E); but under UV irradiation, a photoisomerization occurs, leading to the formation of the β-diketone, AB(K). The latter, which absorbs at 260–280 nm, lacks the photoprotective properties conferred by the excited state intramolecular proton transfer (ESIPT) in AB(E). In addition, the triplet excited state of the β-diketone tautomer is efficiently populated, with Φ_{ISC} close to unity; and it fulfills the requirements to transfer its energy to Thd with $E_T = 292\,\text{kJ mol}^{-1}$ with a bimolecular rate constant $k_{TTET} = 3\times10^8\ \text{M}^{-1}\ \text{s}^{-1}$ [79, 80]. BP-2 (Chart 5.1) is another UVA filter. Phosphorescence in ethanol or methylcyclohexane glass at 77 K showed the typical pattern of BP emission [77], with a E_T of ca. $293\,\text{kJ mol}^{-1}$. Further evidence for the failure of the efficient ESIPT that leads to the undesired population BP-2 triplet excited state was provided by LFP, with the detection of T–T transition at 420 nm [77]. Accordingly with these properties, CPDs were detected by immunofluorescence after irradiation of HaCaT cells in the presence of BP-2 [78].

5.7 Conclusion

Drug photosensitization is a matter of concern due to the large number of compounds involved in photosensitization processes. In this context, a comprehensive knowledge of the photochemical and photophysical properties of drugs is of paramount importance in order to understand and, as far as possible, predict the potential photobiological reactions triggered upon exposure to light. Various biological targets are damaged by photosensitization, but DNA is a key element as the formation of photolesions can provoke a mutation of the original sequence and ends up in the appearance of cancer. In this chapter, the attention has been focused on the ability of drugs to photosensitize the formation of triplet excited state of DNA nucleobases, and especially Pyr ones. Although this reactive species has not been detected directly, the formation of CPDs has provided an indirect indication for its involvement in the photosensitization process. In addition, the photostability of drugs must be considered an important parameter, as clearly demonstrated by the photosensitizing properties of NSAID photoproducts. In any case, it would be useful to include the evaluation of DNA photosensitization in photosafety testing as part of the toxicological assessment of new drug candidates possessing photoactive chromophores prior to their introduction to the market.

References

1 Moore, D.E. (2002). Drug-induced cutaneous photosensitivity. *Drug Safety* 25 (5): 345–372. https://doi.org/10.2165/00002018-200225050-00004.

2 Monteiro, A.F., Rato, M., and Martins, C. (2016). Drug-induced photosensitivity: photoallergic and phototoxic reactions. *Clinics in Dermatology* 34 (5): 571–581. https://doi.org/10.1016/j.clindermatol.2016.05.006.

3 Lhiaubet-Vallet, V. and Miranda, M.A. (2012). Phototoxicity of drugs. In: *CRC Handbook of Organic Photochemistry and Photobiology* (ed. A. Griesbeck, M. Oelgemöller, and F. Ghetti), 1541–1555. Boca Raton: CRC Press https://doi.org/10.1201/9780429100253.

4 Hofmann, G.A. and Weber, B. (2021). Drug-induced photosensitivity: culprit drugs, potential mechanisms and clinical consequences. *Journal der Deutschen Dermatologischen Gesellschaft* 19 (1): 19–29. https://doi.org/10.1111/ddg.14314.

5 Mäkinen, M., Forbes, P.D., and Stenbäck, F. (1997). Quinolone antibacterials: a new class of photochemical carcinogens. *Journal of Photochemistry and Photobiology B: Biology* 37 (3): 182–187. https://doi.org/10.1016/S1011-1344(96)07425-8.

6 Klecak, G., Urbach, F., and Urwyler, H. (1997). Fluoroquinolone antibacterials enhance UVA-induced skin tumors. *Journal of Photochemistry and Photobiology B: Biology* 37 (3): 174–181. https://doi.org/10.1016/S1011-1344(96)07424-6.

7 Johnson, B.E., Gibbs, N.K., and Ferguson, J. (1997). Quinolone antibiotic with potential to photosensitize skin tumorigenesis. *Journal of Photochemistry and Photobiology B: Biology* 37 (3): 171–173. https://doi.org/10.1016/S1011-1344(96)07423-4.

8 Stern, R.S., Liebman, E.J., and Väkevä, L. (1998). Oral psoralen and ultraviolet-A light (PUVA) treatment of psoriasis and persistent risk of non-melanoma Skin cancer. *JNCI Journal of the National Cancer Institute* 90 (17): 1278–1284. https://doi.org/10.1093/jnci/90.17.1278.

9 Gunaydin, G., Gedik, M.E., and Ayan, S. (2021). Photodynamic therapy for the treatment and diagnosis of cancer–a review of the current clinical status. *Frontiers in Chemistry* 9: 686303.

10 Epe, B. (2012). DNA damage spectra induced by photosensitization. *Photochemical & Photobiological Sciences* 11 (1): 98–106. https://doi.org/10.1039/C1PP05190C.

11 Lhiaubet-Vallet, V., Bosca, F., and Miranda, M.A. (2009). Photosensitized DNA damage: the case of fluoroquinolones. *Photochemistry and Photobiology* 85: 861–868. https://doi.org/10.1111/j.1751-1097.2009.00548.x.

12 Baptista, M.S., Cadet, J., Greer, A., and Thomas, A.H. (2021). Photosensitization reactions of biomolecules: definition, targets and mechanisms. *Photochemistry and Photobiology* 97 (6): 1456–1483. https://doi.org/10.1111/php.13470.

13 Miranda, M.A. and Lhiaubet-Vallet. (2021). Triplet channel photochemistry in DNA. In: *DNA Photodamage: From Light Absorption to Cellular Responses and Skin Cancer* (ed. R. Improta and T. Douki), 55–76. Croydon: The Royal Society of Chemistry https://doi.org/10.1039/9781839165580-00055.

14 Cuquerella, M.C., Lhiaubet-Vallet, V., Bosca, F., and Miranda, M.A. (2011). Photosensitised pyrimidine dimerisation in DNA. *Chemical Science* 2 (7): 1219–1232. https://doi.org/10.1039/c1sc00088h.

15 Sandros, K. (1964). Transfer of triplet state energy in fluid solution. *Acta Chemica Scandinavica* 18: 2355–2374.

16 Cuquerella, M.C., Lhiaubet-Vallet, V., Miranda, M.A., and Bosca, F. (2017). Drug–DNA complexation as the key factor in photosensitized thymine dimerization. *Physical Chemistry Chemical Physics* 19 (7): 4951–4955. https://doi.org/10.1039/C6CP08485K.

17 Lhiaubet, V., Paillous, N., and Chouini-Lalanne, N. (2001). Comparison of DNA damage photoinduced by ketoprofen, fenofibric acid and benzophenone via electron and energy transfer. *Photochemistry and Photobiology* 74 (5): 670–678. https://doi.org/10.1562/0031-8655(2001)074<0670:coddpb>2.0.co;2.

18 Moysan, A., Viari, A., Vigny, P. et al. (1991). Formation of cyclobutane thymine dimers photosensitized by pyridopsoralens: quantitative and qualitative distribution within DNA. *Biochemistry* 30 (29): 7080–7088. https://doi.org/10.1021/bi00243a007.

19 Guéron, M., Eisinger, J., and Shulman, R.G. (1967). Excited States of nucleotides and singlet energy transfer in polynucleotides. *The Journal of Chemical Physics* 47 (10): 4077–4091. https://doi.org/10.1063/1.1701580.

20 Lamola, A.A., Gueron, M., Yamane, T. et al. (1967). Triplet state of DNA. *The Journal of Chemical Physics* 47 (7): 2210–2217.

21 Eisinger, J. and Shulman, R.G. (1967). Energy transfer in poly DAT. *Journal of Molecular Biology* 28 (3): 445–449.

22 Eisinger, J. and Shulman, R.G. (1968). Excited electronic states of DNA. *Science* 161: 1311–1319.

23 Helene, C. (1966). Triplet-triplet energy transfer between nucleic acids derivatives in frozen aqueous solutions. *Biochemical and Biophysical Research Communications* 22 (3): 237–242. https://doi.org/10.1016/0006-291X(66)90471-2.

24 Hélène, C. and Montenay-Garestier, T. (1968). Excitation energy transfer in molecular aggregates of nucleic acid derivatives in frozen aqueous solutions. *Chemical Physics Letters* 2 (1): 25–28. https://doi.org/10.1016/0009-2614(68)80137-X.

25 Isenberg, I., Rosenbluth, R., and Baird, S.L. (1967). Comparative phosphorescence quenching of DNA's of different composition. *Biophysical Journal* 7 (4): 365–373. https://doi.org/10.1016/S0006-3495(67)86594-9.

26 Gut, I.G., Wood, P.D., and Redmond, R.W. (1996). Interaction of triplet photosensitizers with nucleotides and DNA in aqueous solution at room temperature.

Journal of the American Chemical Society 118 (10): 2366–2373. https://doi.org/10.1021/ja9519344.

27 Kryukov, P.G., Letokhov, V.S., Nikogosyan, D.N. et al. (1979). Multiquantum photoreactions of nucleic acid components in aqueous solution by powerful ultraviolet picosecond radiation. *Chemical Physics Letters* 61 (2): 375–379. https://doi.org/10.1016/0009-2614(79)80666-1.

28 Oraevsky, A.A. and Nikogosyan, D.N. (1985). Picosecond two-quantum UV photochemistry of thymine in aqueous solution. *Chemical Physics* 100 (3): 429–445. https://doi.org/10.1016/0301-0104(85)87068-3.

29 Nikogosyan, D.N., Angelov, D.A., and Oraevsky, A.A. (1982). Determination of parameters of excited states of DNA and RNA bases by laser falsh photolysis. *Photochemistry and Photobiology* 35 (5): 627–635. https://doi.org/10.1111/j.1751-1097.1982.tb02621.x.

30 Wood, P.D. and Redmond, R.W. (1996). Triplet state interactions between nucleic acid bases in solution at room temperature: intermolecular energy and electron transfer. *Journal of the American Chemical Society* 118 (18): 4256–4263. https://doi.org/10.1021/ja954340+.

31 Lhiaubet-Vallet, V., Cuquerella, M.C., Castell, J.V. et al. (2007). Triplet excited fluoroquinolones as mediators for thymine cyclobutane dimer formation in DNA. *The Journal of Physical Chemistry B* 111 (25): 7409–7414. https://doi.org/10.1021/jp070167f.

32 Bosca, F., Lhiaubet-Vallet, V., Cuquerella, M.C. et al. (2006). The triplet energy of thymine in DNA. *Journal of the American Chemical Society* 128 (19): 6318–6319. https://doi.org/10.1021/ja060651g.

33 Lu, C., Gutierrez-Bayona, N.E., and Taylor, J.S. (2021). The effect of flanking bases on direct and triplet sensitized cyclobutane pyrimidine dimer formation in DNA depends on the dipyrimidine, wavelength and the photosensitizer. *Nucleic Acids Research* 49 (8): 4266–4280. https://doi.org/10.1093/nar/gkab214.

34 Traynor, N.J. and Gibbs, N.K. (1999). The phototumorigenic fluoroquinolone lomefloxacin photosensitizes pyrimidine dimer formation in human keratinocytes In vitro. *Photochemistry and Photobiology* 70 (6): 957–959. https://doi.org/10.1111/j.1751-1097.1999.tb08308.x.

35 Liu, L., Pilles, B.M., Gontcharov, J. et al. (2016). Quantum yield of cyclobutane pyrimidine dimer formation via the triplet channel determined by photosensitization. *The Journal of Physical Chemistry B* 120 (2): 292–298. https://doi.org/10.1021/acs.jpcb.5b08568.

36 Douki, T., Bérard, I., Wack, A., and Andrä, S. (2014). Contribution of cytosine-containing cyclobutane dimers to DNA damage produced by photosensitized triplet-triplet energy transfer. *Chemistry - A European Journal* 20 (19): 5787–5794. https://doi.org/10.1002/chem.201303905.

37 Roca-Sanjuán, D., Olaso-González, G., González-Ramírez, I. et al. (2008). Molecular basis of DNA photodimerization: intrinsic production of cyclobutane cytosine dimers. *Journal of the American Chemical Society* 130 (32): 10768–10779. https://doi.org/10.1021/ja803068n.

38 Zhang, R.B. and Eriksson, L.A. (2006). A triplet mechanism for the formation of cyclobutane pyrimidine dimers in UV-irradiated DNA. *The Journal of Physical Chemistry B* 110 (14): 7556–7562. https://doi.org/10.1021/jp060196a.

39 Gontcharov, J., Liu, L., Pilles, B.M. et al. (2019). Triplet-induced lesion formation at CpT and TpC sites in DNA. *Chemistry - A European Journal* 25 (66): 15164–15172. https://doi.org/10.1002/chem.201903573.

40 Cadet, J., Voituriez, L., Hruska, F.E. et al. (1985). Characterization of thymidine ultraviolet photoproducts. Cyclobutane dimers and 5,6-dihydrothymidines. *Canadian Journal of Chemistry* 63 (11): 2861–2868. https://doi.org/10.1139/v85-477.

41 Douki, T. (2013). The variety of UV-induced pyrimidine dimeric photoproducts in DNA as shown by chromatographic quantification methods. *Photochemical & Photobiological Sciences* 12 (8): 1286–1302. https://doi.org/10.1039/c3pp25451h.

42 Mantel, C., Chandor, A., Gasparutto, D. et al. (2008). Combined NMR and DFT studies for the absolute configuration elucidation of the spore photoproduct, a UV-induced DNA lesion. *Journal of the American Chemical Society* 130 (50): 16978–16984. https://doi.org/10.1021/ja805032r.

43 Douki, T., Court, M., and Cadet, J. (2000). Electrospray–mass spectrometry characterization and measurement of Far-UV-induced thymine photoproducts. *Journal of Photochemistry and Photobiology B: Biology* 54 (2): 145–154. https://doi.org/10.1016/S1011-1344(00)00009-9.

44 Nardi, G., Lineros-Rosa, M., Palumbo, F. et al. (2021). Spectroscopic characterization of dipicolinic acid and its photoproducts as thymine photosensitizers. *Spectrochimica Acta, Part A: Molecular and Biomolecular Spectroscopy* 245: 118898. https://doi.org/10.1016/j.saa.2020.118898.

45 Vendrell-Criado, V., Rodríguez-Muñiz, G.M., Yamaji, M. et al. (2013). Two-photon chemistry from upper triplet states of thymine. *Journal of the American Chemical Society* 135 (44): 16714–16719. https://doi.org/10.1021/ja408997j.

46 Alzueta, O.R., Cadet, J., Cuquerella, M.C., and Miranda, M.A. (2020). Photosensitised biphotonic chemistry of pyrimidine derivatives. *Organic & Biomolecular Chemistry* 18 (12): 2227–2232. https://doi.org/10.1039/d0ob00132e.

47 Kneuttinger, A.C., Kashiwazaki, G., Prill, S. et al. (2014). Formation and direct repair of UV-induced dimeric DNA pyrimidine lesions. *Photochemistry and Photobiology* 90 (1): 1–14. https://doi.org/10.1111/php.12197.

48 Banyasz, A., Douki, T., Improta, R. et al. (2012). Electronic excited states responsible for dimer formation upon UV absorption directly by thymine strands: joint experimental and theoretical study. *Journal of the American Chemical Society* 134 (36): 14834–14845. https://doi.org/10.1021/ja304069f.

49 Bagheri, H., Lhiaubet, V., Montastruc, J.L., and Chouini-Lalanne, N. (2000). Photosensitivity to ketoprofen mechanisms and pharmacoepidemiological data. *Drug Safety* 22 (5): 339–349. https://doi.org/10.2165/00002018-200022050-00002.

50 Boscá, F., Marín, M.L., and Miranda, M.A. (2001). Photoreactivity of the non-steroidal anti-inflammatory 2-arylpropionic acids with photosensitizing side

effects. *Photochemistry and Photobiology* 74 (5): 637–655. https://doi.org/10.1562/0031-8655(2001)0740637POTNAI2.0.CO2.
51 Lhiaubet-Vallet, V., Encinas, S., and Miranda, M.A. (2005). Excited state enantiodifferentiating interactions between a chiral benzophenone derivative and nucleosides. *Journal of the American Chemical Society* 127 (37): 12774–12775. https://doi.org/10.1021/ja053518h.
52 Encinas, S., Belmadoui, N., Climent, M.J. et al. (2004). Photosensitization of thymine nucleobase by benzophenone derivatives as models for photoinduced DNA damage: paterno-büchi vs energy and electron transfer processes. *Chemical Research in Toxicology* 17 (7): 857–862. https://doi.org/10.1021/tx034249g.
53 Costalat, R., Blais, J., Ballini, J.-P. et al. (1990). Formation of cyclobutane thymine dimers photosensitized by pyridopsoralens: a triplet-triplet energy transfer mechanism. *Photochemistry and Photobiology* 51 (3): 255–262. https://doi.org/10.1111/j.1751-1097.1990.tb01709.x.
54 Guillo, L.A., Blais, J., Vigny, P., and Spassky, A. (1995). Selective DNA thymine dimerization during UVA irradiation in the presence of a saturated pyridopsoralen. *Photochemistry and Photobiology* 61 (4): 331–335. https://doi.org/10.1111/j.1751-1097.1995.tb08617.x.
55 Lhiaubet, V., Gutierrez, F., Penaud-Berruyer, F. et al. (2000). Spectroscopic and theoretical studies of the excited states of fenofibric acid and ketoprofen in relation with their photosensitizing properties. *New Journal of Chemistry* 24 (6): 403–410. https://doi.org/10.1039/a909539j.
56 Chouini-Lalanne, N., Defais, M., and Paillous, N. (1998). Nonsteroidal anti-inflammatory drug-photosensitized formation of pyrimidine dimer in DNA. *Biochemical Pharmacology* 55 (4): 441–446. https://doi.org/10.1016/S0006-2952(97)00511-X.
57 Marguery, M.C., Chouini-Lalanne, N., Ader, J.C., and Paillous, N. (1998). Comparison of the DNA damage photoinduced by fenofibrate and ketoprofen, two phototoxic drugs of parent structure. *Photochemistry and Photobiology* 68 (5): 679–684. https://doi.org/10.1111/j.1751-1097.1998.tb02529.x.
58 Michaud, S., Hajj, V., Latapie, L. et al. (2012). Correlations between electrochemical behaviors and DNA photooxidative properties of non-steroïdal anti-inflammatory drugs and their photoproducts. *Journal of Photochemistry and Photobiology B: Biology* 110: 34–42. https://doi.org/10.1016/j.jphotobiol.2012.02.007.
59 Agapakis-Caussé, C., Boscá, F., Castell, J.V. et al. (2000). Tiaprofenic acid-photosensitized damage to nucleic acids: a mechanistic study using complementary in vitro approaches. *Photochemistry and Photobiology* 71 (5): 499–505. https://doi.org/10.1562/0031-8655(2000)071<0499:tapdtn>2.0.co;2.
60 Robinson, K.S., Traynor, N.J., Moseley, H. et al. (2010). Cyclobutane pyrimidine dimers are photosensitised by Carprofen plus UVA in human HaCaT cells. *Toxicology in vitro* 24 (4): 1126–1132. https://doi.org/10.1016/j.tiv.2010.03.007.
61 Trzcionka, J., Lhiaubet-Vallet, V., Paris, C. et al. (2007). Model studies on a carprofen derivative as dual photosensitizer for thymine dimerization and (6-4)

photoproduct repair. *ChemBioChem* 8 (4): 402–407. https://doi.org/10.1002/cbic.200600394.

62 Lhiaubet-Vallet, V., Trzcionka, J., Encinas, S. et al. (2004). The triplet state of a N-phenylphthalimidine with high intersystem crossing efficiency: characterization by transient absorption spectroscopy and DNA sensitization properties. *The Journal of Physical Chemistry B* 108 (37): 14148–14153. https://doi.org/10.1021/jp0498926.

63 Trzcionka, J., Noirot, A., Fabre, P.L., and Chouini-Lalanne, N. (2005). Comparative study of the photophysical properties of indoprofen photoproducts in relation with their DNA photosensitizing properties. *Photochemical & Photobiological Sciences* 4 (3): 298–303. https://doi.org/10.1039/b414654a.

64 Sauvaigo, S., Douki, T., Odin, F. et al. (2001). Analysis of fluoroquinolone-mediated photosensitization of 2′-deoxyguanosine, calf thymus and cellular DNA: determination of type-I, type-II and triplet–triplet Energy transfer mechanism contribution. *Photochemistry and Photobiology* 73 (3): 230. https://doi.org/10.1562/0031-8655(2001)073<0230:aofmpo>2.0.co;2.

65 Tieli, Z., Huichun, Z., Linpei, J., and Shubin, W. (1999). Study of photochemical fluorescence enhancement of the terbium-lomefloxacin complex. *Journal of Photochemistry and Photobiology A: Chemistry* 121 (1): 37–41. https://doi.org/10.1016/S1010-6030(98)00450-X.

66 Itoh, T., Miyauchi-Hashimoto, H., Sugihara, A. et al. (2005). The photocarcinogenesis of antibiotic lomefloxacin and UVA radiation is enhanced in xeroderma pigmentosum group A gene-deficient mice. *The Journal of Investigative Dermatology* 125 (3): 554–559. https://doi.org/10.1111/j.0022-202X.2005.23862.x.

67 Marrot, L., Belaïdi, J.P., Jones, C. et al. (2003). Molecular responses to photogenotoxic stress induced by the antibiotic lomefloxacin in human skin cells: from DNA damage to apoptosis. *The Journal of Investigative Dermatology* 121 (3): 596–606. https://doi.org/10.1046/j.1523-1747.2003.12422.x.

68 Singh, J., Srivastva, A.K., Mandal, P. et al. (2018). Under ambient UVA exposure, pefloxacin exhibits both immunomodulatory and genotoxic effects via multiple mechanisms. *Journal of Photochemistry and Photobiology B: Biology* 178: 593–605. https://doi.org/10.1016/j.jphotobiol.2017.12.014.

69 Lhiaubet-Vallet, V., Sarabia, Z., Hernández, D. et al. (2003). In vitro studies on DNA-photosensitization by different drug stereoisomers. *Toxicology in vitro* 17 (5–6): 651–656. https://doi.org/10.1016/S0887-2333(03)00108-5.

70 Nardi, G., Lhiaubet-Vallet, V., Leandro-Garcia, P., and Miranda, M.A. (2011). Potential phototoxicity of rosuvastatin mediated by its dihydrophenanthrene-like photoproduct. *Chemical Research in Toxicology* 24 (10): 1779–1785. https://doi.org/10.1021/tx200341f.

71 Kunisada, M., Masaki, T., Ono, R. et al. (2013). Hydrochlorothiazide enhances UVA-induced DNA damage. *Photochemistry and Photobiology* 89 (3): 649–654. https://doi.org/10.1111/php.12048.

72 Tonnesen, H.H. and Moore, D.E. (1991). Photochemical stability of biologically active compounds. III. Mefloquine as a photosensitizer. *International Journal of Pharmaceutics* 70 (1): 95–101. https://doi.org/10.1016/0378-5173(91)90168-N.

73 Navaratnam, S., Hamblett, I., and Tonnesen, H.H. (2000). Photoreactivity of biologically active compounds.: XVI. Formation and reactivity of free radicals in Mefloquine. *Journal of Photochemistry and Photobiology B: Biology* 56 (1): 25–38. https://doi.org/10.1016/S1011-1344(00)00056-7.

74 Yadav, N., Dwivedi, A., Mujtaba, S.F. et al. (2014). Photosensitized mefloquine induces ROS-mediated DNA damage and apoptosis in keratinocytes under ambient UVB and sunlight exposure. *Cell Biology and Toxicology* 30 (5): 253–268. https://doi.org/10.1007/s10565-014-9280-7.

75 Wong-Wah-Chung, P., Rafqah, S., Voyard, G., and Sarakha, M. (2007). Photochemical behaviour of Triclosan in aqueous solutions: kinetic and analytical studies. *Journal of Photochemistry and Photobiology A: Chemistry* 191 (2): 201–208. https://doi.org/10.1016/j.jphotochem.2007.04.024.

76 Dubey, D., Srivastav, A.K., Singh, J. et al. (2019). Photoexcited triclosan induced DNA damage and oxidative stress via P38 MAP kinase signaling involving type I radicals under sunlight/UVB exposure. *Ecotoxicology and Environmental Safety* 174: 270–282. https://doi.org/10.1016/j.ecoenv.2019.02.065.

77 Kumasaka, R., Kikuchi, A., and Yagi, M. (2014). Photoexcited states of UV absorbers, benzophenone derivatives. *Photochemistry and Photobiology* 90 (4): 727–733. https://doi.org/10.1111/php.12257.

78 Amar, S.K., Goyal, S., Srivastav, A.K. et al. (2018). Combined effect of benzophenone-2 and ultraviolet radiation promote photogenotoxicity and photocytotoxicity in human keratinocytes. *Regulatory Toxicology and Pharmacology* 95: 298–306. https://doi.org/10.1016/j.yrtph.2018.04.003.

79 Kikuchi, A., Oguchi, N., and Yagi, M. (2009). Optical and electron paramagnetic resonance studies of the excited states of 4-tert-butyl-4′-methoxydibenzoylmethane and 4-tert-butyl-4′-methoxydibenzoylpropane. *The Journal of Physical Chemistry. A* 113 (48): 13492–13497. https://doi.org/10.1021/jp905236m.

80 Paris, C., Lhiaubet-Vallet, V., Jiménez, O. et al. (2009). A blocked diketo form of Avobenzone: photostability, photosensitizing properties and triplet quenching by a triazine-derived UVB-filter. *Photochemistry and Photobiology* 85 (1): 178–184. https://doi.org/10.1111/j.1751-1097.2008.00414.x.

81 Miro, P., Lhiaubet-Vallet, V., Marin, M.L., and Miranda, M.A. (2015). Photosensitized thymine dimerization via delocalized triplet excited states. *Chemistry - A European Journal* 21 (47): 17051–17056. https://doi.org/10.1002/chem.201502719.

82 Pathirana, D., Ormerod, A.D., Saiag, P. et al. (2009). European S3-guidelines on the systemic treatment of psoriasis vulgaris. *Journal of the European Academy of Dermatology and Venereology* 23 (s2): 1–70. https://doi.org/10.1111/j.1468-3083.2009.03389.x.

83 Sidbury, R., Davis, D.M., Cohen, D.E. et al. (2014). Guidelines of care for the management of atopic dermatitis: section 3. Management and treatment with phototherapy and systemic agents. *Journal of the American Academy of Dermatology* 71 (2): 327–349. https://doi.org/10.1016/j.jaad.2014.03.030.

84 Taieb, A., Alomar, A., Böhm, M. et al. (2013). Guidelines for the management of vitiligo: the European dermatology forum consensus. *The British Journal of Dermatology* 168 (1): 5–19. https://doi.org/10.1111/j.1365-2133.2012.11197.x.

85 Bensasson, R.V., Salet, C., Land, E.J., and Rushton, F.A.P. (1980). Triplet excited state of the 4′5' photoadduct of psoralen and thymine. *Photochemistry and Photobiology* 31 (2): 129–133. https://doi.org/10.1111/j.1751-1097.1980.tb03695.x.

86 Diekmann, J., Theves, I., Thom, K.A., and Gilch, P. (2020). Tracing the photoaddition of pharmaceutical psoralens to DNA. *Molecules* 25: 5242. https://doi.org/10.3390/molecules25225242.

87 Jin, P., Wang, X., Jia, M. et al. (2024). Ultrafast and multichannel generation of the triplet state in DNA-intercalated gilvocarcin V for an enhanced photoaddition reaction. *Journal of Physical Chemistry Letters* 15 (10): 2765–2771. https://doi.org/10.1021/acs.jpclett.4c00255.

88 Makovets, S. (ed.) (2013). *DNA Electrophoresis - Methods and Protocols*. Totowa, NJ: Humana Press https://doi.org/10.1007/978-1-62703-565-1.

89 Gordon, L.K. and Haseltine, W.A. (1980). Comparison of the cleavage of pyrimidine dimers by the bacteriophage T4 and micrococcus luteus UV-specific endonucleases. *The Journal of Biological Chemistry* 255 (24): 12047–12050. https://doi.org/10.1016/S0021-9258(19)70242-7.

90 Morikawa, K., Matsumoto, O., Tsujimoto, M. et al. (1992). X-ray structure of T4 endonuclease V: an excision repair enzyme specific for a pyrimidine dimer. *Science* 256: 523–526.

91 Singh, N.P., McCoy, M.T., Tice, R.R., and Schneider, E.L. (1988). A simple technique for quantitation of low levels of DNA damage in individual cells. *Experimental Cell Research* 175 (1): 184–191. https://doi.org/10.1016/0014-4827(88)90265-0.

92 Ostling, O. and Johanson, K.J. (1984). Microelectrophoretic study of radiation-induced DNA damages in individual mammalian cells. *Biochemical and Biophysical Research Communications* 123 (1): 291–298. https://doi.org/10.1016/0006-291X(84)90411-X.

93 Olive, P.L., Banáth, J.P., Durand, R.E., and Banath, J.P. (1990). Heterogeneity in radiation-induced DNA damage and repair in tumor and normal cells measured using the “comet” assay. *Radiation Research* 122 (1): 86–94. https://doi.org/10.2307/3577587.

94 Sauvaigo, S., Serres, C., Signorini, N. et al. (1998). Use of the single-cell gel electrophoresis assay for the immunofluorescent detection of specific DNA damage. *Analytical Biochemistry* 259: 1–7.

95 Helbock, H.J., Beckman, K.B., Shigenaga, M.K. et al. (1998). DNA oxidation matters: the HPLC–electrochemical detection assay of 8-oxo-deoxyguanosine and 8-oxo-guanine. *Proceedings of the National Academy of Sciences* 95 (1): 288–293. https://doi.org/10.1073/pnas.95.1.288.

96 Artuso, T., Bernadou, J., Meunier, B. et al. (1991). Mechanism of DNA cleavage mediated by photoexcited non-steroidal antiinflammatory drugs. *Photochemistry and Photobiology* 54 (2): 205–213. https://doi.org/10.1111/j.1751-1097.1991.tb02008.x.

97 Michaud, S., Bordeau, G., Sartor, V. et al. (2015). Formation of cyclobutane thymine dimers by tiaprofenic acid and its photoproducts: approach to the photosensitizer triplet state energy limit value. *RSC Advances* 5 (84): 68595–68600. https://doi.org/10.1039/c5ra11869g.

98 Cuquerella, M.C., Lhiaubet-vallet, V., Cadet, J., and Miranda, M.A. (2012). Benzophenone photosensitized DNA damage. *Accounts of Chemical Research* 45 (9): 1558–1570. https://doi.org/10.1021/ar300054e.

99 Lhiaubet-Vallet, V., Trzcionka, J., Encinas, S. et al. (2003). Photochemical and photophysical properties of Indoprofen. *Photochemistry and Photobiology* 77 (5): 487–491. https://doi.org/10.1562/0031-8655(2003)0770487PAPPOI2.0.CO2.

100 Trzcionka, J., Lhiaubet-Vallet, V., and Chouini-Lalanne, N. (2004). DNA photosensitization by indoprofen – is DNA damage photoinduced by indoprofen or by its photoproducts? *Photochemical & Photobiological Sciences* 3 (2): 226–230. https://doi.org/10.1039/b307719e.

101 Fief, C.A., Hoang, K.G., Phipps, S.D. et al. (2019). Examining the impact of antimicrobial fluoroquinolones on human DNA topoisomerase IIα and IIβ. *ACS Omega* 4 (2): 4049–4055. https://doi.org/10.1021/acsomega.8b03428.

102 de Vries, H. and Beijersbergen van Henegouwen, G.M.J. (2000). Photochemical decomposition of Lomefloxacin in vitro and in vivo. *Journal of Photochemistry and Photobiology B: Biology* 58 (1): 6–12. https://doi.org/10.1016/S1011-1344(00)00103-2.

6

Triplet Decay in Canonical and Noncanonical DNA/RNA Nucleobases

Siman Chen[1,2]*, Yuhsuan Lee*[1,2]*, and Shuming Bai*[1,2]

[1] *Beijing National Laboratory for Molecular Sciences, State Key Laboratory for Structural Chemistry of Unstable and Stable Species, Institute of Chemistry, Chinese Academy of Sciences, Beijing 100190, China*
[2] *University of Chinese Academy of Sciences, Beijing 100049, China*

6.1 Introduction

DNA and RNA nucleobases exhibit remarkable photostability, which is essential for maintaining genetic integrity upon exposure to ultraviolet (UV) radiation [1–3]. The inherent resistance of nucleic acids to UV-induced photodamage has intrigued researchers for decades, sparking numerous studies aimed at uncovering the photophysical and photochemical mechanisms responsible for their unique properties. This topic has extensive implications across various fields, including photobiology, phototherapy, and the design of photoprotective materials. Historically, studies on nucleobase photostability date back to the mid-20th century when foundational photochemical experiments first described the stability and reactivity patterns of these molecules. Early investigations employed steady-state photochemical methods, revealing basic information on the UV absorption characteristics and general photostability of nucleobases. However, these early studies provided limited mechanistic insights due to their methodological constraints.

Advancements in technology, particularly the development of ultrafast spectroscopy techniques such as transient absorption, femtosecond fluorescence upconversion, and time-resolved infrared spectroscopy, have since revolutionized this field. These techniques allow for direct observation and characterization of excited state dynamics on femtosecond-to-picosecond timescales, significantly enhancing the mechanistic understanding of nucleobase photostability. Concurrently, theoretical developments, including quantum chemical calculations employing methods like Time-Dependent Density Functional Theory (TD-DFT) and Complete Active Space Perturbation Theory (CASPT2), have complemented experimental studies by providing detailed mechanistic insights at the molecular level.

Triplet states, which form after photoexcitation, are longer lived compared to singlet excited states, making them important intermediates in photoinduced

Excited State Dynamics in DNA and RNA: Experimental and Computational Approaches on Triplet State, First Edition. Edited by Jinquan Chen.

processes. Canonical nucleobases—adenine, thymine, guanine, cytosine, and uracil—primarily exhibit rapid internal conversion (IC) to the ground electronic state after photoexcitation, a process that prevents significant accumulation of reactive intermediates such as triplet states. In other words, these nucleobases inhibit efficient intersystem crossing (ISC) and high triplet quantum yields, making them not candidates for photochemical applications.

In contrast, noncanonical nucleobases, such as thiobases (e.g. 4-thiothymine, 6-thioguanine [6tGua]), halogenated nucleobases (e.g. 5-bromouracil), and aza-analogs, exhibit distinctly different photophysical properties, characterized notably by significantly enhanced ISC efficiencies and prolonged triplet state lifetimes [4–7]. These differences come largely from modifications such as heavy atom substitution, which drastically alter the electronic and spin–orbit coupling (SOC) characteristics of these molecules, resulting in distinctly altered photophysical behaviors. Structural modifications in these nucleobases often lead to altered electronic environments, SOC, and triplet state lifetimes, affecting the triplet decay process and the overall photosensitizing efficiency. Therefore, understanding the mechanisms governing triplet state decay is essential for advancing the design of efficient DNA-targeted photosensitizers [8–11].

This chapter systematically synthesizes the accumulated knowledge from both the experimental and theoretical perspectives, emphasizing recent advances and the current state of understanding regarding nucleobase triplet state dynamics. By thoroughly reviewing past and contemporary studies, this chapter sets the stage for an in-depth exploration of nucleobase triplet state dynamics, highlighting the vital role that advanced experimental and computational methodologies have played and continue to play in unraveling these complex photophysical processes. It further outlines critical unresolved questions and identifies promising future research directions. The significance of fully elucidating these mechanisms is underscored by their extensive practical implications, ranging from the development of novel phototherapeutic agents to the design of improved photostabilizers for biomedical and technological applications.

6.2 Low Triplet Yields in Natural DNA/RNA Nucleobases

The intrinsic photostability of DNA and RNA nucleobases—adenine, thymine, guanine, cytosine, and uracil—is essential for protecting genetic information against harmful UV radiation [2, 12]. Despite their significant absorption cross sections in the UV range, these natural nucleobases exhibit exceptionally low triplet quantum yields, typically in the order of 10^{-3} to 10^{-4} or even lower. This distinctive photophysical behavior primarily arises due to ultrafast IC pathways that rapidly channel the absorbed electronic energy back to the electronic ground state, thereby circumventing the formation of long-lived and potentially reactive triplet states [3, 12].

The low triplet yield in canonical nucleobases has been extensively studied using ultrafast spectroscopic techniques and advanced computational methods, as it originates from the efficient nonradiative decay in monomeric nucleobases that occurs predominantly via conical intersections (CIs) [13, 14]. These intersections provide near-barrierless pathways from electronically excited singlet states back to the ground state, facilitating relaxation within femtoseconds to picoseconds. Such rapid and efficient IC substantially reduces the probability of ISC events, thus minimizing triplet formation. According to computational analysis, five canonical nucleobases feature highly accessible CIs, effectively funneling the excited state population rapidly back to the ground state [15]. This unique energetic landscape stands in sharp contrast with structurally related modified nucleobases discussed in the following section, which frequently possess energy barriers on their excited state potential energy surfaces.

From an evolutionary perspective, the selection of these specific nucleobases for genetic encoding is closely linked to their exceptional photostability. By rapidly dissipating electronic excitation energy through IC rather than populating long-lived triplet states, canonical nucleobases significantly decrease the likelihood of photochemical damage and mutagenic events, preserving genomic integrity even under intense solar UV exposure prevalent during Earth's early history. Therefore, natural selection favored these molecules precisely for their robust resistance against photochemical transformations and their superior ability to mitigate photodamage compared to other closely related molecular variants. Consequently, from a biological and genetic integrity perspective, the detailed study of triplet decay in natural nucleobases may hold limited practical significance, as their triplet state populations are negligible under typical physiological conditions.

6.3 Triplet Decay in Thionucleobases

Noncanonical nucleobases, including modified structures such as thiobases, halogenated derivatives, and aza-analogs, exhibit fundamentally different photophysical behaviors compared to canonical nucleobases [16]. These differences primarily manifest as significantly enhanced ISC rates and prolonged triplet state lifetimes, characteristics that have profound implications in photochemistry, biology, and medical science. Electronic transitions between states of different multiplicities are spin-forbidden and typically exhibit low probabilities. However, the presence of heavy atoms such as sulfur, selenium, or halogens significantly enhances SOC, thereby facilitating ISC. SOC enables effective mixing of electronic states with different spin multiplicities, making otherwise spin-forbidden transitions considerably more probable. Thiobases represent one of the most studied groups of noncanonical nucleobases, characterized by sulfur substitution at positions normally occupied by oxygen [17–19]. This seemingly minor substitution dramatically alters electronic configurations, significantly enhancing SOC and ISC efficiencies. Experimental studies employing femtosecond transient absorption spectroscopy have revealed that thiobases, such as 4-thiothymine, 2-thiocytosine, and 6tGua, demonstrate nearly 100% yield of triplet states.

The sulfur substitution in thionucleobases enhances ISC efficiency, which leads to rapid triplet state generation and subsequent singlet oxygen production

[20, 21]. Thionucleobases such as 2-thiothymine (2tThy) and 6-aza-2-thiothymine (6n-2tThy) are well-established models for studying triplet state dynamics due to their nearly unity triplet quantum yields and ultrafast ISC processes. Experimental work by Ichimura and coworkers provided clear transient absorption spectra demonstrating that both 2tThy and 6n-2tThy undergo highly efficient ISC on sub-picosecond timescales, leading to long-lived triplet states [20, 22]. These findings confirmed that sulfur substitution, combined with nitrogen incorporation in the ring, promotes rapid ISC and stabilizes the triplet state, making these molecules excellent benchmarks for understanding triplet decay in both canonical and noncanonical systems. These experimental insights laid the groundwork for subsequent computational studies aiming to dissect the mechanistic details of triplet decay pathways in modified nucleobases.

The impact of glycosylation on triplet state decay in thionucleobases and their nucleoside derivatives has also been investigated [14, 23–28]. This modification can significantly alter the electronic environment and geometry of the nucleobase, influencing activation energy for ISC and triplet state lifetimes. The introduction of sugar moieties into nucleobases, such as 2-Thiothymidine (2tThd) and 6-thioguanosine (6tGuo), significantly influences the triplet decay rate, making these compounds particularly valuable as photosensitizers for photodynamic therapy (PDT) [17]. Glycosylation not only facilitates ISC by enhancing vibronic coupling between the singlet and triplet states but also reduces the activation energy for ISC, leading to faster triplet state decay and more efficient singlet oxygen production [9, 29–31].

Beyond molecular structure, the photophysical properties of nucleobases are substantially impacted by their environment. Solvent polarity, viscosity, and specific solvent–solute interactions critically influence excited state dynamics. Polar solvents can stabilize certain excited states, altering the energetic landscape and pathways available for ISC. High-viscosity environments typically slow down molecular vibrations and rotations, effectively prolonging excited state lifetimes and potentially increasing ISC efficiencies. Specific interactions, such as hydrogen bonding, also profoundly impact ISC pathways by stabilizing or destabilizing particular excited state conformations or intermediates. In addition to glycosylation, Crespo-Hernández and coworkers explored the impact of solvent environments on triplet state decay. They found that polar solvents, such as acetonitrile, could modulate triplet decay rates by altering the electronic interactions between the nucleobase and the solvent molecules. For example, hydrogen bonding between 6tGuo and solvent molecules slowed the ISC rate, highlighting the complex role of solvent in controlling triplet decay [17]. Similarly, Taras-Goslinska et al. confirmed that glycosylation enhances triplet decay rates, specifically noting that 2tThd exhibited a 14 times faster decay than 2tThy. Their study emphasized the role of sugar-induced changes to the T_1 state geometry, particularly the destabilization of the triplet state's potential energy surface, in accelerating triplet decay. Additionally, they found a direct correlation between triplet decay rates and singlet oxygen production, reinforcing the therapeutic potential of glycosylated nucleosides in PDT [29].

In summary, these studies highlight the dual impact of molecular modifications (such as glycosylation and heavy atom substitution) and environmental factors (such as solvent interactions) on triplet decay dynamics in modified nucleobases. Theoretical studies have significantly contributed to understanding the mechanism of triplet decay and the interplay between structural and environmental factors in determining triplet state lifetimes, which paves the way for further research aimed at optimizing photosensitizers for PDT and other photochemical applications.

6.4 Theoretical and Computational Studies of Triplet Decay of Thionucleobases

The observed modification of triplet decay raises an important question: how do structural modifications, such as glycosylation, influence triplet decay dynamics? While experimental methods have provided useful insights, they fall short of fully elucidating the mechanisms behind [29]. These methods have difficulty capturing the detailed excited state dynamics and the intrinsic decay mechanisms of the T_1 state, especially when considering interactions with external species like molecular oxygen. Theoretical approaches are necessary to bridge this gap and provide deeper insights into the underlying mechanisms driving triplet decay, which cannot be fully captured by experimental techniques alone. By combining theoretical and experimental perspectives, a more comprehensive understanding of triplet decay dynamics will be achieved, thereby improving the design of more efficient photosensitizers for photochemical applications. Meanwhile, although significant progress has been made for excited state dynamics through theoretical studies using nonadiabatic dynamics simulations, understanding the intrinsic decay dynamics of the T_1 state is quite different as the lifetime is relatively long.

Computational studies of triplet decay rely on quantum chemical methods to model the excited state dynamics. Key methods include DFT for optimizing ground state geometries and TD-DFT and Multistate Complete Active Space Perturbation Theory (MS-CASPT2) for excited state description [16, 32]. MS-CASPT2 and TD-DFT often can offer more accurate descriptions [16]. The geometry optimization of the T_1 state was performed using CASPT2 to obtain two minimum structures for 2tThy and 6n-2tThy [16]. These structures were characterized by strong out-of-plane sulfur (op-S) displacement and the slight ring distortion ($B_{3,6}$). The CASPT2 method, combined with a well-defined active space and the ANO-RCC-VTZP basis set, was crucial for accurately describing the energy balance of these minima [33–36]. Additional DFT and linear-response TD-DFT calculations were done with B3LYP [37, 38], CAM-B3LYP [39], and oB97X-D [40] functionals based on the 6-311G(d,p) basis set [41]. Also, complementary linear-response algebraic diagrammatic construction to second-order (ADC(2)) [42] and coupled cluster to approximated second-order (CC2) [43] calculations, both with the aug-cc-pVDZ basis set, were done with Turbomole. Singlet/triplet crossing points were optimized at the CASPT2 level to an energy gap smaller than 0.02 eV, with an in-house modified version of the CIOpt program [44]. As shown in Figure 6.1, CC2 and ADC(2) provided very similar results to CASPT2, with minimal error (~0.1 eV), confirming their reliability. In contrast, CASSCF was found to perform poorly,

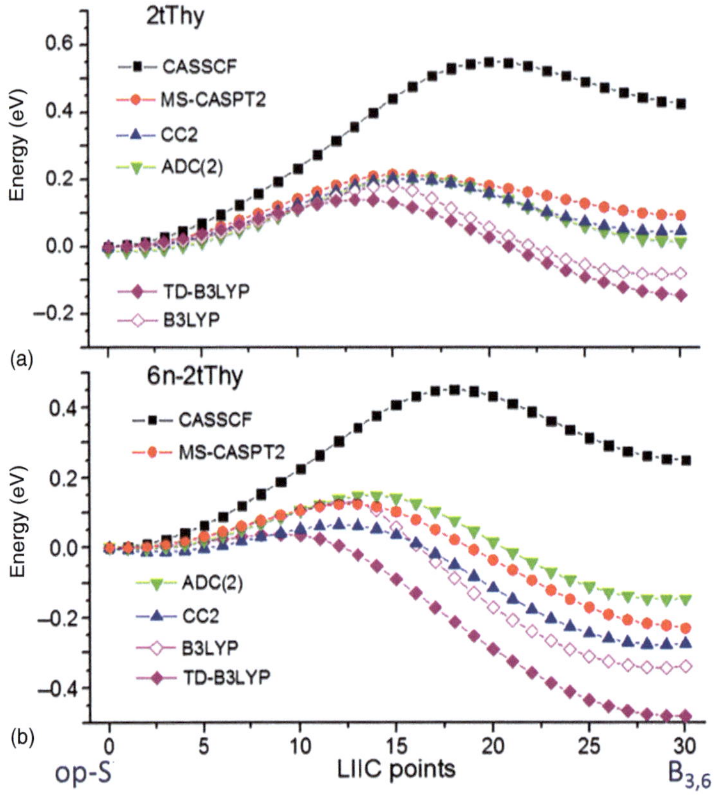

Figure 6.1 T_1 energy profile between the two T_1 minima at different levels. The energy of the op-S minimum was set to zero in all cases. TD-B3LYP, CC2, and ADC(2) calculations are based on the linear response of the closed-shell ground state, while B3LYP is a direct unrestricted DFT calculation of the triplet configuration ([15] / Royal Society of Chemistry).

particularly with the $B_{3,6}$ minimum, which was overestimated by approximately 0.4 eV [16]. These findings underscore the importance of selecting the appropriate computational method for optimizing triplet state geometries and calculating energy differences between minima.

The existence of a double-well potential on the lowest triplet excited state T_1 potential energy surfaces of thionucleobases has been experimentally validated using ultrafast transient vibrational absorption spectroscopy (TVAS) [45]. By exciting 2-Thiouracil (2TU) at 300 nm in deuterated acetonitrile, they observed transient infrared absorption spectra clearly indicating two separate sets of vibrational bands corresponding to two different conformations of the molecule in the T_1 state. Specifically, following photoexcitation, an initial vibrational absorption band at 1643 cm^{-1} was detected, decaying within approximately 7 ps and concurrently giving rise to additional bands at 1626 and 1660 cm^{-1}. These subsequent bands exhibited no decay on the timescale of their measurements (up to 1 ns), strongly suggesting the stabilization of two distinct minima with substantial energy barriers separating them. These experimental observations, such as the extended lifetimes and distinctive vibrational fingerprints, were consistent with the computational prediction [16]. This experimental confirmation reinforces the validity of the theoretical double-well model and solidifies the conceptual framework initially proposed by theorists, underscoring the importance of integrating advanced computational and experimental methodologies to unravel the photodynamics of modified nucleobases.

Then the two-step model to describe the triplet decay was introduced to explain the triplet state decay of thionucleobases through two distinct minima in the T_1 state: the op-S minimum and the ring-distorted ($B_{3,6}$) minimum [16]. According to this model shown in Figure 6.2, the triplet decay occurs via either a slow transition between these minima or a rapid ISC to the S_0 ground state, depending on the energetic balance between the two minima. The transition from the op-S minimum to the S_0 state is particularly important for PDT, as it influences singlet oxygen yield, a key factor in the therapeutic efficacy of photosensitizers. This model was further extended to incorporate the glycosylation effect and solvent effect on triplet decay [46, 47]. Glycosylation destabilizes the T_1 minimum, lowering the activation energy for ISC and accelerating the decay process.

The accurate prediction of triplet decay rates is pivotal for designing photosensitizers with tailored lifetimes. For the nonadiabatic rate in a weak coupling regime, such as triplet/singlet intersystem crossing, the rate can be computed using the quasi-Marcus semi-classical method [48]:

$$k_{\text{isc}} = \frac{2\pi}{\hbar}|\text{H}_{\text{SOC}}|^2 \frac{1}{\sqrt{4\pi\lambda \text{k}_\text{B} T}} \exp\left(-\frac{\Delta G^{\ddagger}_{\text{isc}}}{\text{k}_\text{B} T}\right) \tag{6.1}$$

Here, H_{SOC} represents the SOC at the ISC point; λ is the internal reorganization energy, calculated as the energy difference of the S_0 state between the T_1 minimum and the S_0 minimum; k_B is the Boltzmann constant, T is the absolute temperature, and ΔG^0 is the total Gibbs free energy variation for the transfer. The term in the exponential is a quadratic approximation for the activation energy $\Delta G^{\ddagger}_{\text{isc}}$:

$$\Delta G^{\ddagger}_{\text{isc}} \approx \frac{(\lambda + \Delta G^0)^2}{4\lambda} \tag{6.2}$$

$\Delta G^{\ddagger}_{\text{isc}}$ is the activation energy defined by the energy difference in the T_1 state between the initial minimum and the crossing point. These parameters are shown in Figure 6.1. Regarding λ, in the standard Marcus theory involving two diabatic states, the reorganization energy can be calculated equally from either adiabatic state. However, in real cases, sometimes a third diabatic state may be involved in mixing with

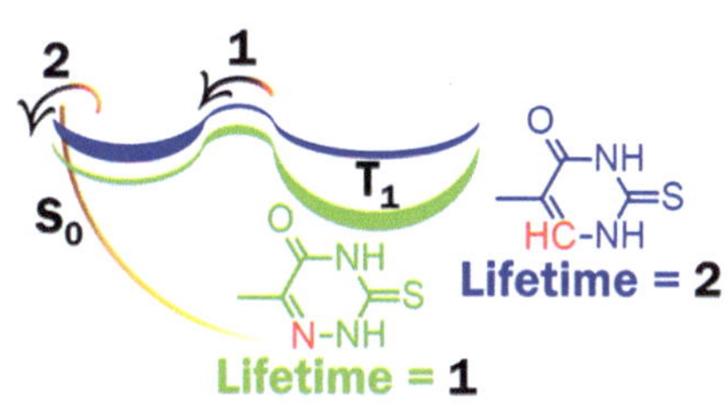

Figure 6.2 The two-step model of triplet decay proposed for thionucleobases ([15] / Royal Society of Chemistry).

T_1. As a result, we used S_0 to calculate the reorganization energy, as it aligns more effectively with the Marcus framework.

It is important to highlight that this rate model is a simplified expression in classical limit [49–51], as it only works if the frequencies defining the Marcus parabolic potentials are much smaller than the thermal energy, which is about 200 cm^{-1} at 300 K. Although these conditions are roughly satisfied for sulfur-substituted nucleobases, there is a significant concern regarding the activation energy, which is incorporated into an exponential term, thus amplifying potential computational errors. More details can be found in the original reference [16].

Similarly, taking into account the energy barrier between the two T_1 minima, we can utilize the Eyring equation [52]:

$$k_{\text{crossing}} = \frac{k_B T}{h} \exp\left(\frac{-\Delta G^{\ddagger}_{\text{cross}}}{k_B T}\right) \tag{6.3}$$

where k_{crossing} is the rate of ISC, k_B is the Boltzmann constant, T is the temperature, h is Planck's constant, and $\Delta G^{\ddagger}_{\text{cross}}$ is the activation free energy barrier for the transition. This equation provides insight into how the energy gap between different T_1 minima influences the decay rate, as well as the effects of conformer transitions in glycosylated nucleosides [46]. We note that by linking the conformational flexibility of nucleosides to the ISC rates, how structural modifications facilitate triplet state decay can be investigated.

6.5 Mechanisms of Chemical Modulation of Triplet Decay

Under the abovementioned two-step model, critical parameters to govern triplet state decay can be determined. First is the SOC. SOC is significantly higher in selenium derivatives (e.g. 6SeGua) compared to their sulfur analogs, such as 6tGua, due to the heavy atom effect [47]. In fact, 6SeGua exhibits SOC values 5–6 times greater than those of its sulfur counterpart [47, 53]. Additionally, the geometry dependence of SOC plays a pivotal role in triplet state decay. SOCs at the T_1/S_0 crossing points with out-of-plane distortions, such as the op-S conformation in 2tThy, are much larger than the ones at the counterpart $B_{3,6}$ conformation [16]. The activation energy ($\Delta G^{\ddagger}$) plays a crucial role in determining the triplet decay rate. Modifications such as glycosylation lower $\Delta G^{\ddagger}$ by inducing steric repulsion between the sugar group and sulfur, facilitating faster ISC. Additionally, hydrogen bonding—particularly the stronger interactions between sulfur and water compared to selenium and water in 6tGua—increases $\Delta G^{\ddagger}$, explaining the slower triplet decay of 6tGua compared to 6SeGua [53, 54]. Another parameter is the reorganization energy (λ), which represents the energy required to reorganize the system during ISC. The solvent environment has a direct impact on λ, with polar solvents stabilizing charge-separated states and reducing λ, leading to a lower energy barrier for ISC.

For the strategy of heavy atom substitution, Fang et al. investigated the excited state dynamics of selenium-substituted guanine derivatives, with a particular focus on the double-well topography of the triplet state (T_1). In their 2019 study, using MS-CASPT2 calculations, they identified the existence of two distinct triplet minima for 6tGua and 6-selenoguanine (6SeGua), analogous to the op-S and $B_{3,6}$ minima found in thionucleobases, such as 2tThy [55]. These minima represent different geometries of the T_1 state, where the sulfur or selenium atoms undergo significant out-of-plane distortions, leading to different energetic profiles and transition paths. Computational results revealed that the introduction of selenium in 6SeGua altered the energy landscape of the T_1 state, making the selenium-substituted T_1 state more stable compared to its sulfur counterpart. This double-well structure plays a crucial role in triplet state formation and decay, as the dynamics between these minima influence the efficiency of ISC and the rate of triplet state relaxation. Specifically, the T_1 state topography of thionucleobases reveals a double-well structure, where two distinct minima correspond to the op-S configuration and the ring-distorted conformation. The energy difference between these two minima governs the rate of ISC and thus the triplet decay process. For 6tGua, the op-S structure is lower in energy than the ring-distorted structure, and ISC primarily occurs from this minimum. In contrast, 6SeGua exhibits a larger energy gap between these minima, which contributes to its faster triplet decay compared to 6tGua. The energy difference between the two minima significantly impacts the decay rate and triplet yield, which is critical for optimizing the photophysical properties of photosensitizers. In 2021, Fang et al. further confirmed these findings experimentally, demonstrating that the presence of selenium in the T_1 state stabilizes one of the minima, enhancing ISC rates and triplet state formation [56]. These results provide valuable insights into how specific structural modifications, such as selenium substitution, influence the topography of the T_1 state, contributing to a better understanding of triplet state decay in thionucleobases and their derivatives.

Besides the atom substitution, the geometry and relative energy of these minima are also influenced by group substitution. We have established that glycosylation enhances the triplet decay rate in thionucleosides compared to their corresponding thionucleobases [46]. The introduction of the sugar group increases vibronic coupling between the singlet and triplet states, facilitating a more efficient ISC. This interaction is driven by both electronic and steric effects between the sugar moiety and the sulfur atom, which destabilize the T_1 minimum, lowering the activation energy for ISC. The destabilization of the T_1 minimum due to glycosylation reduces the activation energy for ISC, making the transition from T_1 to S_0 more rapid and efficient. This reduction is primarily attributed to both geometric distortion and electronic interactions between the sugar moiety and the sulfur atom. In 2tThy, for example, as the molecule transitions from the T_1 minimum to the ISC geometry, the out-of-plane displacement of sulfur decreases, effectively shortening the pathway for ISC and thus making the transition more efficient. As a result, glycosylated nucleosides, such as 2tThd, exhibit significantly enhanced triplet decay rates compared to their parent thionucleobases, such as 2tThy. This shortened ISC pathway results in a lower activation energy, facilitating faster ISC and promoting triplet decay. This model not only explains the faster ISC observed in glycosylated nucleosides but also highlights how structural modifications can be used to tune triplet state dynamics, improving the efficiency of photosensitizers in PDT applications. The effects from the SOC and the reorganization energy (λ) were also examined, but they are secondary factors compared to the effect on the activation energy.

Solvent interactions also play a critical role in triplet state dynamics, particularly in modulating triplet decay and ISC efficiency. Valverde et al. showed that in 6SeGua, the substitution of sulfur with selenium significantly enhances ISC rates in water. Their study indicated that selenium's heavier atomic mass accelerates the ISC process, leading to faster triplet state formation and shorter triplet lifetimes compared to 6tGua [53]. We explored the solvent effects there and found that hydrogen bonding between the solvent and the sulfur atom in 6tGua significantly stabilizes the triplet state, thereby slowing down its decay [47]. This effect was less pronounced in 6SeGua, as selenium forms weaker hydrogen bonds compared to sulfur, leading to faster triplet decay. These effects were modeled using both implicit and explicit solvent models, such as the polarizable continuum model (PCM) and explicit water molecules. The explicit microsolvation model, which included a water molecule interacting with the sulfur/selenium atom, demonstrated that the hydrogen bonding between 6tGua and water stabilized the triplet state and raised the activation energy, thus slowing triplet decay. In contrast, 6SeGua exhibited weaker interactions with water, resulting in a faster decay process. The difference in solvent effect originates from the intermolecular interactions with water molecules. As shown in Figure 6.3, a hydrogen atom from the water molecule near the sulfur/selenium atom

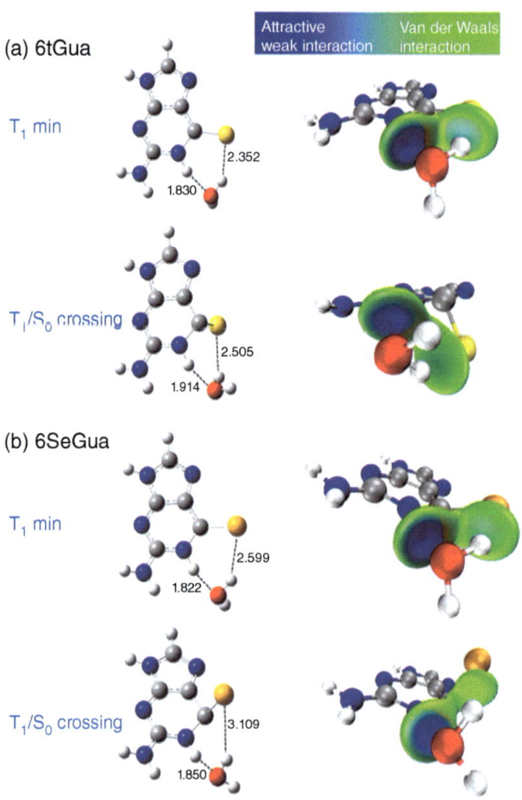

Figure 6.3 Hydrogen bond lengths with the structures of T_1 minima and T_1/S_0 crossing points and their IGMH analysis, including (a) 6tGua monohydrated complex and (b) 6SeGua monohydrated complex. Isovalue: 0.005; bond length unit: Å ([47] / Royal Society of Chemistry).

in hydrated 6tGua/6SeGua, forming an S/Se–H–O structure. This specific structure could form a hydrogen bond and change the activation energies. Marian explored solvent effects on excited state dynamics in noncanonical nucleobases, underscoring the importance of solvent interactions in modifying the electronic structure of the triplet state [57]. For the ultrafast photoinduced processes in complex environments, the roles of solvent polarity and hydrogen bonding were also emphasized in modulating triplet state reactivity [54]. These results highlight how solvent interactions can influence ISC and triplet state reactivities. Understanding how solvents interact with photosensitizers, especially in biological systems, is crucial for the rational design of more efficient and stable photosensitizers. Future research should focus on refining solvent–solute interaction models and investigating their impact in complex biological environments to further optimize the performance of photosensitizers.

6.6 Remaining Challenges and Future Directions

The mechanistic insights into triplet decay dynamics of canonical and noncanonical nucleobases, including glycosylated derivatives and thionucleobases, provide a solid foundation for designing next-generation photosensitizers. These findings are pivotal for enhancing the efficiency of PDT and other therapeutic applications involving reactive oxygen species (ROS). By improving our understanding of the factors influencing triplet decay, such as structural modifications and solvent interactions, future research can optimize photosensitizers for more effective cancer treatments and other therapeutic purposes. However, challenges remain in translating these insights into clinical practice.

While substantial progress has been made in understanding the mechanisms behind enhanced triplet decay in glycosylated nucleosides, important areas remain for further investigation. For instance, the conformational flexibility of glycosylated nucleosides and their impact on ISC needs to be explored in more detail. The sugar moiety's ability to affect the nucleobase's electronic environment and its interaction with the triplet state could significantly alter triplet decay rates. A deeper understanding of these structural variations could lead to more efficient design strategies for photosensitizers with optimized triplet decay rates. Another challenge lies in understanding the biological compatibility of glycosylated thionucleosides. While these compounds show promise in photosensitization, their ability to evade metabolic degradation while maintaining their therapeutic efficacy remains to be determined. This is critical for developing effective PDTs that can selectively target diseased cells without being compromised by rapid metabolic breakdown.

Another challenge lies in the trade-off between the accuracy of explicit microsolvation models and the computational cost associated with them. While explicit models, which incorporate solvent molecules explicitly interacting with the nucleobase, are essential for understanding detailed interactions like hydrogen bonding, they are computationally demanding. Moreover, solvent effects in more complex environments, such as biological systems, must be studied in greater detail [54, 58, 59]. The presence of proteins, ionic strength, and other solution conditions may influence triplet decay, making it essential to explore the interactions between nucleobases and their surroundings. New approaches to multiscale modeling that combine quantum mechanical methods for excited states with molecular dynamics for the solvent or protein environments are gaining traction. Hybrid quantum mechanics/molecular mechanics (QM/MM) approaches offer a potential solution, combining the precision of quantum methods for the nucleobase with the computational efficiency of MM for the solvent [54, 60–68]. Additionally, machine learning (ML) techniques, trained on computational data such as CASPT2 or DFT, could potentially bypass expensive ab initio calculations while predicting triplet state decay behaviors for new molecular derivatives. These methods will provide more accurate and comprehensive predictions of triplet state behavior in biologically relevant systems.

Addressing these challenges requires not only refining computational models but also bridging the gap between theoretical predictions and experimental results. Recent studies on noncanonical nucleobases have expanded our understanding of triplet decay mechanisms and provided valuable insights for tuning triplet lifetimes. Modifications such as glycosylation and atomic substitutions have been shown to regulate the triplet decay rates. Additionally, introducing sugar groups at strategic positions on nucleobases can improve molecular interactions, promoting more efficient transitions from the triplet state to the ground state. Future design strategies could focus on tuning the flexibility of the sugar moiety or incorporating additional functional groups that favor hydrogen bonding, thereby changing the activation energy for ISC.

Furthermore, solvent accessibility to the triplet state should be considered. Future photosensitizer designs should, therefore, account for solvent polarity, hydrogen bonding networks, and dynamic solvation effects to accurately predict in vivo performance. In cellular environments, localized hydration and interactions with biomolecules like DNA bases and proteins further influence triplet decay. By designing nucleobases that are more adaptable to their microenvironment, photosensitizers can be optimized for enhanced performance in complex biological systems.

The principles governing triplet decay dynamics have implications far beyond phototherapy, opening up exciting opportunities in various fields. In photocatalysis, thionucleoside derivatives, which feature long-lived triplet states, could serve as visible-light photocatalysts for driving energy transfer reactions in organic synthesis. Their ability to absorb light efficiently and transfer energy makes them ideal candidates for sustainable chemical processes. In optoelectronics, the tunability of ISC rates presents opportunities for these compounds to be used in advanced organic light-emitting diodes (OLEDs) and photon upconversion systems [57]. Manipulating triplet states could lead to more efficient light-emitting devices and systems that convert low-energy photons into higher energy ones, with potential applications in energy harvesting, displays, and lighting. Additionally, triplet state dynamics' sensitivity to solvent composition can be leveraged for environmental sensing applications. By monitoring variations in triplet decay rates, real-time monitoring of biological systems, such as tracking cellular redox states or detecting pollutants, becomes possible. These capabilities could revolutionize sensing and diagnostic technologies, offering a novel approach for real-time measurements in complex environments.

As the field progresses, it is essential that these theoretical models be validated and refined through experimental methods. Techniques such as transient absorption spectroscopy, time-resolved phosphorescence, and femtosecond-resolved spectroscopy are critical to measuring triplet lifetimes and ISC rates, providing the empirical data needed to further calibrate computational models. These methods will help bridge the gap between computational predictions and experimental reality, ensuring that computational tools can more reliably predict triplet decay in a wide range of nucleobase systems. Therefore, addressing these challenges requires collaborative efforts between computational chemistry, spectroscopy, and molecular biology. Moving forward, theoretical modeling, experimental validation under biologically relevant conditions, and the exploration of nonmedical applications are critical to fully realizing the potential of these versatile compounds in both therapeutic and diagnostic contexts.

In conclusion, the intricate photophysical processes governing the triplet decay mechanisms in canonical and noncanonical nucleobases have profound implications for both fundamental science and applied biomedical research. Canonical nucleobases are optimized through evolutionary selection to rapidly dissipate energy via ultrafast IC, minimizing potentially harmful triplet states and ensuring genetic integrity. Conversely, noncanonical nucleobases, exhibiting enhanced ISC and prolonged triplet lifetimes, provide unique opportunities for phototherapeutic applications, notably in photochemotherapy and photoprotection. The theoretically proposed double-well potential energy surface model, which has been experimentally validated by following spectroscopic studies, has significantly advanced our mechanistic comprehension of these phenomena. However, substantial challenges remain, including refining computational models, understanding complex environmental interactions, and developing nucleobase analogs with precisely controlled photophysical properties. Addressing these challenges requires concerted efforts to integrate experimental investigations and theoretical computations. Continued interdisciplinary research in this vibrant area promises to yield critical insights, enabling novel applications and further advancing our understanding of nucleobase photochemistry.

References

1 Serrano-Andrés, L. and Merchán, M. (2009). Are the five natural DNA/RNA base monomers a good choice from natural selection?: a photochemical perspective. *Journal of Photochemistry and Photobiology C Photochemistry Reviews* 10 (1): 21–32. https://doi.org/10.1016/j.jphotochemrev.2008.12.001.

2 Sagan, C. (1973). Ultraviolet selection pressure on the earliest organisms. *Journal of Theoretical Biology* 39 (1): 195–200. https://doi.org/10.1016/0022-5193(73)90216-6.

3 Callis, P.R. (1983). Electronic states and luminescence of nucleic acid systems. *Annual Review of Physical Chemistry* 34: 329–357. https://doi.org/10.1146/annurev.pc.34.100183.001553.

4 Reelfs, O., Karran, P., and Young, A.R. (2012). 4-Thiothymidine sensitization of DNA to UVA offers potential for a novel photochemotherapy. *Photochemical & Photobiological Sciences* 11 (1): 148–154.

5 Pridgeon, S., Heer, R., Taylor, G. et al. (2011). Thiothymidine combined with UVA as a potential novel therapy for bladder cancer. *British Journal of Cancer* 104 (12): 1869–1876.

6 Elion, G.B. (1989). The purine path to chemotherapy. *in vitro Cellular & Developmental Biology* 25 (4): 321–330. https://doi.org/10.1007/BF02624593.

7 Massey, A., Xu, Y.-Z., and Karran, P. (2001). Photoactivation of DNA thiobases as a potential novel therapeutic option. *Current Biology* 11 (14): 1142–1146. https://doi.org/10.1016/S0960-9822(01)00272-X.

8 Dolmans, D.E., Fukumura, D., and Jain, R.K. (2003). Photodynamic therapy for cancer. *Nature Reviews Cancer* 3 (5): 380–387.

9 Brem, R., Daehn, I., and Karran, P. (2011). Efficient DNA interstrand crosslinking by 6-thioguanine and UVA radiation. *DNA Repair* 10 (8): 869–876.

10 Zhang, X., Jeffs, G., Ren, X. et al. (2007). Novel DNA lesions generated by the interaction between therapeutic thiopurines and UVA light. *DNA Repair* 6 (3): 344–354.

11 Euvrard, S., Kanitakis, J., and Claudy, A. (2003). Skin cancers after organ transplantation. *The New England Journal of Medicine* 348 (17): 1681–1691.

12 Barbatti, M., Aquino, A.J.A., Szymczak, J.J. et al. (2010). Relaxation mechanisms of UV-photoexcited DNA and RNA nucleobases. *Proceedings of the National Academy of Sciences* 107 (50): 21453–21458. https://doi.org/10.1073/pnas.1014982107.

13 Kohler, B. (2010). Nonradiative decay mechanisms in DNA model systems. *Journal of Physical Chemistry Letters* 1 (13): 2047–2053. https://doi.org/10.1021/jz100491x.

14 Jiang, J., Zhang, T.-S., Xue, J.-D. et al. (2015). Short-time dynamics of 2-Thiouracil in the light absorbing S2 ($\pi\pi^*$) state. *The Journal of Chemical Physics* 143 (17): 175103.

15 Serrano-Andrés, L., Merchán, M., and Borin, A.C. (2008). A three-state model for the photophysics of guanine. *Journal of the American Chemical Society* 130 (8): 2473–2484. https://doi.org/10.1021/ja0744450.

16 Bai, S. and Barbatti, M. (2017). On the decay of the triplet state of thionucleobases. *Physical Chemistry Chemical Physics* 19 (20): 12674–12682.

17 Pollum, M., Martinez-Fernandez, L., and Crespo-Hernandez, C.E. (2015). Photochemistry of nucleic acid bases and their thio-and aza-analogues in solution. In: *Photoinduced Phenomena in Nucleic Acids I: Nucleobases in the Gas Phase and in Solvents* (ed. M. Barbatti, A.C. Borin, and S. Ullrich), 245–327. Cham: Springer International Publishing.

18 Matsika, S. (2015). Modified nucleobases. In: *Photoinduced Phenomena in Nucleic Acids I: Nucleobases in the Gas Phase and in Solvents*, 209–243. Cham: Springer International Publishing.

19 Pollum, M., Jockusch, S., and Crespo-Hernández, C.E. (2014). 2,4-Dithiothymine as a potent UVA chemotherapeutic agent. *Journal of the American Chemical Society* 136 (52): 17930–17933. https://doi.org/10.1021/ja510611j.

20 Kuramochi, H., Kobayashi, T., Suzuki, T., and Ichimura, T. (2010). Excited-state dynamics of 6-aza-2-thiothymine and 2-thiothymine: highly efficient intersystem crossing and singlet oxygen photosensitization. *The Journal of Physical Chemistry B* 114 (26): 8782–8789. https://doi.org/10.1021/jp102067t.

21 Schweitzer, C., Mehrdad, Z., Noll, A. et al. (2003). Mechanism of photosensitized generation of singlet oxygen during oxygen quenching of triplet states and the general dependence of the rate constants and efficiencies of O_2 ($^1\Sigma_g^+$), O_2 ($^1\Delta_g$), and O_2 ($^3\Sigma_g^-$) formation on sensitizer triplet state energy and oxidation potential. *The Journal of Physical Chemistry A* 107 (13): 2192–2198.

22 Harada, Y., Okabe, C., Kobayashi, T. et al. (2010). Ultrafast intersystem crossing of 4-Thiothymidine in aqueous solution. *Journal of Physical Chemistry Letters* 1 (2): 480–484.

23 Mai, S., Marquetand, P., and González, L. (2016). Intersystem crossing pathways in the noncanonical nucleobase 2-Thiouracil: a time-dependent picture. *Journal of Physical Chemistry Letters* 7 (11): 1978–1983. https://doi.org/10.1021/acs.jpclett.6b00616.

24 Mai, S., Marquetand, P., and González, L. (2015). A static picture of the relaxation and intersystem crossing mechanisms of photoexcited 2-Thiouracil. *The Journal of Physical Chemistry A* 119 (36): 9524–9533.

25 Cui, G. and Thiel, W. (2014). Intersystem crossing enables 4-Thiothymidine to act as a photosensitizer in photodynamic therapy: an ab initio QM/MM study. *Journal of Physical Chemistry Letters* 5 (15): 2682–2687.

26 Gobbo, J.P. and Borin, A.C. (2014). 2-Thiouracil deactivation pathways and triplet states population. *Computational & Theoretical Chemistry* 1040: 195–201.

27 Cui, G. and Fang, W.-h. (2013). State-specific heavy-atom effect on intersystem crossing processes in 2-thiothymine: a potential photodynamic therapy photosensitizer. *The Journal of Chemical Physics* 138 (4): 044315.

28 Martinez-Fernandez, L., Corral, I., Granucci, G., and Persico, M. (2014). Competing ultrafast intersystem crossing and internal conversion: a time resolved picture for the deactivation of 6-thioguanine. *Chemical Science* 5 (4): 1336–1347.

29 Taras-Goślińska, K., Burdziński, G., and Wenska, G. (2014). Relaxation of the T1 excited state of 2-thiothymine, its riboside and deoxyriboside-enhanced nonradiative decay rate induced by sugar substituent. *Journal of Photochemistry and Photobiology A: Chemistry* 275: 89–95. https://doi.org/10.1016/j.jphotochem.2013.11.003.

30 Ashwood, B., Jockusch, S., and Crespo-Hernández, C.E. (2017). Excited-state dynamics of the thiopurine prodrug 6-thioguanine: can N9-glycosylation affect its phototoxic activity? *Molecules* 22: 379.

31 Reichardt, C., Guo, C., and Crespo-Hernández, C.E. (2011). Excited-state dynamics in 6-thioguanosine from the femtosecond to microsecond time scale. *The Journal of Physical Chemistry B* 115 (12): 3263–3270. https://doi.org/10.1021/jp112018u.

32 Ditchfield, R., Hehre, W.J., Pople, J.A., and Radom, L. (1970). Molecular orbital theory of bond separation. *Chemical Physics Letters* 5 (1): 13–14. https://doi.org/10.1016/0009-2614(70)80116-6.

33 Finley, J., Malmqvist, P.-Å., Roos, B.O., and Serrano-Andrés, L. (1998). The multi-state CASPT2 method. *Chemical Physics Letters* 288 (2): 299–306. https://doi.org/10.1016/S0009-2614(98)00252-8.

34 Ghigo, G., Roos, B.O., and Malmqvist, P.-Å. (2004). A modified definition of the zeroth-order Hamiltonian in multiconfigurational perturbation theory (CASPT2). *Chemical Physics Letters* 396 (1): 142–149. https://doi.org/10.1016/j.cplett.2004.08.032.

35 Malmqvist, P.Å., Roos, B.O., and Schimmelpfennig, B. (2002). The restricted active space (RAS) state interaction approach with spin–orbit coupling. *Chemical Physics Letters* 357 (3): 230–240. https://doi.org/10.1016/S0009-2614(02)00498-0.

36 Roos, B.O., Veryazov, V., and Widmark, P.-O. (2004). Relativistic atomic natural orbital type basis sets for the alkaline and alkaline-earth atoms applied to the ground-state potentials for the corresponding dimers. *Theoretical Chemistry Accounts* 111 (2): 345–351. https://doi.org/10.1007/s00214-003-0537-0.

37 Stephens, P.J., Devlin, F.J., Chabalowski, C.F., and Frisch, M.J. (1994). Ab initio calculation of vibrational absorption and circular dichroism spectra using density functional force fields. *The Journal of Physical Chemistry* 98 (45): 11623–11627. https://doi.org/10.1021/j100096a001.

38 Becke, A.D. (1993). Density-functional thermochemistry. III. The role of exact exchange. *The Journal of Chemical Physics* 98 (7): 5648–5652.

39 Yanai, T., Tew, D.P., and Handy, N.C. (2004). A new hybrid exchange–correlation functional using the coulomb-attenuating method (CAM-B3LYP). *Chemical Physics Letters* 393 (1–3): 51–57.

40 Chai, J.-D. and Head-Gordon, M. (2008). Long-range corrected hybrid density functionals with damped atom–atom dispersion corrections. *Physical Chemistry Chemical Physics* 10 (44): 6615–6620.

41 Hehre, W.J., Ditchfield, R., and Pople, J.A. (1972). Self—consistent molecular orbital methods. XII. Further extensions of Gaussian—type basis sets for use in molecular orbital studies of organic molecules. *The Journal of Chemical Physics* 56 (5): 2257–2261.

42 Trofimov, A.B., Moskovskaya, T.É., Gromov, E.V. et al. (2000). Core-level electronic spectra in ADC(2) approximation for polarization propagator: carbon monoxide and nitrogen molecules. *Journal of Structural Chemistry* 41 (3): 483–494. https://doi.org/10.1007/BF02742009.

43 Christiansen, O., Koch, H., and Jørgensen, P. (1995). The second-order approximate coupled cluster singles and doubles model CC2. *Chemical Physics Letters* 243 (5–6): 409–418.

44 Levine, B.G., Coe, J.D., and Martínez, T.J. (2008). Optimizing conical intersections without derivative coupling vectors: application to multistate multireference second-order perturbation theory (MS-CASPT2). *The Journal of Physical Chemistry B* 112 (2): 405–413.

45 Koyama, D., Milner, M.J., and Orr-Ewing, A.J. (2017). Evidence for a double well in the first triplet excited state of 2-Thiouracil. *The Journal of Physical Chemistry B* 121 (39): 9274–9280. https://doi.org/10.1021/acs.jpcb.7b06917.

46 Bai, S. and Barbatti, M. (2018). Mechanism of enhanced triplet decay of thionucleobase by glycosylation and rate-modulating strategies. *Physical Chemistry Chemical Physics* 20 (24): 16428–16436.

47 Liu, S., Lee, Y., Chen, L. et al. (2024). Unexpected longer T1 lifetime of 6-sulfur guanine than 6-selenium guanine: the solvent effect of hydrogen bonds to brake the triplet decay. *Physical Chemistry Chemical Physics* 26 (18): 13965–13972.

48 Ou, Q. and Subotnik, J.E. (2013). Electronic relaxation in benzaldehyde evaluated via TD-DFT and localized diabatization: intersystem crossings, conical intersections, and phosphorescence. *Journal of Physical Chemistry C* 117 (39): 19839–19849.

49 Yost, S.R., Hontz, E., Yeganeh, S., and Van Voorhis, T. (2012). Triplet vs singlet energy transfer in organic semiconductors: the tortoise and the hare. *Journal of Physical Chemistry C* 116 (33): 17369–17377.

50 (a) Marian, C. (2012). Spin–orbit coupling and intersystem crossing in molecules. *Wiley Interdisciplinary Reviews: Computational Molecular Science* 2: 187–203. (b) Marian, C.M. (2021). *Annual Review of Physical Chemistry* 72: 617–640.

51 Marcus, R. (1984). Nonadiabatic processes involving quantum-like and classical-like coordinates with applications to nonadiabatic electron transfers. *The Journal of Chemical Physics* 81 (10): 4494–4500.

52 Eyring, H. (1935). The activated complex in chemical reactions. *The Journal of Chemical Physics* 3 (2): 107–115.

53 Valverde, D., Mai, S., Canuto, S. et al. (2022). Ultrafast intersystem crossing dynamics of 6-selenoguanine in water. *JACS Au* 2 (7): 1699–1711. https://doi.org/10.1021/jacsau.2c00250.

54 Petrone, A., Perrella, F., Coppola, F. et al. (2022). Ultrafast photo-induced processes in complex environments: the role of accuracy in excited-state energy potentials and initial conditions. *Chemical Physics Reviews* 3 (2): 021307.

55 Fang, Y.-G., Peng, Q., Fang, Q. et al. (2019). MS-CASPT2 studies on the photophysics of selenium-substituted guanine nucleobase. *ACS Omega* 4 (6): 9769–9777. https://doi.org/10.1021/acsomega.9b01276.

56 Fang, Y.-G., Valverde, D., Mai, S. et al. (2021). Excited-state properties and relaxation pathways of selenium-substituted guanine nucleobase in aqueous solution and DNA duplex. *The Journal of Physical Chemistry B* 125 (7): 1778–1789. https://doi.org/10.1021/acs.jpcb.0c10855.

57 Marian, C.M. (2021). Understanding and controlling intersystem crossing in molecules. *Annual Review of Physical Chemistry* 72: 617–640. https://doi.org/10.1146/annurev-physchem-061020-053433.

58 Kohler, G. and Rechthaler, K. (1993). Solvent effects on excited state relaxation phenomena. *Pure and Applied Chemistry* 65 (8): 1647–1652. https://doi.org/10.1351/pac199365081647.

59 Cerezo, J., Petrone, A., Ferrer, F.J.A. et al. (2016). Electronic spectroscopy of a solvatochromic dye in water: comparison of static cluster/implicit and dynamical/explicit solvent models on structures and energies. *Theoretical Chemistry Accounts* 135 (12): 263.

60 Raucci, U., Perrella, F., Donati, G. et al. (2020). Ab-initio molecular dynamics and hybrid explicit-implicit solvation model for aqueous and nonaqueous solvents: GFP chromophore in water and methanol solution as case study. *Journal of Computational Chemistry* 41 (26): 2228–2239.

61 Donati, G., Petrone, A., Caruso, P., and Rega, N. (2018). The mechanism of a green fluorescent protein proton shuttle unveiled in the time-resolved frequency domain by excited state ab initio dynamics. *Chemical Science* 9 (5): 1126–1135.

62 Rega, N., Brancato, G., Petrone, A. et al. (2011). Vibrational analysis of x-ray absorption fine structure thermal factors by ab initio molecular dynamics: the Zn (II) ion in aqueous solution as a case study. *The Journal of Chemical Physics* 134 (7): 074504.

63 Coppola, F., Perrella, F., Petrone, A. et al. (2020). A not obvious correlation between the structure of green fluorescent protein chromophore pocket and hydrogen bond dynamics: a choreography from ab initio molecular dynamics. *Frontiers in Molecular Biosciences* 7: 569990.

64 Chiariello, M.G. and Rega, N. (2018). Exploring nuclear photorelaxation of pyranine in aqueous solution: an integrated ab-initio molecular dynamics and time resolved vibrational analysis approach. *The Journal of Physical Chemistry A* 122 (11): 2884–2893.

65 Battista, E., Scognamiglio, P.L., Di Luise, N. et al. (2018). Turn-on fluorescence detection of protein by molecularly imprinted hydrogels based on supramolecular assembly of peptide multi-functional blocks. *Journal of Materials Chemistry B* 6 (8): 1207–1215.

66 Perrella, F., Raucci, U., Chiariello, M.G. et al. (2018). Unveiling the structure of a novel artificial heme-enzyme with peroxidase-like activity: a theoretical investigation. *Biopolymers* 109 (10): e23225.

67 Petrone, A., Donati, G., Caruso, P., and Rega, N. (2014). Understanding THz and IR signals beneath time-resolved fluorescence from excited-state ab initio dynamics. *Journal of the American Chemical Society* 136 (42): 14866–14874.

68 Perrella, F., Petrone, A., and Rega, N. (2021). Direct observation of the solvent organization and nuclear vibrations of [Ru(dcbpy) 2 (NCS) 2] 4−, [dcbpy = (4,4′-dicarboxy-2,2′-bipyridine)], via ab initio molecular dynamics. *Physical Chemistry Chemical Physics* 23 (40): 22885–22896.

7

Triplet State Dynamics in DNA/RNA Nucleobases and DNA by Ab Initio Static and Surface-Hopping Dynamics Simulations

Marta Ibanez-deLuis[1] *and Inés Corral*[1,2]

[1] *Universidad Autónoma de Madrid, Ciudad Universitaria de Cantoblanco, Departamento de Química, Av. Francisco Tomás y Valiente 7, Cantoblanco, 28049, Spain*
[2] *Universidad Autónoma de Madrid, Ciudad Universitaria de Cantoblanco, IADCHEM, Institute for Advanced Research in Chemistry, Av. Francisco Tomás y Valiente 7, Cantoblanco, 28049, Spain*

7.1 Introduction

DNA plays a pivotal role in biology as the carrier of genetic information. Hence, damage to this biomolecule, as UV light exposure can trigger, can lead to mutations or carcinogenesis. The potential damaging effects of solar UV light are reduced by the photostability of DNA, based on the conversion of electronic excitation energy into vibrational motion in the ground state.

IC is at the heart of photostability of DNA monomers (Figure 7.1). IC occurs in the (sub)picosecond timescale and is mediated by conical intersections. This ultrafast dissipation of the absorbed energy can be considered a key factor for the evolution of the DNA biomolecule until its current form. In fact, light, and more particularly UV light, can be considered one of the strongest evolutionary force during the prebiotic era, and the ability of nucleobases to withstand these radiative extreme conditions is believed to have provided these chromophores a significant superiority, favoring their incorporation into the genetic biomolecule [1].

The mechanism of photostability in the DNA macromolecule, where several bases are present, is, however, much more complex. In DNA, besides localized excitations, excitons and charge-transfer states, as a consequence of base-stacking and base-pairing interactions, can also be populated. Thus, for single- and double-stranded DNA, additional transients in the range of tens or hundreds of picoseconds and even nanoseconds have been registered [2–4].

Despite IC being the major deactivation process in natural nucleobases, the photophysics of DNA monomers is intricate. Nucleobases exhibit a multiexponential behavior, implying that several deactivation pathways may compete [5–10].

In DNA monomers, IC competes with photon emission, other nonradiative processes such ISC, or photoproduct formation. The quantum yields for fluorescence in

Excited State Dynamics in DNA and RNA: Experimental and Computational Approaches on Triplet State, First Edition. Edited by Jinquan Chen.

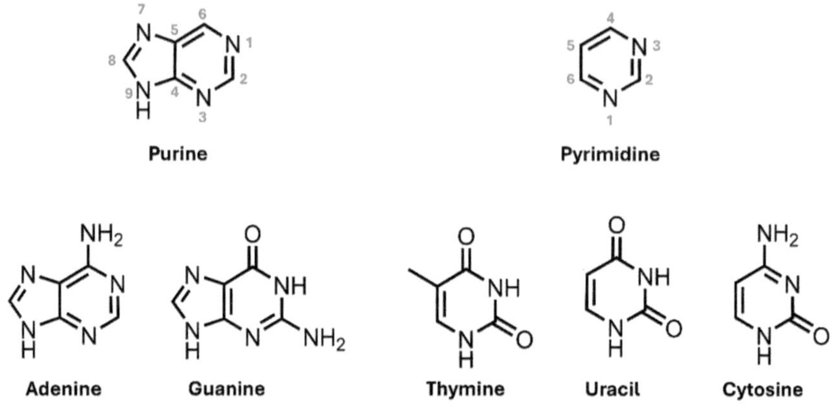

Figure 7.1 Core purine and pyrimidine atom labelling (above) and natural nucleobases (below).

DNA monomers are small (ca. 10^{-4}) [11–13]. Emission usually takes place from ππ* excited states and occurs on a picosecond timescale [5].

Triplet quantum yields in nucleobases are also small in water (ca. 10^{-2}) but grow larger in aprotic and apolar solvents [14]. Although triplets could potentially be populated from both ππ* and nπ* singlet states, in their original work on 1-cyclohexyluracil, Hare et al. signal dark nπ* states as the mediators for population transfer to the triplet manifold [15]. On the one hand, the constant yield for direct IC from the bright ππ* state, independent of the solvent considered, in contrast with the yield of nπ*, which was found to be solvent correlated and proportional to the triplet yield, suggested these dark states as door states to the triplet manifold.

Besides ISC, triplet states in nucleobases in DNA can be populated following triplet–triplet energy transfer from excited triplet photosensitizers, following ISC from singly excited states from the latter [16–19]. This process is especially relevant for thymine, as it is a nucleobase with the lowest lying triplet state.

Remarkably, triplet quantum yields in nucleobases have been demonstrated to be extraordinarily sensitive to the position and nature of the substituents along the core heterocycles of purine and pyrimidine. In fact, even minor modifications such as replacing the natural oxo substituent with a sulfur atom or removing the native substituents of nucleobases, leading to free purine and pyrimidine cores, have been shown to enhance the quantum yield by up to two orders of magnitude, approaching near-unity triplet yields. From this detaches that the precise nature and position of the substituents along purine and pyrimidine are responsible for the photostability of DNA monomers, and ultimately of DNA, and that these structural and electronic changes profoundly affect the topographies of the potential energy surfaces (PESs), blocking some deactivation routes while activating or enhancing others.

The fate of these triplets varies depending on whether the excited triplet is formed in solution or if the chromophore is embedded in a DNA duplex or strand. The detection of pyrimidine–pyrimidine dimers upon DNA photosensitization [18, 19] following triplet–triplet energy transfer from an excited photosensitizer or the resemblance of time-resolved fluorescence and transient absorption spectra for thymine monomer and oligonucleotides [20, 21] suggests that triplet states may act as precursors for these types of lesions. Furthermore, the detection of a biradical intermediate, identified as a triplet state, along with the formation of cyclobutane pyrimidine dimer (CPD) by nanosecond infrared (IR) spectroscopy experiments, provides additional support for the participation of the triplet manifold in the origin of these photoproducts [22]. Some other studies, however, indicate that thymine–thymine dimers are formed on an ultrafast timescale (around 1 ps) from singlet excimers [23, 24], competing with nonradiative decay along the singlet pathway, consistent with the lack of triplet signal in the transient spectra when thymine dimers are formed from the direct excitation of $(dT)_{20}$ [25].

To reconcile these experimental observations, computational studies have thoroughly investigated the photoinduced formation mechanisms of both the minor pyrimidine–pyrimidone (6–4) adducts and the major CPDs of thymine, uracil, and cytosine, along both singlet and triplet manifolds. These studies provide molecular support to the experimental findings [26–29].

The triplet states of the DNA monomers have also been identified as the precursors of singlet oxygen in O_2-saturated solutions [30, 31].

In the case of nonnatural nucleobases, and more particularly in the case of thionated analogs of natural nucleobases, their triplet states have also shown a photoreactive profile. In solution, for 4-thiothyimidine and 2,4-dithiothymidine, singlet oxygen yields of 0.42 and 0.46 were, respectively, recorded [32, 33]. A slightly lower singlet oxygen yield has been registered for 2-thiothymidine [34]. However, once incorporated into a DNA strand, these nucleobases exhibit a different reactivity, which is independent of molecular oxygen. In fact, the irradiation of a single-strand DNA oligonucleotide incorporating 4-thiothymidine leads to the formation of (6-4) photoadducts. This photoproduct would have its origin in the triplet state of the thiated nucleobase that would undergo a photocycloaddition reaction with an adjacent thymine, leading to a triplet thietane intermediate that would finally decay to the ground state via ISCs [35]. The potential of these systems as agents in photodynamic therapy relies on these two photoreactions.

Within this frame, the purpose of this chapter is to provide an overview of the mechanisms behind triplet formation in natural nucleobases and in a selection of structurally similar derivatives. Due to the vast amount of information available, we focus on the most recent studies that prioritize the CASPT2 or CASPT2//CASSCF approaches, as well as density functional theory (DFT), under which most of the research has been conducted. Our overview covers from absorption spectra to the mechanistic insight of the photophysical and photochemical processes from a static and time-resolved perspective. The chapter covers five natural nucleobases that conform DNA and RNA, along with some close derivatives, for which high triplet quantum yields have been recorded. Each nucleobase section follows the same scheme, including a brief discussion of its optical properties, an overview of the PES topography (including both multiplicities), and, when available, a summary of dynamic studies incorporating both multiplicities.

A synergistic interplay between theory and experiment is essential for exhaustively examining both the triplet formation mechanisms in nucleobases and their subsequent evolution. Exploring the PESs beyond the FC region, by locating critical stationary points such as minima, transition states, and surface crossings between states of the same or different multiplicity, can provide useful insight. Furthermore, the excited state nonadiabatic dynamics simulations, which capture simultaneously temporal and spatial dimensions, can provide a better understanding of the competition between different deactivation processes. Here, quantum dynamics, when feasible, or "on-the-fly" nonadiabatic dynamics based on Tully's surface-hopping (TSH) approach, which incorporates the full dimensionality of the system, can be considered a good compromise between accuracy and computational cost. However, the dynamical modeling of triplet states presents some challenges, due to the long propagation times required.

The level of theory used in simulations is crucial for providing useful mechanistic information at the molecular level and for assisting in the interpretation of experimental results. It has been demonstrated that dynamical correlation is often essential to accurately capture key details of the excited state PES topography for many systems. The complex electronic structure of these compounds may sometimes necessitate highly correlated methods, such as (MS-)CASPT2, to avoid misinterpretations of the mechanism, which is a challenge itself. Another critical factor when investigating excited state mechanisms is the biological or solvation environment. Due to the higher computational cost and technical challenges of setting up these simulations, many studies in the literature have been performed in vacuo. This should be considered with caution when analyzing experimental results.

7.2 Adenine

Adenine is the closest nucleobase to the *mother* chromophore purine, being the only difference an amino substituent in position C_6. However, in contrast with raw purine, adenine is able to deactivate in an ultrafast manner, as stated by different works, which situate the deactivation component (τ_2) within 0.75–2 ps, postulated as related to the lowest triplet state, with a faster transient (τ_1) shorter than 100 fs [36–42].

Serrano-Andrés et al. [43] employed ab initio CASPT2//CASSCF methods to study the photophysics of biologically relevant 9H-adenine, as well as the 7H-tautomer. They found that the bright state for adenine is S_3, $\pi\pi^*$ L_a, located at 5.35 eV. Below it, an nπ^* S_1 state is located at 4.96 eV and in between a rather dark $\pi\pi^*$ state, L_b, is placed at 5.16 eV [43]. This ordering of states seems to be replicable when employing DFT/MRCI level of theory [44], having a slight shift on the bright state (5.15 eV). In the very same year as Serrano-Andrés et al., Blancafort published a work in which he also performed CASPT2//CASSCF calculations on 9H-adenine [45]. With very similar results to those obtained by Serrano-Andrés et al., Blancafort found the same ordering of states in the FC region, with the energies being a bit shifted, presumably due to Blancafort including one less virtual orbital in the active space (AS) and employing a slightly less flexible basis set. Similarly, Conti et al. performed a parallel mapping of the PES of the nucleobase, in this case at the CASPT2//CASSCF(12,11) level of theory [46]. With an AS including σ orbitals, they were able to locate the same ordering for the three lowest lying states at the FC region.

When undertaking minimum energy path (MEP) calculations from the FC region, in order to map the PES of adenine, Serrano-Andrés et al. found a barrierless path connecting the bright state and a conical intersection (CI) able to deactivate the population to the ground state [43], as is depicted in Figure 7.2. This absence of a minimum in the surface of the bright state accounts for the aforementioned ultrafast decay. The S_1/S_0 CI is characterized by an out-of-plane movement of the hydrogen atom bonded to C_2, ending up perpendicular to the plane of purine. For completion, MEP calculations for the secondary paths were also performed. They found a $^1(\pi\pi^*$ $Lb)_{min}$ at around 4.9 eV that could be populated as well via direct absorption, despite its small oscillator strength. Upon surpassing a small barrier of 0.14 eV, the population could reach a crossing point with the nπ* surface, from where a $^1(n\pi^*)_{min}$ at 4.53 eV can be accessed. This minimum is characterized by a twisting of the amino group, and upon a further rotation of the N_1–C_6 bond (recall Figure 7.1). A small barrier of 0.08 eV separates this minimum from a CI that deactivates to the ground state. Blancafort found roughly the same potential energy landscape, with the bright ππ* L_a state being able to decay in a barrierless manner, and the two darker ππ* L_b and

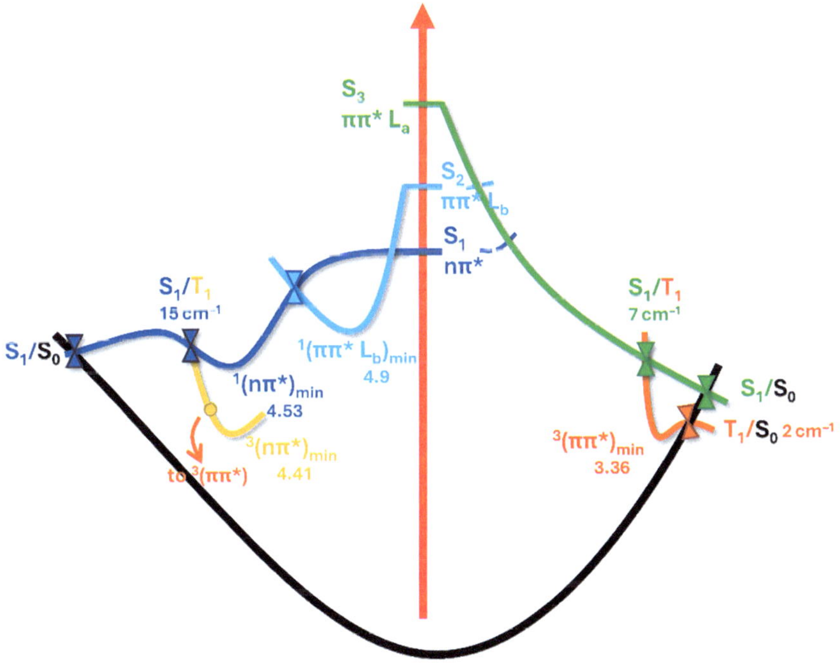

Figure 7.2 PES of adenine as described by Serrano-Andrés et al. [43] and Blancafort [45]. Singlets represented in cold colors, while triplets as described by González-Luque et al. [48] are represented in warm colors. Energies in eV and referred to S_0.

nπ^* states presenting respective minima in the S_1 surface [45]. The decay coordinates are also equivalent in both works. In addition, Blancafort considers the possibility of ISC by computing the difference in energy between the longest lived minimum, $^1(n\pi^*)_{min}$, and the closest triplet states, as well as the SOC values. He found two degenerate triplet states 0.04 eV below the S_1, of $\pi\pi^*$ and nπ^* characters. The computed SOC values amount to 13 and 4 cm^{-1}, respectively, in line with El-Sayed rules [47]. Thus, Blancafort proposes the possibility of crossing to the $^3(\pi\pi^*)$ state after population of the $^1(n\pi^*)_{min}$, even though triplet minima were not optimized and characterized in this work. (In Figure 7.2, this crossing is instead with $^3(n\pi^*)$, since triplets states are depicted as described by González-Luque et al. [48]) Conti et al. found the same excited state minima, $^1(\pi\pi^*\ L_b)_{min}$ and $^1(n\pi^*)_{min}$, with the same energies along the S_1 and S_2 MEPs [46] and were also able to locate a crossing point between the $^1(n\pi^*)$ and $^3(\pi\pi^*)$ states with an SOC of 11 cm^{-1}, which is similar to that computed by Blancafort et al.

In the aforementioned work, Serrano-Andrés et al. studied the nonbiological 7H-tautomer of adenine [43]. They found that this tautomer had a quite similar PES to the 9H, except for the minimum found along the $\pi\pi^*$ L_a state. From the computed energy barrier of 0.22 eV necessary to access the S_1/S_0 CI region, they postulated that this tautomer is responsible for the fluorescence emission recorded for adenine in solution [5, 49–51].

In order to try to rationalize the aforementioned experimental time constants, several works performing surface-hopping (SH) dynamics at different levels of theory, such as DFT [52, 53], semiempirical [53, 54] and multiconfigurational methods [55], have been published. Barbatti et al. performed singlet SH dynamics at the MR-CIS(6,4)/CASSCF(12,10) level from the first $\pi\pi^*$ state, trying to gain some insight into the competing deactivation routes, decay rates of nucleobases, and monitoring of the decay coordinates inferred from the dynamics [55]. Adenine showed an ultrafast decay, with the population reaching the ground state in 0.63 ps. Moreover, they report that 100% of the propagated trajectories underwent deactivation via a CI characterized by a puckering of the C_2 position. Some years later, Plasser et al. developed a linear vibronic coupling model within SH approaches [56]. In order to assess the performance of the method in describing ultrafast deactivation, they resorted to adenine and 2-aminopurine, two structurally similar molecules with very different photophysical profiles. Just as Barbatti et al. [55], they were able to observe an ultrafast decay to the ground state, from the population of the S_2, which subsequently injects population into the S_1. Nonetheless, they were able to find several S_1/S_0 hopping structures, characterized by the out-of-plane distortion of the aromatic ring. These results differ from those shown in the study by Barbatti et al. [55], who could only observe IC via the C_2-puckered CI.

González-Luque et al. published a thorough study in which they performed CASPT2//CASSCF ab initio calculations for all canonical nucleobases, trying to shed some light on the mechanisms for ISC [48], for which an experimental quantum yield of $0.23{\cdot}10^{-2}$ was found for adenine [57]. They corroborated the results previously reported by Blancafort [45], calculating an SOC of 15 cm^{-1} at the minimum energy crossing point (MECP) with the $^3(\pi\pi^*)$ state close to $^1(n\pi^*)_{min}$.

Later, the population crosses to the $^3(n\pi^*)$ state, and after computing the MEP from $^3(\pi\pi^*)/^3(n\pi^*)$, they were able to find a critical point $^3(n\pi^*)_{min}$, located at 4.41 eV. Moreover, González-Luque et al. located another relevant S/T MECP, in this case, along the barrierless path connecting L_a in FC with the S_1/S_0 CI, closer to the latter. Even though the crossing is of $^1(\pi\pi^*)/^3(\pi\pi^*)$ character, they argue that it has enough SOC for the transfer of population to occur (7 cm^{-1}) due to the nonplanarity of the structure in this region. Moreover, they propose that due to the characteristic geometry of the S_1/S_0 CI, the population could get trapped in this region of the PES instead of deactivating swiftly, hence providing the chance of undergoing ISC at this MECP. From this crossing point, population could evolve to the $^3(\pi\pi^*)_{min}$ located adiabatically at 3.36 eV, which can also be reached from the previous $^3(\pi\pi^*)/^3(n\pi^*)$ crossing. This minimum is characterized by the tilting of the hydrogen in position C_8. The population trapped in $^3(\pi\pi^*)_{min}$ would need to surpass a high barrier, of 0.6 eV, in order to reach the T_1/S_0 MECP, which is—in addition—characterized by a 2 cm^{-1} SOC value, somewhat low. Hence, González-Luque et al. propose that $^3(\pi\pi^*)_{min}$ is responsible for the experimental phosphorescence band of adenine [58].

7.3 Guanine

Guanine is the heaviest of the purine nucleobases. Contrary to adenine, guanine is a disubstituted chromophore with both keto and amino functional groups. However, this does not seem to affect the decay timescales, since both purine nucleobases seem to relax swiftly, in the sub-picosecond range. More specifically, experiments found biexponential decays for guanine, with τ_1 ca. 0.15 ps and τ_2 ranging between 0.36 and 0.80 ps [37, 40].

Serrano-Andrés et al. employed ab initio CASPT2//CASSCF to characterize the photophysics of this nucleobase [59]. Their calculations suggest that upon UV light absorption, S_1 and S_3 could get populated. Their characterization of the absorption spectrum establishes S_1 and S_3 as $\pi\pi^*$ states, from here on L_a and L_b, respectively. S_2, on the other hand, seems to be a dark $n_O\pi^*$ state. S_1, S_2, and S_3 were calculated at 4.93, 5.54, and 5.77 eV, respectively [59]. When employing TD-DFT, the energies of L_a seem to be well reproduced (TD-B3LYP/6-311++G(d,p), 4.64 eV; TD-B3LYP/6-311(3+,3+)G(df,pd), 4.88 eV), but the oscillator strength associated with L_a and L_b is swiped [60].

Serrano-Andrés et al. attribute the ultrafast decay of guanine to the barrierless character of the path connecting the FC L_a state with the crossing seam with the ground state (see Figure 7.3) [59]. This S_1/S_0 conical intersection is characterized by a puckering of the C_2 position as well as the twisting of the C_2–N_3 bond (recall Figure 7.1). Moreover, when computing the MEP for S_2, Serrano-Andrés et al. were able to find a minimum $^1(n_O\pi^*)_{min}$ lying 1.4 eV below the FC region, with the N_1–C_6 bond slightly rotated. This minimum is isoenergetic with an S_1/S_0 CI that can be accessed after surpassing a small barrier of 0.07 eV. When following the L_b state, yet another minimum, in this case much shallower, $^1(\pi\pi^*)_{min}$ could be found. However, the population of the S_3 L_b minimum would only gain prominence at high excitation

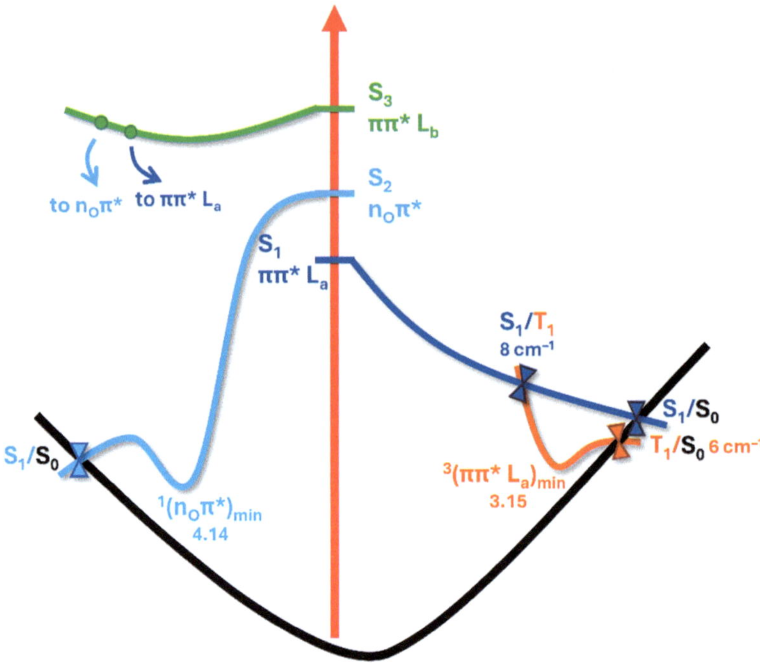

Figure 7.3 PES of guanine as described by Serrano-Andrés et al. [59] Singlets represented in cold colors, while triplets as described by González-Luque et al. [48] are represented in warm colors. Energies in eV and referred to S_0.

energies. The latter $^1(\pi\pi^*)_{min}$ is connected to both lower lying states via different seams, allowing the decay of the population.

Just as for adenine, Barbatti et al. performed SH dynamics at the MR-CIS(10,7)/CASSCF(12,9) level in guanine [55], through which they confirm the results by Serrano-Andrés et al. In fact, Barbatti et al. observed a clear tendency of the $\pi\pi^*$ state to deactivate via an S_1/S_0 CI of this character. They report that all the trajectories from the pool run for guanine jump at geometries puckered at the C_2 position, which is in agreement with the static calculations. They also report a lifetime of deactivation of 0.28 ps, which lies in the sub-picosecond window. A year later, the same group publishes a work in which they revisit their dynamics but end up reporting a slightly shorter lifetime of 224 fs [61]. Moreover, some works propagating semiempirical SH dynamics can also be found in the literature [62].

Even though guanine seems to be the only canonical nucleobase for which no phosphorescence or triplet-state formation data is available, González-Luque et al. explored the location of potential singlet–triplet crossings, aiming at understanding the rate of ISC [48]. First, they computed the energy of the lowest lying triplet states at the FC region, observing T_1 ($\pi\pi^*$, L_a) and T_2 ($\pi\pi^*$) below S_1, at 4.11 and 4.76 eV, respectively. They also computed T_3 ($\pi\pi^*$) slightly above S_1, being located at 5.14 eV, but 0.4 eV below the second singlet excited state. This information agrees with that obtained by Petke et al., who calculated T_1 and T_2 to be below S_1 at the MRCI level of theory [63]. Later on, González-Luque et al. explored the path connecting S_1 L_a in FC with the S_1/S_0 CI and encountered an S_1/T_2 crossing of $^1(\pi\pi^*)/^3(\pi\pi^*)$ character near the FC region (not shown in Figure 7.3). However, the computed SOC for this crossing was smaller than 0.1 cm^{-1}, not being enough to efficiently populate the triplet manifold. Nonetheless, close to the S_1/S_0 region, they were able to find an S_1/T_1 crossing involving states of the same character. In this case, an SOC of 8 cm^{-1} hinted at the possibility of populating the lowest lying triplet state. This MECP was connected to a minimum on the triplet surface, almost 1.2 eV lower in energy. From the T_1 minimum, González-Luque et al. were able to locate a T_1/S_0 MECP with enough SOC to return population to the ground state. However, a large barrier of 0.85 eV would hinder deactivation, favoring the trapping of population at the $^3(\pi\pi^*)_{min}$. They were able to locate another S/T MECP, in this case of $^1(\pi\pi^*)/^3(n_O\pi^*)$ character, but they discarded the possibility of population transfer to the triplet manifold via this funnel due to its unfavorable accessibility, being placed far from the main MEP region. Hence, González-Luque et al. conclude that guanine, when compared to adenine, presents less favorable crossing to the triplet manifold, which may be the reason for the aforementioned lack of experimental data involving the triplet excited states of guanine.

7.4 Uracil

The low energy region of the gas-phase absorption spectrum of uracil is dominated by an absorption band with a maximum at 5.08 eV [64]. This absorption has been ascribed to a $\pi\pi^*$ electronic state, which has been vertically computed at 4.92–5.18 eV at the (MS-)CASPT2 level of theory [65–68]. The main experimental absorption band slightly shifts to the red in water solution (4.78 eV) and so consistently does the calculated vertical energy for this state, which shifts from 5.25 to 5.14 eV upon inclusion of bulk and explicit solvent–solute interactions, according to DFT-PBE0 combined with a large basis set [69].

As reported by the available computational works, this bright state would lie close to a dark $n_O\pi^*$ transition situated below the $\pi\pi^*$ state in the gas phase, according to both multiconfigurational and DFT approaches (4.5–4.93 and 4.8 eV), or above it when incorporating solvent effects with DFT (5.27 eV).

The decay of uracil has been found to be governed by a multiexponential behavior, dominated by two excited state lifetimes of 130 fs and 1.05 ps in the gas phase [40] and of 1.9 and 24 ps in aqueous solution [70]. CASPT2//CASSCF MEP calculations starting from the bright $\pi\pi^*$ excited state revealed a direct and barrierless connection between the FC region and an S_1/S_0 ethene-like CI (dashed blue line in Figure 7.4), compatible with the femtosecond lifetime of the multiexponential decay. Along this path, the electronically excited population would return to the ground state in ultrafast timescales [67]. This funnel presents a quite distorted geometry with a stretched and twisted ethylenic bond, and the H at C_5 describes a 120° dihedral. A planar minimum of $\pi\pi^*$ character along the S_1 PES was located using a reduced AS and basis set and was postulated to participate in the decay of the system

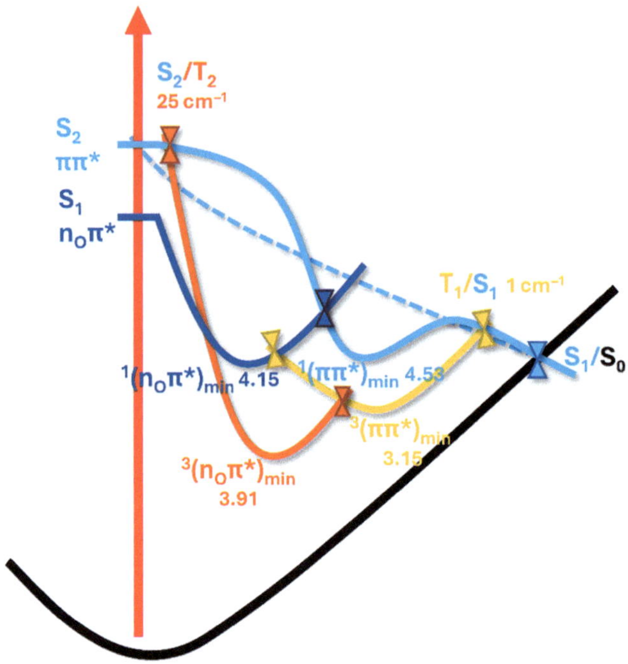

Figure 7.4 PES of uracil (Adapted from References [65, 68]). Singlets represented in cold colors, while triplets are represented in warm colors. Energies in eV and referred to S_0.

in relation to the longer picosecond component of the fit (Figure 7.4, solid blue line). Despite the 0.7 eV energy difference between the calculated vertical energy and the experimental band origin (4.6 eV) [69], the TD-DFT S_1 minimum was also proposed to contribute to the weak fluorescence observed experimentally in water for this system (ϕ_F ~10^{-4}) [69]. Within a gas-phase TD-DFT frame, however, the former planar structure was identified as a transition state, which upon relaxation of the planar constraints, evolves into a crossing with the more stable nπ* state. Instead, a boat-like S_1 $\pi\pi$* minimum was alternatively optimized either at TD-DFT considering bulk solvent effects [69] or at the MS-CASPT2 level of theory [65]. This minimum was predicted to be separated by small energy barriers from the S_0/S_1 conical intersection [65, 69]. Later works propose the population of an nπ* minimum [65, 71], pyramidalized at the N_3, along the relaxation of the system toward the S_1/S_0 conical intersection that would be responsible for the experimental longer component of the decay of uracil. The gap between the nπ* minimum and the S_2/S_1 CI was determined to be small when considering explicit water uracil interactions or a high level of correlation what would allow the recrossing of the system to the $\pi\pi$* state and the decay along the S_1/S_0 CI [65].

The role of the triplets in the deactivation of uracil has also been thoroughly explored in the literature. In fact, the kinetics of dimerization of uracil to form cyclobutane derivatives in different oxygen concentrations suggested triplet states as the precursors of uracil dimers [20, 72]. This was supported by the generation of dimers via photosensitization of uracil that would lead to the population of triplet states through triplet-triplet energy transfer without involvement of singlet levels [73]. Also interesting, the quantum yield for ISC was found to present a wavelength-dependent behavior, increasing by an order of magnitude (1.4×10^{-3}–1.6×10^{-2}) when switching from 280 to 230 nm light [72]. From the computational point of view, two different funnels were located for the transfer of population to the triplet manifold [68]. These two crossings lie along the main deactivation pathway of photoexcited uracil along the $^1(\pi\pi^*)$ state. The first crossing with the T_2 $^3(n_O\pi^*)$ state, which lies in the vicinity of the FC region, is expected to be rather efficient due to the large calculated SOC (ca. 25 cm^{-1}). The other crossing with the $^3(\pi\pi^*)$ state would be located near the S_1/S_0 CI and, as expected, presents a smaller SOC of ca. 1 cm^{-1}. These two crossings would, respectively, populate the $^3(n_O\pi^*)$ and $^3(\pi\pi^*)$ minima. Population relaxation from the former to the latter would be possible after surpassing a small energy barrier of 0.01 eV. Climent et al. also succeeded in locating a funnel for depopulation of the triplet manifold via a T/S_0 crossing (data not shown in Figure 7.4). Also for this crossing, a small SOC amounting to 1 cm^{-1} was calculated. An additional ISC funnel was proposed by Etinski et al. [74], this time involving the S_1 $(n_O\pi^*)$ and the T_1 $^3(\pi\pi^*)$, who identify the S_1 $(n\pi^*)$ excited state as the long-lived picosecond dark state, in agreement with Hare et al. [70] Nonetheless, discrepancies between the quantum yield determined for this long-lived state (0.28) by transient absorption spectroscopy and more recent time-resolved photoelectron spectroscopy experiments, which assign a quantum yield of 0.06 for the $^1(n_O\pi^*)$ of uracil, call for more experiments and computational work on this issue [75].

The time-resolved deactivation of uracil has been explored by several groups [55, 76–82]. For the sake of simplicity, we constrain to the revision of the most recent works employing multiconfigurational CASSCF methods with different ASs and TSH approach to account for nonadiabatic effects.

The work by Nachtigallová et al. explore the singlet-only dynamics of uracil, employing a (10,8) AS, including one n, four π, and three π^* orbitals [55, 82]. This work explored two different excitation energies, i.e. 6.56 eV (59 trajectories) and 7.06 eV (31 trajectories), slightly above and below the computed maximum of absorption, respectively. Three different decay mechanisms were unraveled by this study. The predominant mechanism, both for the two excitation energies, corresponds to the decay of the population excited to the $\pi\pi^*$ state mediated by the dark $n\pi^*$ state. Population following this route gets trapped in the S_2 state and contributes to the slower relaxation (τ = 1.5 ps), corresponding to the longest experimental time constant. The second most important pathway corresponds to the direct relaxation $\pi\pi^* \rightarrow S_0$ that would contribute to the calculated time constant of 0.7 ps. Decay to the ground state takes place through different conformations of the ethylenic conical intersection, consistent with the static works described above. Finally, a third minority (10% of the trajectories with higher excitation energies and 2.5% started from the lower energy) mechanism that would contribute to the shorter time constant was pointed out in this study. Trajectories following this mechanism do not recover the original FC minimum but instead lead to an open ring photoproduct, where the N_3–C_4 dissociates (recall Figure 7.1). This route was found to be mediated by a σ(nπ)π* state, with the ππ orbital being an antibonding combination of a *p* orbital on N_3 and a nonbonding orbital localized on O_8.

A later work by Fingerhut et al. monitored the decay of 44 trajectories along the singlet manifold and employed a slightly larger AS (14,10), including one more oxygen lone pair and an extra π orbital [76]. Along this work, the two main deactivation mechanisms depicted by Nachtigallová et al. were only observed. Here, the primary relaxation mechanism involves $\pi\pi^* \rightarrow n_O\pi^*$ IC. Population transfer to the $n_O\pi^*$ state leads to trapping, and only residual relaxation (two trajectories) to the S_0 was observed for propagation time exceeding 1 ps. The second most relevant decay pathway would retain the diabatic ππ* character, allowing the trajectories to return to the ground state in less than one picosecond (Figure 7.5).

In contrast with the previous two works, the study by the group of González activates the possibility of ISC upon simultaneously tackling nonadiabatic and SOC (Figure 7.5) [77]. In this work, the authors employ a (14,10) AS, including 8 π/π* orbitals and 2 lone pair orbitals located on the oxo groups of uracil, and 7 electronic states, 4 singlets and 3 triplets, with the triplets being considered explicitly, which makes a total of 13 states. This work considers three different sets of propagations: (i) singlet-only dynamics and singlet–triplet dynamics, (ii) considering initial conditions across the full spectral range, and (iii) aligned with the experimental irradiation energy, with an energy window centered at 6.59 eV. In all three cases, the starting electronic states are the S_2 and S_3, and both ππ* character and the dynamics were run for 1 ps.

Similar to the work of Fingerhut, for the three ensembles, the authors observe a very quick deactivation of the S_3 in favor of the S_2 state. From 40 fs onward, the S_2 undergoes a very slow deactivation to the S_1 (mostly nπ*). From this state, some trajectories reach the S_0 in the first 200 fs via the open ring conical intersection, attributed to the σ(nπ)π* state, already identified in the dynamics of Nachtigallová et al. Simultaneously to ground state recovery, the S_1 also injects population to the triplet manifold through the T_2 (of ππ* character), which finally relaxes to the T_1 via IC. The blockage of the ISC channel in the singlet dynamics logically leads to the accumulation of population in the S_1 state, which surpasses the population of the S_2 after 140 fs, while the constraints in the starting energy matching the experimental

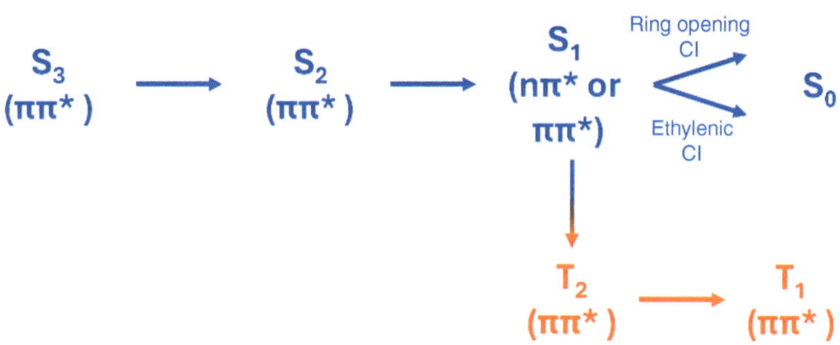

Figure 7.5 Deactivation mechanisms as proposed by References [76, 77, 82] in which ISC processes are described by gradient-colored arrows and IC mechanisms with static-colored arrows.

setup reduce the ISC yield in favor to ground state recovery. Interestingly, along the dynamics, the authors were able to identify several funnels not characterized by static or previous dynamical works. As a matter of fact, the authors identify a second S_2/S_1 CI, differing from the planar structure of previous works and characterized by a strong distortion of the aromatic ring. This work also points out that IC to the ground state occurs in early times via the open ring S_1/S_0 CI, while after 200 fs, it is the ethylenic S_1/S_0, which facilitates ground state deactivation. Finally, ISC was found to require the strong elongation of the oxo group, while T_2/T_1 IC is mediated by the asymmetric stretching of the two CO bonds, concomitant to the out-of-plane distortion of the oxo group at C_4. Altogether, the authors identify a biexponential decay with the fastest component (30 fs) ascribed to the change of $\pi\pi^*$ to $n\pi^*$ character, whereas the slowest one (2.4 ps) corresponds to competing IC and ISC processes.

7.5 Thymine

Due to their structural resemblance, the absorption of thymine and uracil in solution is similar, with the maximum of the first absorption band of thymine peaking at 4.68 eV [69]; that is, slightly red shifted compared to uracil. The TD-PBE0 vertical absorption for the first bright state ($\pi\pi^*$, S_1) of solvated thymine, including four explicit water molecules and embedded in a polarizable continuum model (PCM) water continuum, was calculated at 4.97 eV. Similar to uracil, this first bright $\pi\pi^*$ state is followed by a closely lying dark $n\pi^*$ state at 5.28 eV, which becomes stabilized and switches order with the $\pi\pi^*$ in the gas phase (4.87 vs 5.06 eV). These energies are quite close to the CASPT2 gas-phase results predicting the $n\pi^*$ and $\pi\pi^*$ states at 4.39–5.11 and 4.68–5.38 eV, respectively, depending on the size of the AS employed [65, 66, 83–85]. It is worth noting that with the (12,9) AS, the $n\pi^*$ is predicted slightly above the $\pi\pi^*$ state [65].

Two components contribute to the decay of photoexcited thymine both in the gas phase and solution: the first 105 fs (0.15–2.8 ps) and the second 5.12 ps (0.76–30 ps), demonstrating that at least two mechanisms contribute to the deactivation of this system [21, 40, 70]. Remarkably, thymine is a nucleobase with the longest excited state lifetime.

Analogously to uracil, two double-well profiles were proposed for the S_1 and T_1 states, based on the location of stationary points at the CASPT2 level of theory (Figure 7.6). Also, equivalent S_2/S_1, T_2/T_1, S_1/S_0, and T_1/S_0 interstate funnels were localized. Small adiabatic energy differences for these stationary points were, however, identified due to the presence of the extra methyl group compared to uracil. More specifically, a stabilization by 0.1 eV of the $\pi\pi^*$ PES, i.e. minimum, TS to the CI and S_1/S_0 were detected. In contrast, a destabilization by 0.1 eV was calculated for the energy of the S_1 $n\pi^*$ minimum. Similar energy barriers connecting the $\pi\pi^*$ and $n\pi^*$ S_1 minima in the forward and reverse directions were predicted for uracil and its methylated analog [65]. Similar to absorption, fluorescence of thymine was found to be red shifted compared to uracil. Also interestingly, fourfold smaller fluorescence lifetimes were observed for thymine [69]. Two additional

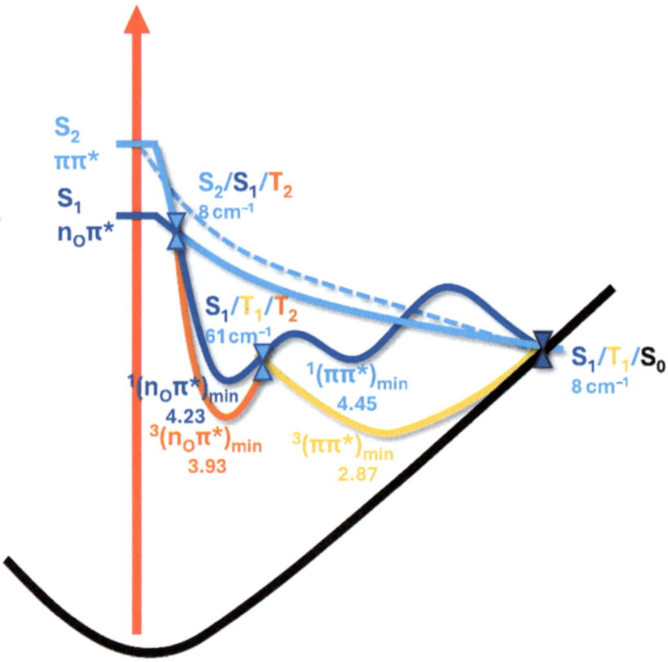

Figure 7.6 PES of thymine as described by References [65, 87]. Singlets represented in cold colors, while triplets are represented in warm colors. Energies in eV and referred to S_0.

$n\pi^*/S_0$ crossings for thymine, which were shown to lie above the vertical energy of this state, were also localized in the frame of the CASSCF/MRCI method by Blancafort, Domcke, and Barbatti [84–86]. One of these crossings is characterized by a strongly stretched and out of plane distorted CO bond, while the other shows a boat-like deformed structure. Altogether, and similar to uracil, four different possible deactivation paths were scrutinized for this system: (i) direct deactivation along the $\pi\pi^*$ state from the FC structure to the $S_1(\pi\pi^*)/S_0$; (ii) deactivation mediated by the trapping of population at the $\pi\pi^*$ minimum; and two more routes consisting of (iii) IC to the $n\pi^*$ state followed by direct deactivation to the S_0 via the $S_1(n\pi^*)/S_0$ or (iv) alternatively through the $S_1(\pi\pi^*)/S_0$ funnel after recrossing to the $\pi\pi^*$ state. MS-CASPT2 calculations predict similar slopes for decays (i) and (ii), suggesting the competition between both paths following photoexcitation [86]. Calculations for solvated thymine, however, reveal that accessing the $\pi\pi^*/S_0$ CI region is more energetically expensive for thymine, with a methyl group in the C_5 position that gets distorted at the S_1/S_0 degeneracy region [69]. Reaction paths based on geometry interpolations reveal that paths (iii) and (iv) involve the surpassing of an energy barrier [84, 86]. The combination of these two facts would justify the longer excited state lifetimes registered for thymine.

The triplet yield for thymine amounts to 6% [14]. Similar to uracil, three different funnels were characterized to leak population to the triplet manifold (Figure 7.6) [87]. The first two funnels were found along the barrierless direct path for ground state population recovery and would transfer population from the bright $^1(\pi\pi^*)$ state to the $^3(n_O\pi^*)$ and $^3(\pi\pi^*)$ states, respectively. These two crossings lie in different regions of the pathway connecting the FC region and the $\pi\pi^*/S_0$ CI. While the funnel to the $^3(n_O\pi^*)$ state is located at the beginning of the pathway, the crossing with the $^3(\pi\pi^*)$ lies close to the S_1/S_0 degeneracy region. Thus, while the first is characterized by a rather planar geometry, the second presents a screw boat-like structure, consistent with the ethylenic CI. The third funnel would derive population from the dark $^1(n_O\pi^*)$ intermediate to the $^3(\pi\pi^*)$ state (Figure 7.6).

It is worth noting that greater SOC terms were computed for thymine, compared to uracil. This is especially relevant for the $^1(n_O\pi^*)/^3(\pi\pi^*)$ crossing where the SOC amounts to ca. 60 cm^{-1}. Moreover, thymine presents parallel singlet and triplet $n_O\pi^*$ PES, paving the way for efficient ISC.

Different electronic structure methods and approaches for the dynamics have been considered for modeling the time-resolved decay of thymine, including CASSCF [55, 78, 86, 88, 89], semiempirical methods [80, 90], TD-DFT/PBE0 [91], ADC(2) [92], and CASPT2 [93].

CASSCF ab initio multiple spawning (AIMS) [78] and SH [88] dynamics considering (10,8) or smaller AS report inefficient $\pi\pi^* \rightarrow n\pi^*$ IC in the first stages of the dynamics ($t < 500$ fs) and assign the sub-picosecond decay of thymine (and uracil) to the relaxation of the excited population to a minimum in the S_2 ($\pi_{CO}\pi^*$) potential (not shown in Figure 7.6). The preferred population diversion toward the minimum S_2 is ascribed to the greater gradient connecting the FC region to this minimum, compared to the S_2/S_1 IC route. A subsequent work by the same group, this time employing ADC(2) as an electronic structure method, highlights the importance of nondynamic correlation and reports a much reduced trapping in the $S_2(\pi_{CO}\pi^*)$ minimum compared to the one predicted employing CASSCF surfaces [92]. This minimum, which presents a C_6 O_8 biradical character, would correlate with the higher lying excited states S_3 and S_4. Thymine remains in the S_2 minimum for ca. 2.5 ps. Getting out of this minimum requires a change in the state character to $\pi\pi^*$ through the puckering of the C_6 position. Relaxation along this coordinate destabilizes the S_1 state and facilitates the access of the S_2 $\pi\pi^*/S_1$ $n\pi^*$ degeneracy region. Further relaxation involves the access of the S_1 $n\pi^*$ minimum followed by direct deactivation via $n\pi^*/S_0$ funnel or recrossing and IC from the $\pi\pi^*$ state [55, 88]. Other dynamical studies also addressed this mechanism, allowing a more detailed analysis of the underlying relaxation pathways. The on the fly SH dynamics with CASSCF(8,6) starting from the TS along the $\pi\pi^*$ path, separating the $(\pi\pi^*)_{min}$ and the ethylenic CI, predict a branching between a direct decay via the $\pi\pi^*$ state and an indirect mechanism populating the $(\pi\pi^*)_{min}$ and evolving through a crossing with the $n\pi^*$ state towards the $\pi\pi^*/S_0$ CI [86]. In a later work by Asturiol et al. [89] employing quantum dynamics propagated from the S_2/S_1 CI and CASSCF(8,6) PESs, the deactivation pathway is shown to depend on the region of the conical intersection accessed by the nuclear wavepacket. When the wavepacket is directed toward the peaked region of the intersection, relaxation to S_0 occurs from the $\pi\pi^*$ state, whereas when it evolves toward the sloped region, the population is funneled into the $n\pi^*$ state. Singlet–triplet SH dynamics with a larger (12,9) AS registers fast $S_2 \rightarrow S_1$ IC ($\tau = 160$ fs) at the first stages of the dynamics ($t < 70$ fs), which slows down for longer propagation times [94]. The authors propose two different S_2/S_1 IC funnels, of which one of them gets blocked at early propagation times. This crossing, accessed by 20% of the trajectories, presents

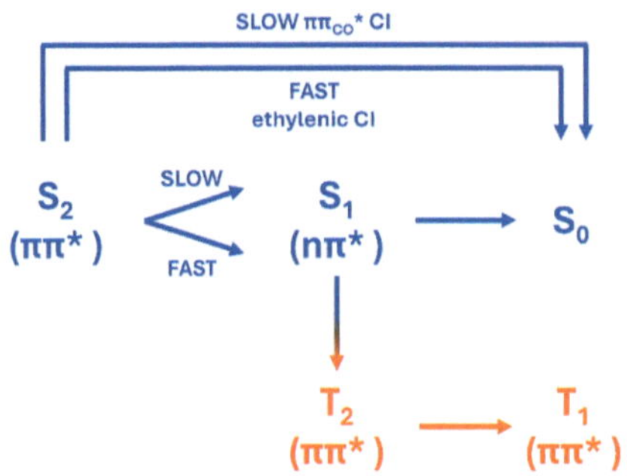

Figure 7.7 Deactivation mechanisms as proposed by References [93, 94] in which ISC processes are described by gradient-colored arrow, and IC mechanisms with static-colored arrows. Upper part as described in Nakayama et al. [93] and lower part as described in Mai et al. [94].

a reinforced C_4–O distance and is quite close to the FC, consistent with its early inhibition. The trajectories that skip this crossing evolve to the S_2 minimum and slowly decay to the nπ* state in a longer timescale ($\tau = 3.8$ ps) through a crossing with a more stretched C_4–O bond distance. This work proposes two different mechanisms for the subsequent decay of the nπ* state, namely, direct ground state recovery along the singlet manifold or mediated by ISC (Figure 7.7). The fitted time constants for these two processes amount to $\tau_{S1S0IC} = 5.2$ ps and $\tau_{ISC} = 900$ fs, respectively. ISC promoted by the reinforcement of the C_5–C_6 bond length was found to mainly occur via the S_1(nπ*)/T_2($\pi\pi$*) funnel. Only residual ^{1}(nπ*) → ^{3}(nπ*) and $\pi\pi$* → triplet states processes were registered along the singlet–triplet dynamics. Most of the trajectories would finally decay to the most stable triplet T_1($\pi\pi$*). These dynamics highlight the insignificant role played by the other ISC channels identified in the static works with the ^{1}($\pi\pi$*) state as injector of population to the triplet manifold.

Considering the remarkable differences between the CASSCF and CASPT2 PES, especially with reduced ASs, the deactivation dynamics of photoexcited thymine was also explored with CASPT2 in both vacuum and solution environments (Figure 7.7) [93]. Due to the high computational cost of these simulations, only 10 trajectories and reduced (2,2) ASs for the CASSCF reference wavefunctions were employed. Nonetheless, the reliability of the CASPT2(2,2) and CASPT2(4,4) potential energies was previously assessed against CASPT2(12,9) calculations. For all the monitored gas-phase trajectories, and in contrast with the CASSCF dynamics, the authors observe the access of the ethylenic CI in less than 1 ps due to the disappearance of the $\pi\pi$* minimum (or at least its shallowing) [65] after considering dynamic correlation. Interestingly, for the CASPT2(4,4)/MM molecular dynamics, 7 out of 10 trajectories reach the ethylenic CI in a range of 0.16–0.65 ps, while the other 3 trajectories access a $\pi\pi_{CO}$*/S_0 facilitated by the stretching and out-of-plane distortion of the carbonyl group. In fact, hydrogen bonds between thymine and water molecules were found to stabilize the biradical path. Quite unexpectedly, the lowest lying nπ* state was found to lie remarkably high in energy along the whole trajectories CASPT2(12,9) energies, revealing its much secondary role both in gas phase and solution.

7.6 Cytosine

As the other nucleobases, cytosine presents a multiexponential decay signal. Experimental studies have found up to three different time constants: τ_3 of several picoseconds and τ_2 ranging between 0.24 and 1.2 ps [37]. τ_1 was not found in all studies, but when recorded is very fast, below 200 fs [39, 42].

The vertical absorption spectrum of cytosine has been studied employing a variety of methods, such as DFT/MRCI [44], TD-DFT [60], LR-CC2 [95], and CASPT2 [83]. All of these works situate the lowest lying state of cytosine, of $\pi\pi^*$ character, at around 4.6 eV. More specifically, CASPT2 calculations performed by Schreiber et al. calculate the S_1 at 4.68 eV and the dark S_2 $n\pi^*$ state at 5.12 eV [83].

According to Blancafort and Robb, the slow lifetimes recorded can be ascribed to recrossing processes [96]. This statement was seconded by their ab initio CASPT2//CASSCF calculations, in which they gained insight into the crossing regions present along the singlet PES of cytosine. The S_1 FC region is directly connected to a $^1(\pi\pi^*)_{min}$ excited state minimum in the S_1 surface, which, in turn, is connected to an energetically accessible S_1/S_0 conical intersection, explaining the ultrafast decay of the nucleobase. This seam involves mainly the pyramidalization of C_6 (recall Figure 7.1). They found, however, that this S_1/S_0 CI, in turn, degenerates with the S_2/S_1 seam, geometrically located between the $^1(\pi\pi^*)_{min}$ and S_1/S_0 CI (Figure 7.8). This CI could allow the nonadiabatic trapping of the population in the S_2 state, of $n_O\pi^*$ character, thus explaining the slower decay components of cytosine.

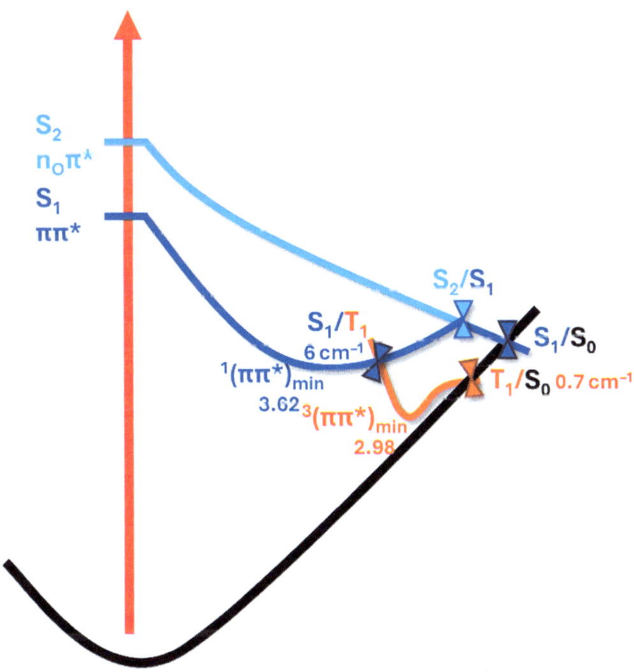

Figure 7.8 PES of cytosine as described by Blancafort and Robb [96] and Merchán and Serrano-Andrés [98]. Singlets represented in cold colors, while triplets as described by González-Luque et al. [48] are represented in warm colors. Energies given by González-Luque et al. in eV and referred to S_0.

Nonetheless, the photophysics of cytosine has been subject to a strong debate. The shape of the PES of cytosine seems to be strongly dependent on the chosen electronic structure method. For example, Ismali et al. propose the bright state to be connected in a barrierless manner to a $^1(n_O\pi^*)_{min}$. These calculations were performed at the CASSCF/6-31G* level of theory [97]. When performing CASPT2 energy corrections on CASSCF structures, including an additional *n* orbital to the AS used in Blancafort and Robb [96], Merchán and Serrano-Andrés proposed a different decay mechanism for cytosine [98]. Their calculations corroborate the existence of the $^1(\pi\pi^*)_{min}$ connected directly to the FC region, but rule out the possibility of recrossing to S_2. Instead, they suggest $^1(\pi\pi^*)_{min}$ to be connected to an S_1/S_0 CI, accessible after surpassing a small barrier of ca. 0.1 eV.

González-Luque et al. argued that the presence of the $^1(\pi\pi^*)_{min}$ minimum sets the difference between the rest of the nucleobases and cytosine, increasing the probability of ISC [48]. They locate an MECP almost degenerate with the S_1/S_0 CI, which is also very close to the $^1(\pi\pi^*)_{min}$. At this S_1/T_1 MECP, they calculate an SOC of 6 cm^{-1}, which is relatively large for a molecule with the characteristics of cytosine, lacking heavy atoms. From the S_1/T_1 MECP, they were able to locate a minimum on the triplet manifold, hereafter $^3(\pi\pi^*)_{min}$, lying less than 0.7 eV below the MECP. From $^3(\pi\pi^*)_{min}$, they were able to locate yet another MECP, in this case with the S_0, with a much smaller associated SOC value. In contrast with other nucleobases, cytosine was postulated to only undergo ISC via the low energy channel. This is, as hypothesized, due to the nonmixing of S_1 and S_2, of $\pi\pi^*$ and $n_O\pi^*$ characters, respectively. This is in line with the work by Merchán and Serrano-Andrés, who ruled out recrossing processes [98], but in disagreement with the results of Blancafort and Robb, who suggested a strong mixing of the $\pi\pi^*$ and $n_O\pi^*$ electronic states along the deactivation coordinate due to the presence of a three-state CI [96].

Richter et al. performed singlet–triplet SH dynamics on cytosine, employing CASSCF as the electronic structure method [99]. They focused on the biologically relevant *keto* tautomer of the nucleobase. They report ISC happening even in the first few tens of femtoseconds, with triplet population transfer happening via different routes. For the dynamic calculations, they propagated a pool of trajectories that started mainly from S_2 to S_1, with a minor contribution of S_3. As per their calculations, the trajectories that started in S_3 were able to relax to S_2 in a swift manner (25 fs) or slower directly to S_1 via a three-state CI. From S_2, the population can decay to the ground state by either a cascade of CIs connecting S_2 and S_0 or an $S_2/S_1/S_0$ three-state CI as proposed by Blancafort and Robb [96]. However, from S_2, ~3% of the population was found to cross to the triplet manifold (T_3 state). Balancing the income of population decaying from higher lying excited states and the deactivation to S_0, they report S_1 to have nearly 30% of the population after a propagation time of 100 fs. Moreover, they computed S_1/T_2 to be the principal ISC mechanism, with more than 10% of the population crossing to the triplet manifold coming from the S_1. Following radiationless decay from higher lying triplet excited, Richter et al. conclude that ~10% of the population remain stable in the T_1 state for more than 1 ps without crossing back to the singlet manifold and decaying to the ground state. A sum up of the deactivation is depicted in Figure 7.9. When

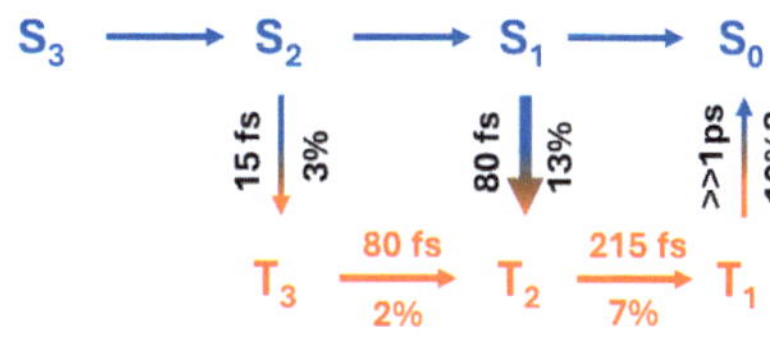

Figure 7.9 Deactivation mechanisms as proposed by Richter et al. [99]. in which ISC processes are described by gradient-colored arrows and IC mechanisms with static-colored arrows. The thickness of the arrows indicates the importance of the pathway.

studying the geometrical parameters, Richter et al. proposed that the singlet–triplet energy gap becomes reduced upon simultaneous NH_2 group pyramidalization and rotation, including a strong out-of-plane motion of the N–H in *para* to NH_2.

However, a later study by Mai et al., resorting to the same level of theory but employing a molecular Coulomb Hamiltonian (MCH), introduced some important clarifications regarding the dynamics of cytosine [100]. Just as in the work by Richter et al., they propose $S_1(n_O\pi^*) \rightarrow T_2 \rightarrow T_1$ to be the main ISC path, even if they clarify that this process occurs preferably via a three-state near-degeneracy point involving the three states. Moreover, the study details that direct ISC from S_1 to T_1 is possible and actually almost mechanistically identical, due to the aforementioned $S_1(n_O\pi^*)/T_2/T_1$ near-degeneracy seam. In fact, while Richter and coworkers report crossing to the triplet manifold after a few hundred femtoseconds, Mai et al. propose a mechanism that reduces this timescale 10-fold. Mai et al. argue that this difference arises because their calculations were propagated in the MCH basis, where the algorithm remains highly stable for weakly coupled surfaces. Moreover, Mai et al. also argue that the NH_2 pyramidalization proposed to reach the ISC crossing is certainly accidental, since their exhaustive analysis of the singlet–triplet hops at the dynamics indicates that the most significant changes in the geometries are an elongated C_2–O bond and a shortening of the adjacent C_2–N_3 bond.

Besides performing MCH SH to *keto* cytosine, Mai et al. also studied the *enol* tautomer of the nucleobase, explaining that even if the *keto* conformation is the biologically predominant one, the *enol* is the most stable tautomer in the aqueous phase [100]. Even though the dynamic calculations of this tautomer seem to differ greatly from those of the *keto* one, they were able to extract some similarities in the ISC mechanism. Crossing to the triplet manifold in the *enol* tautomer seems to happen scarcely, but through the S_1/T_2 funnel, with an associated SOC value of around $10\,cm^{-1}$. Subsequent IC to the lowest lying triplet state T_1 is to follow the ISC. Hence, they concluded that the ISC mechanism for both *keto* and *enol* tautomer takes place predominantly via $S_1 \rightarrow T_2 \rightarrow T_1$. Mai et al. calculate for the *enol* tautomer a smaller yield and slower ISC than for its counterpart, with a 5% of the trajectories crossing to the triplets vs the ~25% of the *keto* tautomer. The study proposes the lack of the keto group, linked to a decrease in the values of the SOC, as one of the reasons for the *enol* tautomer to depict less ISC.

7.7 Nonnatural Nucleobases

In this section, we provide a brief overview of the mechanism for triplet population in some nucleobase derivatives showing significant ISC yields. We mainly focus on thionated nucleobases arising for the substitution of oxo by thio groups in natural nucleobases, or the purine and pyrimidine raw heterocycles.

7.7.1 Thionucleobases

Oxo by sulfur substitution triggers important changes in the absorption spectrum. More specifically, thionated nucleobases absorption spectra undergo a significant red shift (up to ca. 100 nm) and an increase in the molar absorption coefficients compared to their oxo analogs [101].

The S heavy atom also affects the intersystem quantum yields, which approach unity in these systems (0.2 -> 0.9) [33, 102–104]. Moreover, these systems present some of the fastest ISC processes, which lie in the sub-picosecond regime (180–830 fs) [33, 102–104].

These photophysical properties can be rationalized based on quantum mechanical (QM) calculations. An overview of the ISC mechanisms at the molecular level extracted from molecular simulations is explained in this section.

7.7.1.1 6-Thioguanine

Even though guanine seems to scarcely undergo ISC to the triplet manifold, the literature hints at the substitution of the carbonyl for a thiocarbonyl as a dramatic enhancer of this phenomenon. Martínez-Fernández et al. performed ab initio CASPT2//CASSCF(14,12) calculations to map the excited states of 6-thioguanine [105]. The vertical energies in the FC region showed the lowest lying state to be a dark $n_S\pi^*$ state, whereas S_2 is a $\pi\pi^*$ state. This information was corroborated by Pirillo et al., who performed TD-DFT calculations, both at the B3LYP/6-31+G* and M062X/6-31+G* levels of theory [106]. Moreover, Martínez-Fernández et al. performed an extensive mapping of the PES of 6-thioguanine, deeming ISC and IC to the ground state as two potential competitive processes. They established that, upon population of the bright state and deactivation to S_1 via the $^1(\pi\pi^*)/^1(n_S\pi^*)$, the system could populate two S_1 minima of $n\pi^*$ and $\pi\pi^*$ characters. The population that reaches $^1(\pi\pi^*)_{min}$ should be more prone to deactivate through the almost-degenerate $S_1(\pi\pi^*)/S_0$ CI. On the other hand, since the $S_1(n_S\pi^*)/S_0$ CI mediating IC for the population reaching the $n_S\pi^*$ minimum is energetically inaccessible, the population trapped in $^1(n_S\pi^*)_{min}$ could undergo ISC and cross to the triplet manifold easily, since the computed value for the SOC exceeds 200 cm^{-1}. They also argue that the high density of states near $^1(n_S\pi^*)_{min}$ could also increase the possibility of crossing to the triplet manifold. Hence, they propose a kinetic model in which the direct population of the S_2, upon irradiation, would allow the transfer of population to the dark S_1 state, from where ISC to T_2 is more probable. From here, the population could deactivate to the T_1 state. This kinetic model is similar to the one proposed by Pirillo et al. [106], who suggest that while this deactivation route is plausible, the $S_2 \rightarrow S_1 \rightarrow T_1$ route is more likely due to the significantly higher computed SOC values for the S_1/T_1 (190 vs 17 cm^{-1}) at the B3LYP/6-31+G* level of theory.

Martínez-Fernández et al. subsequently performed FOMO-CI SH dynamics with a reparameterized semiempirical Hamiltonian for 6-thioguanine [107]. For their

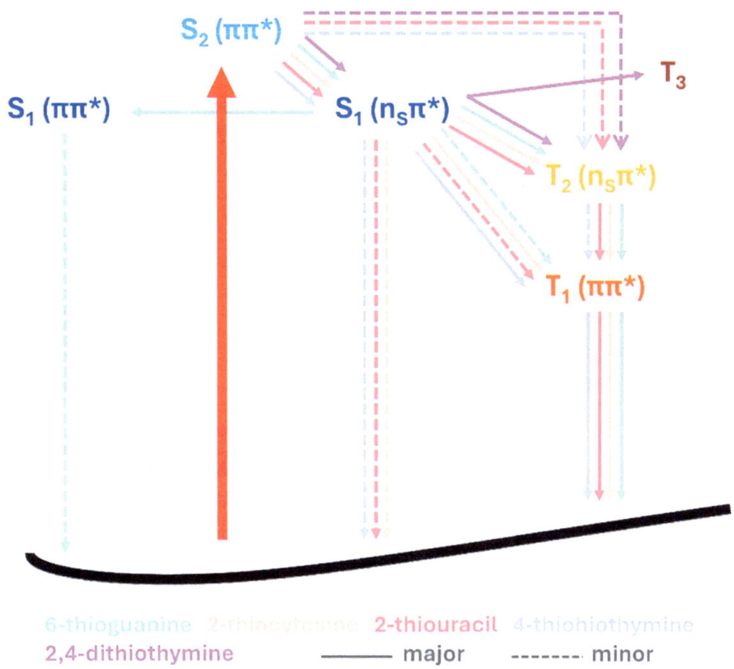

Figure 7.10 Summary of the kinetic model for the different thiobases studied in the literature.

singlet–triplet dynamics, a pool of 377 trajectories were propagated for a span of 10 ps after excitation of the system to the S_2. Deactivation to the dark S_1 state allows the population of the $^1(n_S\pi^*)_{min}$ and then further ISC to either T_1 or T_2 is proposed (Figure 7.10). Moreover, they additionally observe minority IC from $^1(n_S\pi^*)_{min}$ to $S_1(\pi\pi^*)$ (corresponding to the FC S_3), followed by IC to the ground state, revealing a pathway not considered in the first static work. This deactivation mechanism was found to be residual, since they observe that after 0.2 ps, T_1 is the state carrying most of the population. A fitting of the rise of triplet population reveals a time constant of $\tau(T) = 122$ fs [107].

7.7.1.2 2-Thiocytosine

Mai et al. propose thionation as an efficient tool for increasing the triplet quantum yield in cytosine [103]. In their work, they study the influence of substituting the keto group with a thioketo one by employing ab initio and experimental methods. According to their MS-CASPT2//CASSCF(14,10)/ANO-S-VDZP calculations, the vertical absorption energies are 3.65 eV for $S_1(n_S\pi^*)$ and 3.73 eV for S_2, the bright $\pi\pi^*$ state. Other works at different levels of theory reproduced the ordering of states but showed a slight shift in the energies. The shift might be small, as the one calculated for S_2 at 3.47 eV with MS-CASPT2(14,10)/cc-pVDZ calculations [108], or larger, as is the case of TD-B3LYP that locates the bright state at 3.08 eV [109]. Mai et al. locate two different singlet excited state minima for the thiobase, of $n_S\pi^*$ and $\pi\pi^*$

character. These minima get populated after excitation to S_2 and subsequent branching of the population at a peaked S_2/S_1 CI. As per their calculations, the S_1/S_0 CIs are located almost 1 eV higher in energy than the minima, ruling them inaccessible. Nonetheless, two MECPs with the triplets were located less than 0.15 eV away from the minima. The energetic accessibility plus the high values of the SOC, which were calculated to be higher than 150 cm^{-1}, hint to an efficient crossing to the triplet manifold. Mai et al. propose a crossing from S_1 to both T_2 and T_1, deeming the former faster and proposing consecutive IC to T_1 (Figure 7.10). Their SH calculations average an ISC lifetime of 250 fs, which agrees with the data obtained experimentally. Their calculations estimate more than 70% of the population to have undergone transfer to the triplet manifold after 1 ps, with a 10% that cross back and decay to the ground state. Nonetheless, they argued that this recrossing may be an artefact of the method employed for the dynamics, MRCIS and that ISC is to happen with a yield greater than 90%. As a possible explanation for the great increment in the ISC, besides the inherent increment in the SOC that introducing a sulfur atom carries, Mai et al. propose the localization of the electronic excitation. They hypothesize that whenever an electronic excitation is localized on the thioketo moiety, considering the difference in energy it takes to excite an electron from an n_S orbital to a π^* one, rather than from a n_O one, the stabilization of the electronic state should be greater. This excludes the $\pi\pi^*$ states and CIs, which, in the case of canonical cytosine, seem to be the main deactivation funnels. The significant stabilization of the minima, while the conical intersections retain their energetic positions as in canonical nucleobases, along with the substantial increase in SOC values, leads Mai et al. to suggest that thionation of cytosine—and nucleobases in general—greatly enhances the probability of ISC.

7.7.1.3 2-Thiouracil and 2-Thiothymine

The absorption spectra of 2-thiouracil (2-TU) and 2-thiothymine (2-TT) registered in aqueous solution present maximal absorption at 4.52 eV [33, 104]. According to gas-phase MS-CASPT2 calculations, considering CASSCF(12,9) and (14,10) reference wavefunctions, the first bright state for these two systems (S_2) is predicted at 4.30 eV (2-TU) and 4.34 (2-TT) eV with slightly red-shifted absorptions when reducing or augmenting the AS [110–112]. This state with $\pi_S\pi^*$ character is preceded by a $n_S\pi^*$ transition (S_1), involving the lone pair of the sulfur atom. Consistent with the experimental spectra, both S_1 and S_2 transitions appear significantly red shifted compared to their natural counterparts.

According to MS-CASPT2 [110], the S_2 and T_1 states for 2-TU present double-well topographies with the two minima carrying $\pi_S\pi^*$ character, correlating with the FC S_2 (S_2min and T_1min) and S_4 states (S_2min' and T_1min') and the equivalent triplets. For the S_1 state, a single minimum with $n_S\pi^*$ character was also localized for this system. Interpolated pathways between the above stationary points reveal an almost barrierless connection between the FC S_2 state, the S_2 minimum preserving the same character (S_2min) and the S_1 minimum along the C_6 pyramidalization coordinate. A small barrier (ca. 0.3 eV) was found for the equivalent pathway leading access to the S_2min', which requires C_2 pyramidalization instead.

The authors find a straightforward pathway with negligible barriers connecting the $n_S\pi^*$minimum with the $S_1(n_S\pi^*)/T_2(n_S\pi^*)$ crossing, the $T_2(n_S\pi^*)/T_1(\pi_S\pi^*)$ crossing, and the T_1 min, which is separated from the second T_1min' by a barrier of 0.10 eV. At difference, CASPT2//CASSCF studies with a similar AS or slightly enlarged propose direct $S_1 \rightarrow T_1$ ISC through planar geometries or $S_2 \rightarrow T_2$ ISC for the population of the triplet manifold [111, 113]. The large SOC values calculated at the $S_1(n_S\pi^*)/T_2(n_S\pi^*)$ crossing above 100 cm^{-1} support the efficiency of the ISC event. Consistent with the static results, molecular dynamics with CASSCF(16,11) reveals $S_1 \rightarrow T_1$ ISC as the predominant channel and shows $S_2 \rightarrow T_3$ ISC as minoritary [114]. MS-CASPT2(12,9) dynamics reveals three different routes for ISC in 2-TU: $S_2 \rightarrow T_{2,3}$, $S_1 \rightarrow T_{2,3}$, and $S_1 \rightarrow T_1$ with relative frequencies of 25%, 40%, and 35%, respectively (Figure 7.10) [115]. The dynamics also reveals a 90% triplet yield in good agreement with the experimental result (75%) and time constants of $\tau = 540$ and 250 fs for $S_1 \rightarrow T_1$ and $S_2 \rightarrow T_{2,3}$ in line with the experimental τ for ISC, which amounts to 350–340 fs for water and acetonitrile solution [116]. The dynamics also reveals the population of minimum S_2min with very few trajectories accessing the S_2min'.

7.7.1.4 4-Thiouracil and 4-Thiothymine

4-Thiouracil and 4-thiothymine present a maximum of absorption at 3.78 and at 3.70 eV in water solution, respectively [117, 118].

In 4-thiothymine, this maximum absorption has been ascribed to the combination of two $\pi\pi^*$ excited states. These two transitions come after a $n\pi^*$ state, with all the orbitals involved, including contributions from the sulfur [118]. Two different mechanisms for S_2/S_1 IC and ISC were identified with help of MS-CASPT2//CASSCF calculations [118]. The triplet degeneracy region $S_2/S_1/T_2$, functioning at the same time as IC and ISC, can actually be accessed via two very different S_2 minima: the first and higher lying with a twisted structure around the C_5–C_6 bond, and the second with a planar geometry and lying lower in energy. The large SOC calculated at the $S_2/S_1/T_2$ crossing supports potential efficient ISC to the T_2 state at this region of the PES. Alternatively, the excited population would internally convert to the $S_1(n\pi^*)$ minimum with the T_2 and T_1 states lying energetically close. Excited state deactivation of 4-thiothymidine was also considered in a water MM environment within a QM(CASPT2//CASSCF)/MM approach [119]. In this work, the PES considering solution was found to be rather similar to that evaluated in vacuo, except for the energies of the S_2 state, which seem to be quite sensitive to the environment. The time-resolved evolution of the decay of 4-thiothymine has been investigated resorting to a TSH approach combined with two different electronic structure methods, namely, a CASPT2/CASSCF semiempirical reparametrized Hamiltonian [118] and Linear Response TD-DFT (LR-TDDFT) [120]. Both studies explore the effect of the water environment on the dynamics.

The vacuum TD-DFT dynamics reports relaxation of the excited population to a S_2 C-pyramidalized minimum, followed by IC to the S_1 within ca. 350 fs driven by C_2-pyramidalization and the stretching of the C–S and C_5–C_6 bond distances. Following this same global coordinate, this work detects S_1/T_2 ISC as the main route for the transfer of population to the triplet manifold. The incorporation of the interaction of the thiobase with an MM water solvation environment activates a second ISC channel, namely, $S_2 \rightarrow T_3 \rightarrow T_2 \rightarrow T_1$. The dynamics employing the parameterized semiempirical Hamiltonian supports deactivation of 4-thiothymine in water following the sequence $S_2 \rightarrow S_1 \rightarrow T_1$ for most of the trajectories (87%) (Figure 7.10) [118]. The access of the S_1 takes place from a twisted conformation, although switches between the planar and distorted conformations are observed until the system abandons the S_2 state.

7.7.1.5 2,4-Dithiouracil and 2,4-Dithiothymine

2,4-Dithiouracil and 2,4-dithiothymine present a maximum of absorption at ca. 3.44 eV [33, 104], which has been ascribed to a $\pi\pi^*$ transition that promotes an electron from one of the sulfurs to the other [121]. Introducing an additional S atom adds complexity to the deactivation mechanism of these systems. Gas-phase CASSCF and CASSCF/MM static calculations on 2,4-dithiothymine reveal the stabilization of the two lowest lying $n\pi^*$ localized on each of the two sulfur atoms along the global deactivation coordinate, which adds to the identification of several high-order degeneracies, multiplying the number of pathways for IC and ISC [121]. From the complex playground of IC and ISC routes derived from the static calculations, TD-DFT SH dynamics identifies two main routes for relaxation of 2,4-dithiothymine [120]. The primary route involves IC to the FC S_1 ($n\pi^*$) followed by ISC to the T_2 and T_3 in vacuum and to T_1, T_2, and T_3 in water solution, while the secondary route involves direct ISC from the S_2 (Figure 7.10).

7.7.2 Purine and Pyrimidine

The photophysical studies performed on purine and pyridine underpin that these heterocyclic cores, common to the natural nucleobases, decay along the triplet manifold. In fact, ISC yields close to unity have been registered for the two systems in water [122, 123]. From this, it can be concluded that specific substitution (both in terms of nature and position of the substituents) of these cores strongly tunes the topographies of the purine and pyrimidine cores, rendering triplet state population as a residual process while turning accessible high energy IC funnels [122, 124].

In the case of purine (N9H-tautomer), two different pathways for the population of the triplet manifold were identified (Figure 7.11) [122]. The first route accounts for ISC following the trapping of the population at the position of the S_2 min ($\pi\pi^*$), straightforwardly accessed from the first bright state at the FC region. In fact, the T_2 ($n\pi^*$) state is energetically degenerate at the minimum position and couples the S_2 with an SOC amounting to $14\,cm^{-1}$. The second pathway for the injecting population into the triplet manifold is mediated by the S_1 ($n\pi^*$) state, accessed via IC from the S_2. Similar to the S_2 min region, the T_2 ($\pi\pi^*$) state is energetically close and coupled to the S_1, suggesting efficient ISC. The population gathered by the T_2 ($\pi\pi^*$) and T_2 ($n\pi^*$) states would be directed to the most stable T_1 ($\pi\pi^*$) minimum through the T_2/T_1 CI. Ab initio SH molecular dynamics simulations considering CASSCF as an electronic structure method predict two main decay pathways for the excited N9H purine tautomer, i.e. ground state recovery via the $S_1(n\pi^*)/S_0$ CI or ISC to the T_2 ($\pi\pi^*$) state via the $S_1(n\pi^*)/T_2(\pi\pi^*)$ crossing. The population of the S_1 ($n\pi^*$) minimum

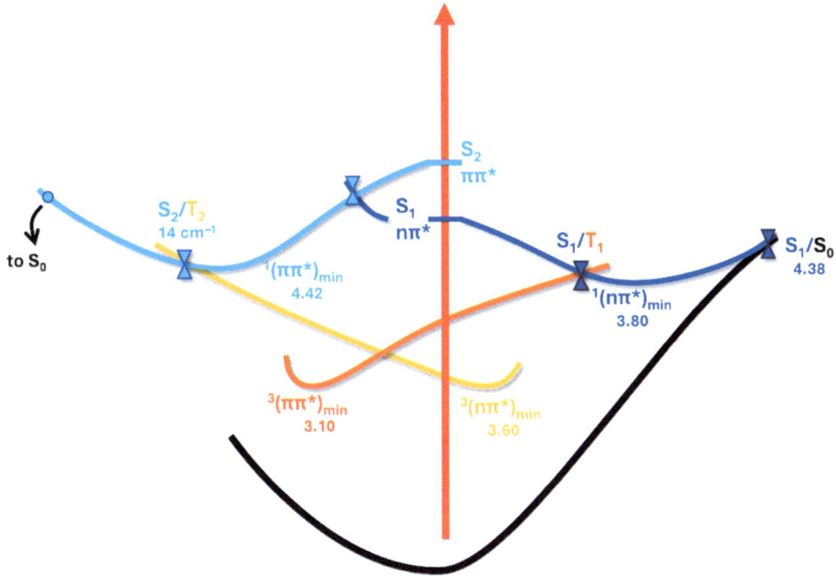

Figure 7.11 PES of the purine chromophore as described by Crespo-Hernández et al. [122] Singlets represented in cold colors, while triplets are represented in warm colors. Energies in eV and referred to S_0.

is achieved through the $S_2(\pi\pi^*)/S_1(n\pi^*)$ crossing (66%) or via a $S_2(n\pi^*)/S_1(n\pi^*)$ crossing not identified in the static calculations. According to the simulations, the $S_2 \rightarrow S_1$ relaxation would be governed by a 25 fs time constant, while the ISC process could be fitted to a 5 ps lifetime. The amplification of $S_2 \rightarrow S_1 \rightarrow S_0$ path is ascribed to the underestimation of the energy barrier to access the S_1/S_0 CI by CASSCF.

Three different time constants were recorded for free purine base, which in water amount to $\tau_1 = 0.20$ ps, $\tau_2 = 8$ ps, and $\tau_3 = 645$ ps. From the results of the static and dynamic calculations, these are attributed to the initial relaxation of the wave packet to the S_1 minimum, vibrational relaxation of the S_1 $(n\pi^*)$ state, and the transfer of population to the triplet manifold followed by conformational relaxation to the T_1 $(\pi\pi^*)$ minimum.

In the case of pyrimidine, two different lifetimes were recorded employing acetonitrile as a solvent: $\tau_1 = 7.8$ ps and $\tau_2 = 1.5$ ns [124]. Based on static calculations considering critical points of the PESs and the simulation of the transient absorption spectra, these time constants have been assigned to ISC to the triplet manifold and relaxation to the triplet minima position, respectively [124].

The S_1 and T_1 PESs of pyrimidine are rather complex in their topographies with several minima of similar energies (Figure 7.12) [124]. Similar to purine, S_1/S_0 IC funnels either lie rather high in energy (ca. 0.9 eV from the closest minima) or are separated from the minima by very high energy barriers precluding relaxation to the ground state. Thus, following excitation to the S_1, the population of the three S_1 $(n\pi^*)$ minima is plausible. The inaccessibility of the S_1/S_0 IC funnels forces the system to decay along an S_1/T_2 ISC funnel, which, despite presenting weak S/T coupling (tens

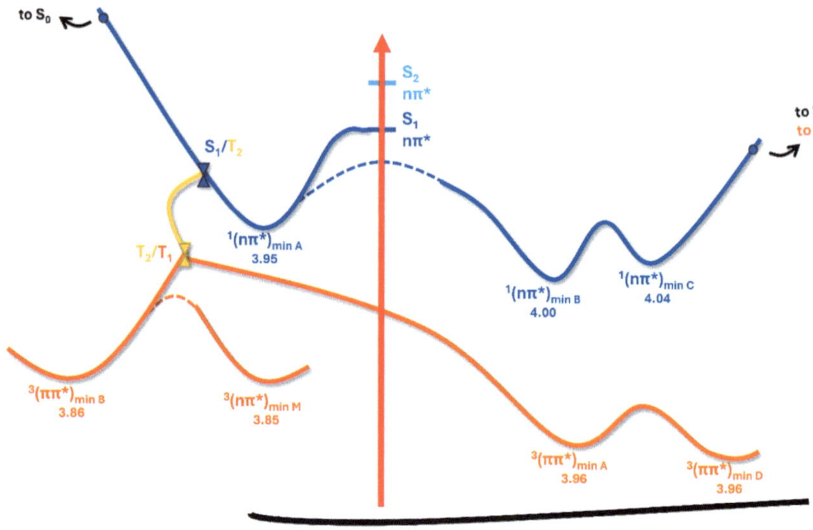

Figure 7.12 Scheme for the PES of the pyrimidine chromophore as described by Arpa et al. [124] Singlets represented in cold colors, while triplets are represented in warm colors. Energies in eV and referred to S_0.

of cm^{-1}), requires small structural arrangements and lies close to the position of the lowest energy S_1 minimum. Subsequent T_2/T_1 IC would allow the population of the T_1 state, with several minima as well.

7.8 Pyrimidine–Pyrimidine Dimer Formation Processes

7.8.1 Cyclobutane Pyrimidine Dimers

The major photolesion identified in cellular DNA is CPD [57, 125]. In spite of this, it is only formed in small percentages [23], since the nucleobases must adopt specific reactive conformations, which are limited [126].

These lesions have been characterized as rather local by MD simulations, mainly affecting the dimer spot, triggering the rolling and tilting of the nucleobases participating and the weakening of the hydrogen-bonding pattern of the 5′ base pair [127].

As already mentioned in the introduction, the formation of CPD has been conceived to occur along both the singlet and triplet manifolds. Direct irradiation of single-stranded oligomers has been postulated to lead to CPDs along the singlet manifold [23], via very short-lived triplet states [21, 25] or photosensitization of DNA [18, 19, 125].

The singlet and triplet mechanisms for CPD formation have been studied within different computational frameworks [27–29, 128–132].

In order to rationalize the mechanism for pyrimidine–pyrimidine dimer formation, the minimum potential energy profiles of the triplet and singlet excimers

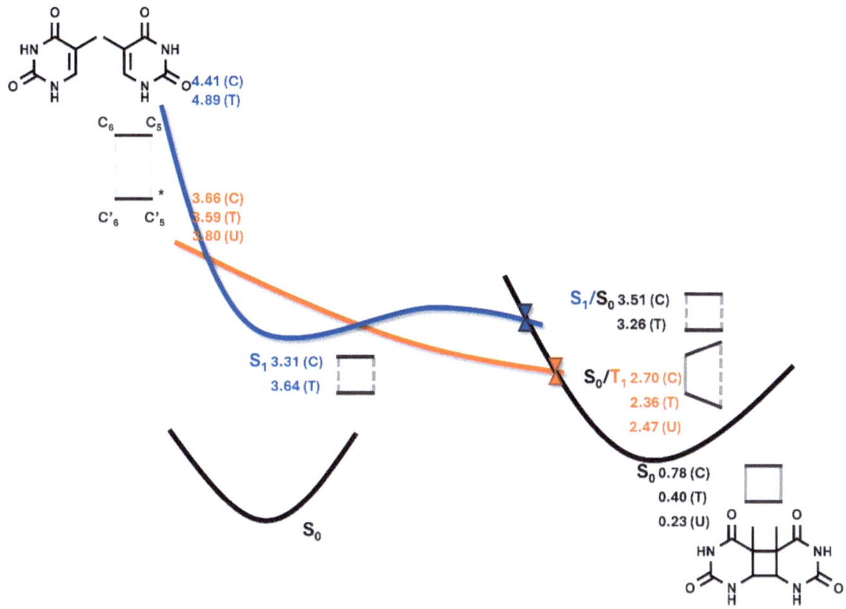

Figure 7.13 Singlet and triplet photochemical paths leading to the formation of CPD (Adapted from References [27–29]). Energies in eV and referred to S_0.

along the coordinate that approaches two π-stacked uracil, cytosine, and thymine molecules were investigated at the CASPT2 level of theory [27–29]. The mechanism optimized for CPD formation along the triplet manifold is of a stepwise nature. The MEP along the T_1 state leads to a degeneracy with the ground state, corresponding to an intermediate biradical structure (see Figure 7.13). Here, the SOC is estimated to amount to a few cm^{-1} in gas phase. Geometrically, at this point, C_6–$C_{6\prime}$ atoms are covalently bonded, while the C_5 and $C_{5\prime}$ centers, holding the two electrons, are kept separated at a longer distance. Interestingly, energetically speaking, the ISC crossing is close to the triplet energy threshold observed for potential DNA photosensitizers. This geometry can actually be accessed from many different initial arrangements of the two nucleobases. The shape of the potential energy profile along this coordinate anticipates a fast deactivation, provided the conditions for efficient ISC are met.

In contrast with the triplet mechanism, the mechanism along the singlet manifold takes place in a concerted manner. MEP from the locally excited S_1 state proceeds barrierless toward a relaxed excimer structure. Depending on the nucleobase, either the excimer lies below the CI (cytosine) and then the system needs to surpass an energy barrier to access the S_1/S_0 conical intersection or no excimer below the CI was found (thymine) and the population would proceed barrierless to the CI (see Figure 7.13) [29]. This would explain, at least in part, why CPD lesions have been found to be more efficient for thymine than for cytosine. The peaked topology of the S_1/S_0 conical intersection would branch the population between the original stacked arrangement of the nucleobases and the CPD product [130].

The dimerization mechanism from excitons delocalized over two thymine nucleobases has been studied at the TD-M052X both in gas phase and in solution.[24] This work predicts a straightforward connection between the FC region and a crossing with the ground state responsible for the formation of thymine-thymine dimers or an exciton $\pi\pi^*$ minimum, separated by small barriers from the former crossing, depending of the starting conformation.

Later, CASPT2//CASSCF/MM MEP calculations along the charge transfer states of a thymine oligomer allowed the identification of a three-state degeneracy $^1(n\pi^*)/^3(n\pi^*)/^3(\pi\pi^*)$ crossing that would allow the leak of population from the $^1(n\pi^*)$ to a triplet minimum [132]. From this minimum, an energy barrier of 0.16 eV to access the T_1/S_0 crossing has been calculated. Overall, Wang and Chen [132] underpin two competitive decay pathways for the decay of the excited oligomer besides the monomer-like deactivation: (i) a singlet concerted downhill pathway involving important structural deformation and (ii) a second route along the triplet $^3(\pi\pi^*)$ state mediated by the intermediate population of the $^1(n\pi^*)$ state. A very complete computational review covering different levels of theory for the dimerization mechanism has been covered by Barbatti [133].

Dynamical effects have been taken into account, to the best of our knowledge, by two different works modeling CPD formation using nonadiabatic dynamics simulations. The first dynamical study resorted to a BLYP/MM approach and employed a B-DNA oligomer, with 10 base pairs surrounded by a water solvent environment [134]. This dynamical study was undertaken on precomputed free energy maps for the S_1 and S_0 surfaces along the distance between the centers of the two C_5–C_6 bonds and the angle defined by the $C_{5'}$–$C_{6'}$–C_6 atoms. The authors attribute the low yield for the photolesion to the greater gradient of the PES in the direction of the standard DNA conformations. The second study addressed CPD formation in the gas phase resorting to the CASSCF(12,10) method [135]. In this study, the trajectories were run starting from different families of electronic states, namely, doubly excited singlet states $^1DE(\pi\pi^*)$, and two single excited states with $^1(\pi\pi^*)$ and $^1(n\pi^*)$ characters. Half of the trajectories started from the doubly excited state, which would become eventually populated from singlet states, showed dimerization in ca. 150 fs. $S_1 \rightarrow S_0$ IC for these trajectories proceeds via the conical intersection reported in a study [130]. The trajectories that do not evolve to the dimer were found to decay to the $^1(n\pi^*)$ state instead. No dimerization was observed from any of the trajectories initiated in the $^1(\pi\pi^*)$ and $^1(n\pi^*)$ states. Likewise, none of the trajectories were found to access the triplet manifold.

7.8.2 (6-4) Photoadduct Formation

The second most common lesion caused by UV light in DNA is the formation of (6-4) adducts. This photolesion consists of the formation of a covalent bond between two adjacent pyrimidine nucleobases, tying positions C_6 and $C_{4'}$ together (recall Figure 7.1). This photoreaction is believed to take place via an oxetane intermediate [25], in which a four-membered ring heterocycle is formed between positions C_5, C_6, $C_{4'}$, and the adjacent O′. Experimental work performed by Marguet and Markovitsi situates the timescales for the formation of (6-4) adducts in the milliseconds, with a peak at ca. 4 ms. [25] From the computational perspective, several works have been published addressing the mechanism of the formation of these adducts. The literature includes studies at classical or QM/MM level [136–138], TD-DFT [137, 139, 140], and even CASPT2//CASSCF calculations [26, 131].

Giussani et al. postulate that the experimental lifetime could be ascribed to the ISC processes required for the formation of the adduct [26]. In their work, they performed CASPT2//CASSCF(6,6) and CASPT2//CASSCF(12,12) calculations, considering one-thymine and two-thymine system, respectively, and emphasize that, in order for the photoreaction to happen, the excitation needs to be localized on a nucleobase, whereas the adjoint thymine needs to remain in the ground state. Hence, Giussani et al. propose that after undergoing ISC and populating a $^3(n\pi^*)$ state, a minimum of the same character is reached (Figure 7.14). This minimum seems to be key for the photoreaction, due to the elongated $C_{4'}$–O′ distance, allowing the reaction with the C_5–C_6 bond of the adjacent thymine. This $^3(n\pi^*)_{min}$ is connected to a $^3(n\pi^*)/^3(\pi\pi^*)$ crossing, which, in turn, eventually leads to another crossing point; in this case, with the ground state, already with the four-membered ring heterocycle formed. On a similar note, Yang et al. conclude that the photoreaction can only happen in a triplet–singlet manner, i.e. with one thymine as a triplet and the other as a singlet [139]. When optimizing the lowest lying triplet state at the TD-B3LYP/6-31G(d,p) level of theory, they observe an elongation of both $C_{4'}$–O′ and C_5–C_6 bonds and computed the reaction path toward the formation of the adduct. From this path, they concluded that the reaction was mediated by a diradical intermediate, and based on the computed barriers, they are able to rationalize the higher experimental quantum yields for CPD formation. From these results, they also demonstrate that the (6-4) adduct is inaccessible in a triplet–triplet reaction. Similar conclusions were extracted by Wang and Yu at the M062X/def2-SVP level of theory, but they also noted that the backbone is key to the mechanism, since it seems to play a nonnegligible role on the computed triplet

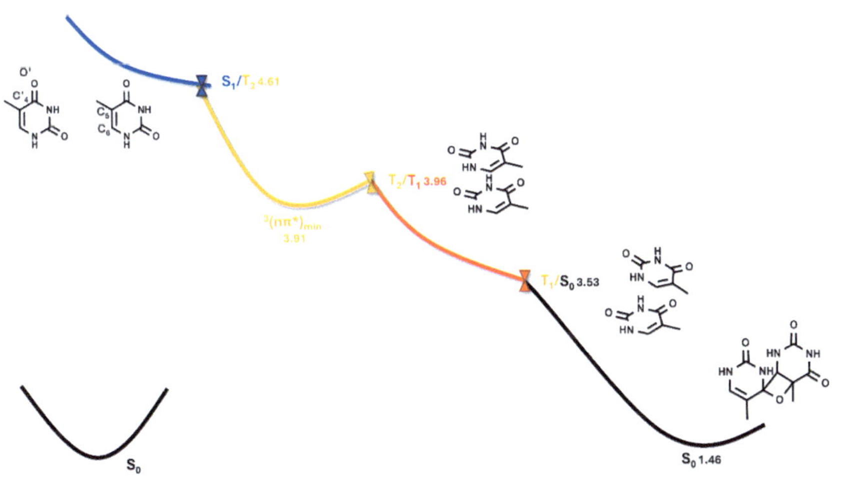

Figure 7.14 Photochemical path leading to the oxetane intermediate in the formation of (6-4) thymine dimers (Giussani et al. [26] / with permission ofAmerican Chemical Society).

pathways [140]. Nonetheless, the work by Conti et al. rules out the necessity of undergoing ISC, determining that the reaction may instead happen via a reactive charge transfer state [137].

7.9 Conclusions

The study of triplet states in nucleobases and their derivatives provides important insight into nucleic acid photophysics and thus into their photostability and photoreactivity. Triplet states are relevant because their long lifetimes allow for subsequent photochemical reactions, including dimer formation or the generation of singlet oxygen, depending on the environment where they are embedded. These processes, which are key for understanding UV-induced DNA damage, also play a central role in other contexts, such as photobiology and photochemotherapy.

Natural nucleobases exhibit triplet quantum yields that, although small, are not negligible, ranging between 10^{-3} and 10^{-2} depending on the excitation wavelength and the solvent. Although relatively inefficient, triplet mechanisms are still relevant when considering the fate of these excited states.

Photophysical studies on the unsubstituted purine and pyrimidine heterocycles, the common scaffolds to the natural nucleobases, reveal that both heterocycles decay along the triplet manifold, with ISC yields close to unity observed in water. Both purine and pyrimidine share a common ISC pathway via the S_1 ($n\pi^*$) state, which leads to efficient triplet population despite weak SOC. This efficiency is due to the inaccessibility of IC channels, which are energetically high or separated by significant barriers from the S_1 minima positions, preventing IC, but favoring ISC. In purine, an additional direct ISC pathway occurs, where the population is transferred directly from the S_2 ($\pi\pi^*$) state to the triplet manifold.

The specific substitution pattern (in terms of the nature and positions of the substituents) in purine and pyrimidine bases significantly modifies the PESs of the raw purine and pyrimidine cores, altering the accessibility of conical intersections and ISC channels [122, 124]. In this sense, the nucleobases conforming DNA and RNA at present can be considered the outcome of selection processes, due to the exposure of the extant protonucleobases to UV light in primordial Earth.

The population of the triplet state in nucleobases has been commonly thought to occur through dark $n\pi^*$ states, which are not directly populated by absorption but are instead accessed through IC from the initially excited bright $\pi\pi^*$ states. Traditionally, ISC in these molecules was considered inefficient due to the weak SOC typical of organic molecules. However, recent studies have demonstrated that, in some cases, ISC can occur on similar timescales as IC processes. In fact, the five natural nucleobases exhibit significant differences in their triplet state formation mechanisms, although they share common features.

A common feature of uracil and thymine is their ability to access the triplet manifold through singlet–triplet crossings, lying in the vicinity of $n\pi^*$ minima, which are populated along the deactivation of the bright $\pi\pi^*$. These two pyrimidines exhibit relatively efficient $\pi\pi^* \rightarrow n\pi^*$ population diversion as shown by molecular dynamics simulations. Despite the weak coupling between the singlet and triplet state, typical of organic molecules consisting of second row elements, the momentarily trapping of the population at the nπ* minimum position assists the ISC process. In contrast, purine nucleobases, adenine and guanine, have been mainly proposed to deactivate along the singlet $\pi\pi$* surface, making triplet state formation less favorable, despite ISC funnels along the deactivation route of the bright state were also located. These funnels were, however, located near the S_1/S_0 funnels, where the system is not expected to remain long enough to undergo ISC processes. This might in part explain the lack of experimental data involving the triplet excited states of guanine. In the case of cytosine, the main ISC funnel also involves the bright $\pi\pi$* state. Overall, statistically significant singlet–triplet molecular dynamics simulations using highly correlated electronic structure methods are scarce in the gas phase, solution, and biological environments, and this hinders the comprehensive understanding of the triplet formation mechanism.

The accessibility of triplet states is significantly altered when even small modifications are introduced into the natural nucleobase structure. One of the examples is the substitution of an oxo group with sulfur, leading to thionucleobases. The incorporation of sulfur changes the excited state landscape of the molecule, leading to much higher triplet quantum yields, which can approach unity. This contrast with natural nucleobases is primarily due to the significant stabilization of singlet excited state minima, particularly the nπ* and $\pi\pi$* states involving sulfur, compared with their oxygen-containing counterparts. Population trapping at the singlet minima and the presence of multiple triplet states with energies similar to those of the singlet state minima allow for efficient ISC. Importantly, stabilization does not affect, to the same extent, the most accessible conical intersections, which remain energetically constant upon thionation. This, combined with the heavy-atom effect of sulfur, enhances SOC and accelerates ISC [103]. These characteristics turn thionucleobases into excellent candidates for applications in photodynamic therapy, where triplet state formation can be exploited for medical treatments. The study of the mechanisms for ISC processes not only enlarges our understanding of DNA photostability but also provides insights into the design of novel photosensitizers. The synergy between experimental and theoretical approaches is fundamental to this purpose. Simulations exploring PESs and molecular dynamics are essential tools to understand the mechanisms of light-induced processes at the molecular level. These computational studies play a crucial role in interpreting experimental results and help unravel the details of nucleobase photophysics.

References

1 Rios, A.C. and Tor, Y. (2013). *Israel Journal of Chemistry* 53: 469–483.

2 Crespo-Hernández, C.E., Cohen, B., and Kohler, B. (2005). *Nature* 436 (7054): 1141–1144.

3 Buchvarov, I., Wang, Q., Raytchev, M. et al. (2007). *PNAS* 104: 4794–4797.

4 Doorley, G.W., McGovern, D.A., George, M.W. et al. (2009). *Angewandte Chemie* 48: 123–127.
5 Crespo-Hernández, C.E., Cohen, B., Hare, P.M., and Kohler, B. (2004). *Chemical Reviews* 104: 1977–2019.
6 Chen, J., Zhang, Y., and Kohler, B. (2015). *Excited States in DNA Strands Investigated by Ultrafast Laser Spectroscopy*. In: Barbatti, M., Borin, A., Ullrich, S. (eds) Photoinduced Phenomena in Nucleic Acids II. Topics in Current Chemistry, vol. 356. Springer.
7 Middleton, C.T., De La Harpe, K., Su, C. et al. (2009). *Annual Review of Physical Chemistry* 60: 217–239.
8 Schreier, W.J., Gilch, P., and Zinth, W. (2015). *Annual Review of Physical Chemistry* 66: 497–519.
9 Kleinermanns, K., Nachtigallová, D., and de Vries, M.S. (2013). *International Reviews in Physical Chemistry* 32: 308–342.
10 Markovitsi, D., Gustavsson, T., and Vayä, I. (2010). *Journal of Physical Chemistry Letters* 1: 3271–3276.
11 Daniels, M. and Hauswirth, W. (1971). *Science* 171: 675–677.
12 Onidas, D., Markovitsi, D., Marguet, S. et al. (2002). *The Journal of Physical Chemistry. B* 106: 11367–11374.
13 Vigny, P. (1971). *C. R. Acad. Sci. Ser. D* 272: 3206–3209
14 Salet, C., Bensasson, R., and Becker, R.S. (1975). *Photochemistry and Photobiology* 22: 231–235.
15 Hare, P.M., Crespo-Hernández, C.E., and Kohler, B. (2006). *The Journal of Physical Chemistry. B* 110: 18641–18650.
16 Cuquerella, M.C., Lhiaubet-Vallet, V., Cadet, J., and Miranda, M.A. (2012). *Accounts of Chemical Research* 45: 1558–1570.
17 Dumont, E. and Monari, A. (2015). *Frontiers in Chemistry* 3: 145566.
18 Wood, P.D. and Redmond, R.W. (1996). *Journal of the American Chemical Society* 118: 4256–4263.
19 Bosca, F., Lhiaubet-Vallet, V., Cuquerella, M.C. et al. (2006). *Journal of the American Chemical Society* 128: 6318–6319.
20 Lamola, A.A. and Mittal, J.P. (1966). *Science* 154: 1560–1561.
21 Kwok, W.M., Ma, C., and Phillips, D.L. (2008). *Journal of the American Chemical Society* 130: 5131–5139.
22 Pilles, B.M., Bucher, D.B., Liu, L. et al. (2014). *Journal of Physical Chemistry Letters* 5: 1616–1622.
23 Schreier, W.J., Schrader, T.E., Koller, F.O. et al. (2007). *Science* 315: 625–629.
24 Banyasz, A., Douki, T., Improta, R. et al. (2012). *Journal of the American Chemical Society* 134: 14834–14845.
25 Marguet, S. and Markovitsi, D. (2005). *Journal of the American Chemical Society* 127: 5780–5781.
26 Giussani, A., Serrano-Andrés, L., Merchán, M. et al. (2013). *The Journal of Physical Chemistry. B* 117: 1999–2004.
27 Roca-Sanjuán, D., Olaso-González, G., González-Ramírez, I. et al. (2008). *Journal of the American Chemical Society* 130: 10768–10779.

28 Climent, T., González-Ramírez, I., González-Luque, R. et al. (2010). *Journal of Physical Chemistry Letters* 1: 2072–2076.
29 Serrano-Pérez, J.J., González-Ramírez, I., Coto, P.B. et al. (2008). *The Journal of Physical Chemistry. B* 112: 14096–14098.
30 Bishop, S.M., Malone, M., Phillips, D. et al. (1994). *Journal of the Chemical Society, Chemical Communications* (7): 871–872.
31 Kobayashi, T., Kuramochi, H., Suzuki, T., and Ichimura, T. (2010). *Physical Chemistry Chemical Physics* 12: 5140–5148.
32 Harada, Y., Suzuki, T., Ichimura, T., and Xu, Y.Z. (2007). *The Journal of Physical Chemistry. B* 111: 5518–5524.
33 Pollum, M., Jockusch, S., and Crespo-Hernández, C.E. (2014). *Journal of the American Chemical Society* 136: 17930–17933.
34 Kuramochi, H., Kobayashi, T., Suzuki, T., and Ichimura, T. (2010). *The Journal of Physical Chemistry. B* 114: 8782–8789.
35 Ortiz-Rodríguez, L.A., Reichardt, C., Hoehn, S.J. et al. (2020). *Nature Communications* 11: 1–9.
36 Kang, H., Jung, B., and Kim, S.K. (2003). *The Journal of Chemical Physics* 118: 6717–6719.
37 Kang, H., Lee, K.T., Jung, B. et al. (2002). *Journal of the American Chemical Society* 124: 12958–12959.
38 Ullrich, S., Schultz, T., Zgierski, M.Z., and Stolow, A. (2004). *Journal of the American Chemical Society* 126: 2262–2263.
39 Ullrich, S., Schultz, T., Zgierski, M.Z., and Stolow, A. (2004). *Physical Chemistry Chemical Physics* 6: 2796–2801.
40 Canuel, C., Mons, M., Piuzzi, F. et al. (2005). *The Journal of Chemical Physics* 122: 74316.
41 Satzger, H., Townsend, D., Zgierski, M.Z. et al. (2006). *PNAS* 103: 10196–10201.
42 Kotur, M., Weinacht, T.C., Zhou, C., and Matsika, S. (2012). *IEEE Journal on Selected Topics in Quantum Electronics* 18: 187–194.
43 Serrano-Andrés, L., Merchán, M., and Borin, A.C. (2006). *PNAS* 103: 8691–8696.
44 Silva-Junior, M.R., Schreiber, M., Sauer, S.P.A., and Thiel, W. (2008). *The Journal of Chemical Physics* 129: 4.
45 Blancafort, L. (2006). *Journal of the American Chemical Society* 128: 210–219.
46 Conti, I., Garavelli, M., and Orlandi, G. (2009). *Journal of the American Chemical Society* 131: 16108–16118.
47 El-Sayed, M.A. (1963). *The Journal of Chemical Physics* 38: 2834–2838.
48 González-Luque, R., Climent, T., González-Ramírez, I. et al. (2010). *Journal of Chemical Theory and Computation* 6: 2103–2114.
49 Callis, P.R. (1979). *Chemical Physics Letters* 61: 568–570.
50 Cohen, B., Hare, P.M., and Kohler, B. (2003). *Journal of the American Chemical Society* 125: 13594–13601.
51 Morgan, J.P. and Daniels, M. (1979). *Chemical Physics Letters* 67: 533–537.
52 Mitrić, R., Werner, U., Wohlgemuth, M. et al. (2009). *The Journal of Physical Chemistry. A* 113: 12700–12705.

53 Barbatti, M., Lan, Z., Crespo-Otero, R. et al. (2012). *The Journal of Chemical Physics* 137: 22–503.

54 Fabiano, E. and Thiel, W. (2008). *The Journal of Physical Chemistry. A* 112: 6859–6863.

55 Barbatti, M., Aquino, A.J.A., Szymczak, J.J. et al. (2010). *PNAS* 107: 21453–21458.

56 Plasser, F., Gómez, S., Menger, M.F.S.J. et al. (2018). *Physical Chemistry Chemical Physics* 21: 57–69.

57 Cadet, J. and Vigny, P. (1990). *Bioorganic Photochemistry*, vol. 1. New York: Wiley.

58 Cohen, B.J. and Goodman, L. (1965). *Journal of the American Chemical Society* 87: 5487–5490.

59 Serrano-Andrés, L., Merchán, M., and Borin, A.C. (2008). *Journal of the American Chemical Society* 130: 2473–2484.

60 Shukla, M.K. and Leszczynski, J. (2004). *Journal of Computational Chemistry* 25: 768–778.

61 Barbatti, M., Szymczak, J.J., Aquino, A.J.A. et al. (2011). *The Journal of Chemical Physics* 134: 14304.

62 Lan, Z., Fabiano, E., and Thiel, W. (2009). *ChemPhysChem* 10: 1225–1229.

63 Petke, J.D., Maggiora, G.M., and Christoffersent, R.E. (1990). *Journal of the American Chemical Society* 112: 5452–5460.

64 Clark, L.B., Peschel, G.G., and Tinoco, I. (1965). *Journal of Physical Chemistry* 69: 3615–3618.

65 Yamazaki, S. and Taketsugu, T. (2012). *The Journal of Physical Chemistry. A* 116: 491–503.

66 Lorentzon, J., Fulscher, M.P., and Roos, B.O. (1995). *Journal of the American Chemical Society* 117: 9265–9273.

67 Merchán, M., González-Luque, R., Climent, T. et al. (2006). *The Journal of Physical Chemistry. B* 110: 26471–26476.

68 Climent, T., González-Luque, R., Merchán, M., and Serrano-Andrés, L. (2007). *Chemical Physics Letters* 441: 327–331.

69 Gustavsson, T., Bányász, Á., Lazzarotto, E. et al. (2006). *Journal of the American Chemical Society* 128: 607–619.

70 Hare, P.M., Crespo-Hernández, C.E., and Kohler, B. (2007). *PNAS* 104: 435–440.

71 Mercier, Y., Santoro, F., Reguero, M., and Improta, R. (2008). *The Journal of Physical Chemistry. B* 112: 10769–10772.

72 Brown, I.H. and Johns, H.E. (1968). *Photochemistry and Photobiology* 8: 273–286.

73 Jennings, B.H., Pastra-Landis, S., and Lerman, J.W. (1972). *Photochemistry and Photobiology* 15: 479–491.

74 Etinski, M., Fleig, T., and Marian, C.M. (2009). *The Journal of Physical Chemistry. A* 113: 11809–11816.

75 Miura, Y., Yamamoto, Y.I., Karashima, S. et al. (2023). *Journal of the American Chemical Society* 145: 3369–3381.

76 Fingerhut, B.P., Dorfman, K.E., and Mukamel, S. (2013). *Journal of Physical Chemistry Letters* 4: 1933–1942.

77 Richter, M., Mai, S., Marquetand, P., and González, L. (2014). *Physical Chemistry Chemical Physics* 16: 24423–24436.

78 Hudock, H.R., Levine, B.G., Thompson, A.L. et al. (2007). *The Journal of Physical Chemistry. A* 111: 8500–8508.

79 Doltsinis, N.L., Markwick, P.R.L., Nieber, H., and Langer, H. (2008). *Ultrafast Radiationless Decay in Nucleic Acids: Insights from Nonadiabatic Ab Initio Molecular Dynamics. In: Shukla, M.K., Leszczynski, J. (eds) Radiation Induced Molecular Phenomena in Nucleic Acids. Challenges and Advances In Computational Chemistry and Physics*, vol. 5. Springer. Dordrecht.

80 Lan, Z., Fabiano, E., and Thiel, W. (2009). *The Journal of Physical Chemistry. B* 113: 3548–3555.

81 Nieber, H. and Doltsinis, N.L. (2008). *Chemical Physics* 347: 405–412.

82 Nachtigallová, D., Aquino, A.J.A., Szymczak, J.J. et al. (2011). *The Journal of Physical Chemistry. A* 115: 5247–5255.

83 Schreiber, M., Silva-Junior, M.R., Sauer, S.P.A., and Thiel, W. (2008) *The Journal of Chemical Physics* 128: 134110.

84 Zechmann, G. and Barbatti, M. (2008). *The Journal of Physical Chemistry. A* 112: 8273–8279.

85 Perun, S., Sobolewski, A.L., and Domcke, W. (2006). *The Journal of Physical Chemistry. A* 110: 13238–13244.

86 Asturiol, D., Lasorne, B., Robb, M.A., and Blancafort, L. (2009). *The Journal of Physical Chemistry. A* 113: 10211–10218.

87 Serrano-Pérez, J.J., González-Luque, R., Merchán, M., and Serrano-Andrés, L. (2007). *Journal of Physical Chemistry B* 111: 11880–11883.

88 Szymczak, J.J., Barbatti, M., Soo Hoo, J.T. et al. (2009). *The Journal of Physical Chemistry. A* 113: 12686–12693.

89 Asturiol, D., Lasorne, B., Worth, G.A. et al. (2010). *Physical Chemistry Chemical Physics* 12: 4949–4958.

90 Alexandrova, A.N., Tully, J.C., and Granucci, G. (2010). *The Journal of Physical Chemistry. B* 114: 12116–12128.

91 Picconi, D., Barone, V., Lami, A. et al. (2011). *ChemPhysChem* 12: 1957–1968.

92 Stojanović, L., Bai, S., Nagesh, J. et al. (2016). *Molecules* 21: 1603.

93 Nakayama, A., Arai, G., Yamazaki, S., and Taketsugu, T. (2013). *The Journal of Chemical Physics* 139: 214304.

94 Mai, S., Richter, M., Marquetand, P., and González, L. (2017). *Chemical Physics* 482: 9–15.

95 Fleig, T., Knecht, S., and Hättig, C. (2007). *The Journal of Physical Chemistry. A* 111: 5482–5491.

96 Blancafort, L. and Robb, M.A. (2004). *The Journal of Physical Chemistry. A* 108: 10609–10614.

97 Ismail, N., Blancafort, L., Olivucci, M. et al. (2002). *Journal of the American Chemical Society* 124: 6818–6819.

98 Merchán, M. and Serrano-Andrés, L. (2003). *Journal of the American Chemical Society* 125: 8108–8109.
99 Richter, M., Marquetand, P., González-Vázquez, J. et al. (2012). *Journal of Physical Chemistry Letters* 3: 3090–3095.
100 Mai, S., Marquetand, P., Richter, M. et al. (2013). *ChemPhysChem* 14: 2920–2931.
101 Ashwood, B., Pollum, M., and Crespo-Hernández, C.E. (2019). *Photochemistry and Photobiology* 95: 33–58.
102 Ashwood, B., Jockusch, S., and Crespo-Hernández, C.E. (2017). *Molecules* 22: 379.
103 Mai, S., Pollum, M., Martínez-Fernández, L. et al. (2016). *Nature Communications* 7: 1–8.
104 Pollum, M., Jockusch, S., and Crespo-Hernández, C.E. (2015). *Physical Chemistry Chemical Physics* 17: 27851–27861.
105 Martínez-Fernández, L., González, L., and Corral, I. (2012) *ChemComm* 48: 2134–2136.
106 Pirillo, J., Mazzone, G., Russo, N., and Bertini, L. (2017). *Journal of Chemical Information and Modeling* 57: 234–242.
107 Martínez-Fernández, L., Corral, I., Granucci, G., and Persico, M. (2014). *Chemical Science* 5: 1336–1347.
108 Mai, S., Ashwood, B., Marquetand, P. et al. (2017). *The Journal of Physical Chemistry. B* 121: 5187–5196.
109 Gomzi, V. (2011). *Journal of Theoretical and Computational Chemistry* 8: 71–83.
110 Mai, S., Marquetand, P., and González, L. (2015). *The Journal of Physical Chemistry. A* 119: 9524–9533.
111 Gobbo, J.P. and Borin, A.C. (2014). *Computational & Theoretical Chemistry* 1040–1041: 195–201.
112 Bai, S. and Barbatti, M. (2016). *The Journal of Physical Chemistry. A* 120: 6342–6350.
113 Cui, G. and Fang, W.H. (2013). *The Journal of Chemical Physics* 138: 44315.
114 Jiang, J., Zhang, T.S., Xue, J.D. et al. (2015). *The Journal of Chemical Physics* 143: 175103.
115 Mai, S., Marquetand, P., and González, L. (2016). *Journal of Physical Chemistry Letters* 7: 1978–1983.
116 Pollum, M. and Crespo-Hernández, C.E. (2014). *The Journal of Chemical Physics* 140: 71101.
117 Igarashi-Yamamoto, N., Tajiri, A., Hatano, M. et al. (1981). *Biochimica et Biophysica Acta (BBA) - Nucleic Acids and Protein Synthesis* 656: 1–15.
118 Martínez-Fernández, L., Granucci, G., Pollum, M. et al. (2017). *Chemistry - A European Journal* 23: 2619–2627.
119 Cui, G. and Thiel, W. (2014). *Journal of Physical Chemistry Letters* 5: 2682–2687.
120 Cao, J. and Chen, D.C. (2020). *Physical Chemistry Chemical Physics* 22: 10924–10933.
121 Bin Xie, B., Wang, Q., Guo, W.W., and Cui, G. (2017). *Physical Chemistry Chemical Physics* 19: 7689–7698.

122 Crespo-Hernández, C.E., Martínez-Fernández, L., Rauer, C. et al. (2015). *Journal of the American Chemical Society* 137: 4368–4381.
123 Bent, D.V., Hayon, E., and Moorthy, P.N. (1975). *Journal of the American Chemical Society* 97: 5065–5071.
124 Arpa, E.M., Brister, M.M., Hoehn, S.J. et al. (2020). *Journal of Physical Chemistry Letters* 11: 5156–5161.
125 Douki, T. and Cadet, J. (2001). *Biochemistry* 40: 2495–2501.
126 Law, Y.K., Azadi, J., Crespo-Hernández, C.E. et al. (2008). *Biophysical Journal* 94: 3590–3600.
127 Miaskiewicz, K., Miller, J., Cooney, M., and Osman, R. (1996). *Journal of the American Chemical Society* 118: 9156–9163.
128 Zhang, R.B. and Eriksson, L.A. (2006). *The Journal of Physical Chemistry. B* 110: 7556–7562.
129 Durbeej, B. and Eriksson, L.A. (2002). *Journal of Photochemistry and Photobiology A: Chemistry* 152: 95–101.
130 Boggio-Pasqua, M., Groenhof, G., Schäfer, L.V. et al. (2007). *Journal of the American Chemical Society* 129: 10996–10997.
131 Blancafort, L. and Migani, A. (2007). *Journal of the American Chemical Society* 129: 14540–14541.
132 Wang, H. and Chen, X. (2017). *Molecules* 22: 60.
133 Barbatti, M. (2014). *ChemPhysChem* 15: 3342–3354.
134 Mendieta-Moreno, J.I., Trabada, D.G., Mendieta, J. et al. (2016). *Journal of Physical Chemistry Letters* 7: 4391–4397.
135 Rauer, C., Nogueira, J.J., Marquetand, P. and González, L. (2016) *Journal of the American Chemical Society* 138: 15911–15916.
136 Bin Xie, B. and Cui, C.X. (2019). *Physical Chemistry Chemical Physics* 21: 2006–2016.
137 Conti, I., Martínez-Fernández, L., Esposito, L. et al. (2017). *Chemistry - A European Journal* 23: 15177–15188.
138 Jin, P., Wang, J.N., Wang, X. et al. (2023). *The Journal of Physical Chemistry. B* 127: 6878–6886.
139 Yang, Z.B., Zhang, R.B., and Eriksson, L.A. (2011). *Physical Chemistry Chemical Physics* 13: 8961–8966.
140 Wang, X. and Yu, H. (2018). *Journal of Molecular Modeling* 24: 1–9.

8

The Photophysics and Photochemistry of Epigenetics and "Natural" Noncanonical Bases: Insights from Quantum Mechanical Calculations

Lara Martínez-Fernández[1] and Roberto Improta[2]

[1] Instituto de Química Física Blas Cabrera, CSIC (Consejo Superior de Investigaciones Científicas), Departamento de Química Física de Materiales, Calle de Serrano 119, Madrid 28006, Spain
[2] Istituto di Biostrutture e Bioimmagini-CNR (IBB-CNR), Via De Amicis 95, Napoli 80145, Italy

8.1 Introduction

Absorption of ultraviolet (UV) light by DNA can start several oxidative reactions, damaging the genetic code and ultimately leading to pathological processes, from apoptosis to carcinogenesis [1, 2]. Consequently, the photophysics and photochemistry of polynucleotides have been the focus of an intense research activity, both experimental [3–8] and computational [8–14], which has allowed significant advances in our understanding of the complex interplay between different excited state processes that give account of the high photostability of nucleic acids (NAs) or, on the other side, of the oxidative damage. Particular attention has been devoted to the study of the five canonical nucleobases (uracil/thymine, cytosine, guanine, and adenine), which are responsible for the absorption of NAs in the UVC–UVA. Time-resolved (TR) experiments have shown that the excited state lifetime of the five nucleobases decays on the sub-picosecond timescale, in line with their very small ($\sim10^{-4}$) fluorescence quantum yield (QY) [3–8]. This outcome has been rationalized by quantum mechanical (QM) calculations, which reveal that very effective, almost barrierless, deactivation paths exist on the potential energy surface (PES) of the spectroscopic state(s) connecting the Franck–Condon (FC) region with a conical intersection (CI) with the ground electronic state (S_0) [8–14]. These deactivation routes have also been shown to be active in oligonucleotides and are one of the factors giving account of the photostability of NAs [9, 10]. At the same time, despite the ultrafast decay of the bright states, experiments and calculations show that the photophysics of the nucleobases is complex; many excited states, both bright and dark, can be transiently populated, and many features are still a matter of a lively debate [9, 10].

However, any attempt to obtain a complete picture of the photophysics of the DNA/RNA constituents should consider that "canonical" nucleobases (those encoded in the genes) are not the only bases "naturally" present in NAs. A century

Excited State Dynamics in DNA and RNA: Experimental and Computational Approaches on Triplet State, First Edition. Edited by Jinquan Chen.

ago [15, 16], indeed, a methylated form of cytosine (5-methyl-cytosine [5mC]) was isolated, bearing a methyl substituent, instead of a hydrogen atom, at C5 (see Scheme 8.1). This discovery has been one of the landmarks for the birth of a new discipline, epigenetics, focusing, inter alia, on all the functional changes occurring in genome that do not include changes in the sequence [17, 18]. For example, 5mC has a critical role in gene transcription and is involved in a plethora of biological, functional, and/or pathological processes [15, 19–28]. Several epigenetic bases have been more recently discovered (an exhaustive review of all the studies in the field is here impossible), as N6-methyl-adenine (relatively common also in RNA), and other cytosine derivatives (see Scheme 8.1), as 5-hydroxymethyl-cytosine (5hmC), 5-formyl-cytosine (5fC), and 5-carboxyl-cytosine (5caC) [29–33]. They are usually intermediate for the synthesis of 5mC from C, and, for some of them, their precise epigenetic role is still a matter of debate.

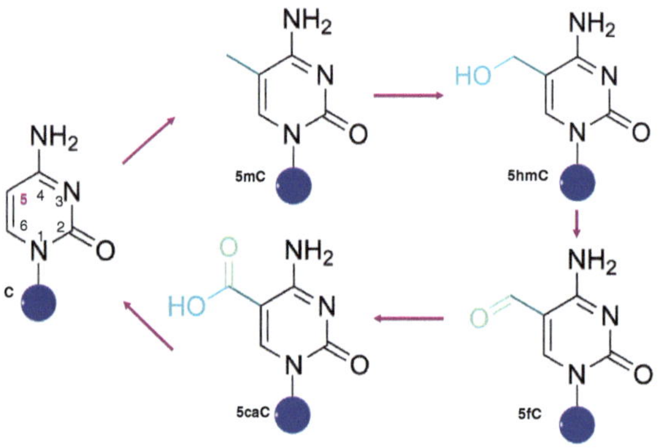

Scheme 8.1 Cytosine (atom labeling included) and its derivatives with the different substituents at position 5.

Although the absolute ratio of epigenetic bases is not large, it has to be highlighted that their location in the genome is not uniform. For example, even if "only" ~5% of cytosine in the human genome are "on the average" methylated, there are sequences (such as the CpG repeats) that are much richer in 5mC [34, 35]. This is a crucial point: even less common epigenetic bases are more frequent in certain DNA tracts, whose photophysical and photochemical behaviors could thus be strongly affected by the presence of the epigenetic bases. Consequently, many studies, especially in the past years, tackled the description of the photophysics/photochemistry of epigenetic bases, either isolated or included in oligonucleotides (see, for instance, References [36–44].

Epigenetic bases are not the only "noncanonical" nucleobases that can be isolated in NAs. Many other "natural" derivatives can indeed be found, produced by possible oxidative processes occurring in living cells. The most important one is likely 8-oxoguanine which is produced by the oxidation of guanine, i.e. the most readily oxidizable base [45–47]. Etheno adducts, as N4-etheno-2′-deoxycytidine (εdC) or 1,N2-etheno-2′-deoxyguanosine (εdG), can be produced not only by exogenous oxidants (as vinyl chloride) but also by highly reactive aldehydes produced during inflammation-induced lipid peroxidation [48–54]. All these molecules are important oxidative markers and some of them have been proposed to be mutagenic, motivating a fairly intense research activity on their photoactivated behavior, by both experimental and computational techniques (as detailed later). The computational study of the photophysics and photochemistry of noncanonical nucleobases, in addition to contributing to the interpretation of TR experiments and to disentangling their excited state dynamics, also has an intrinsic modeling and methodological interest. Indeed, considering the large number of experimental and computational studies existing on canonical nucleobases [10], the comparison of the behaviors of noncanonical epigenetic bases can give interesting information on the interplay of chemical effects determining the photophysics of heteroatomic organic molecules and the accuracy of the different computational methods adopted in their study. For example, studying how the 5-methyl substituent can affect the photophysics of cytosine provides useful and general insights into the substituent effect, and, at the same time, it is an important test to examine the ability of the different methods in reproducing and explaining these effects. Finally, it is possible to obtain indirect, but useful, information on canonical nucleobases (in this example, cytosine). In this chapter, we thus concisely review the outcome of the computational studies addressing natural noncanonical nucleobases. We focus in particular on the papers published in the past 10 years, as quite comprehensive reviews of older studies can be found in the literature [10, 12, 55].

8.2 Epigenetic Bases

We first discuss the epigenetic derivatives of cytosine, which bear a substituent at position 5 (see Scheme 8.1), whose epigenetic role is better assessed and whose photophysics has been investigated more thoroughly.

8.2.1 Cytosine Derivatives

8.2.1.1 5-Methyl-Cytosine

5mC can be considered the most important epigenetic base, and due to this relevance, it has been defined as the fifth "dynamic" letter of the genetic code [15, 25, 28]. As anticipated earlier, it was the first one to be discovered, the most frequently found by sequencing DNA, and it has a key epigenetic role. On the other side, it is a well-known mutational hotspot, which has been connected to carcinogenesis [20, 22, 24, 26, 27]. As a focus of this chapter, 5mC has been correlated with an increase in the formation of cyclobutane pyrimidine dimers (CPDs) [19, 21, 23, 27], explaining the several studies (both experimental and theoretical) devoted to elucidating the photophysics/photochemistry of 5mC and of sequences containing 5mC. A fairly recent review of its effect on CPD formation can be found in Martinez-Fernandez et al. [56].

In this chapter, we focus on the main keto-neutral tautomer of 5mC, the one that is biologically relevant. Readers are referred to studies [57–60] concerning other tautomers and to the study [61] about how methylation affects the photophysics of C cationic species.

A first basic effect of 5-methylation is the red-shift of the absorption spectrum of C, which has also been considered one of the factors explaining its larger photoreactivity [62]. The absorption spectrum of 5mC has, thus, been investigated using different methods and environmental conditions. In the gas phase, the bright $\pi\pi^*$ state corresponds to the first excited state (in the following, S_1) and is placed at 4.41 (MS-CASPT2/DZP), 4.48 (RI-CC2/aug-ccc-pvtz), 4.78 (TD-B3LYP/6-31G(d)), and 4.59 eV (TD-B3LYP/TZVP) [57, 63, 64]. Computational studies in a solution reproduce the differences in the absorption spectrum of C and 5mC and the solvent effects [36, 65]. In Martínez-Fernández et al. [65], it has been shown that the 5-methyl substituent reduces the HOMO/LUMO gap (and therefore red-shifts the lowest energy band) by providing an antibonding contribution to the HOMO (which mainly corresponds to the C5=C6 double bond). Moreover, the absorption spectrum was simulated in solvents of different polarity, by means of TD-CAM-B3LYP/6-31G(d) calculations, in which the solvent was considered either implicitly or explicitly (both with QM or molecular mechanics [MM] water molecules). In short, the absorption spectrum of 5mC in water has three bright $\pi\pi^*$ transitions ($S_1(\pi\pi_1^*)$, $S_4(\pi\pi_2^*)$, and $S_6(\pi\pi_3^*)$) that are responsible for the three main peaks at 241 (5.13), 205 (6.03), and 189 nm (6.54 eV). The low energy band is red-shifted in tetrahydrofuran and acetonitrile, which also strongly relatively stabilize the dark ($n_N\pi^*$) and ($n_O\pi^*$) states [65]. At the QM(CASPT2)/MM, the energy of the S_1 state is in the 4.46 eV [36]. When comparing, at the same level of theory, with the energy of the $S_1(\pi\pi_1^*)$ of cytosine, it can be concluded that methylation induces a red-shift of ~0.1 eV either in gas phase or in water at the RI-CC2 and Time-Dependent Density Functional Theory (TD-DFT) levels of theory [36, 63, 64, 66]. CASPT2/MM in water is the only approach that provides a larger shift of ~0.3 eV [36].

The PES of $\pi\pi_1^*$, the lowest energy bright state of 5mC, has been mapped both in the gas phase [57] and in solution [36, 38]. All these studies find a stable minimum, which is separated by a fairly large barrier from the lowest energy CI with the ground state (GS; S_0), involving puckering of the ring at the C4 or C5 position (see Figure 8.1 and Scheme 8.1 for atom labeling) [36, 57].

In the gas phase, CASPT2//CAS(14,10)/6-31G(d)-LIIC concluded that for $\pi\pi_1^*$, the most accessible CI with GS has a C4 ring-puckered structure (labelled as Bend-CI), which can be accessed by overcoming an energy barrier of ~0.4 eV [57]. This barrier is slightly smaller than that estimated to be necessary to access the lowest energy triplet state. In solution, a combined TD-CAMB3LYP and CASPT2/MM study characterized the photophysics of 5mC and C on the same foot [36]. For 5mC, it was found that $\pi\pi_1^*$, after passing an energy plateau where the ring is almost planar, reaches a slightly twisted minimum, separated by a barrier (0.08 eV at CAMB3LYP and 0.27 eV at CASPT2 level) from a CI with S_0, exhibiting an out-of-plane distortion at the C5 atom (ethylenic CI). These calculations provided Stokes shifts very close to their experimental counterpart and accounted for the significant increase in the excited state lifetime and fluorescence QY induced by 5-methylation. These effects have also been proven by experimental studies using two-color resonant two-photon

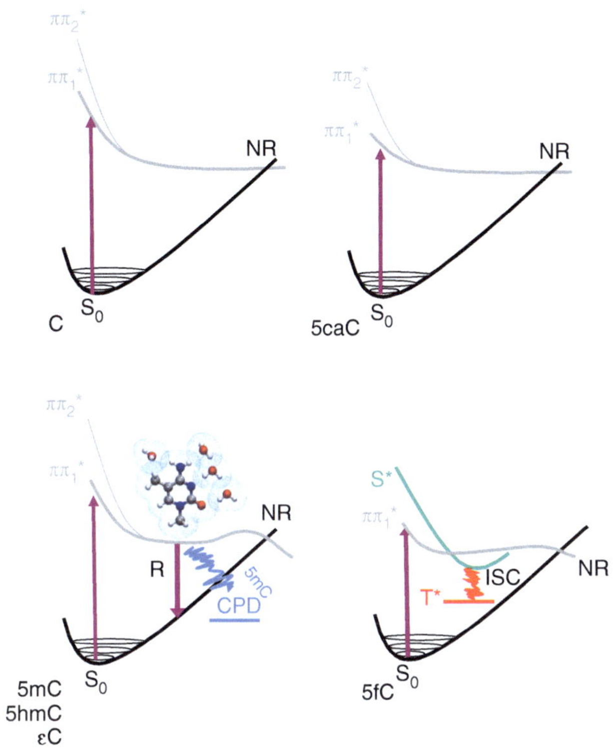

Figure 8.1 Scheme of the main probable excited state mechanisms extracted from calculations for C derivatives.

ionization, combined with calculations [64], femtosecond transient absorption spectroscopy [67], and fluorescence measurements [36]. As schematically depicted in Figure 8.1, in 5mC, the energy barrier to reach the CI is indeed larger than that computed for C at the same levels of theory. The decay of the second bright state, $\pi\pi_2$*, has also been studied, being predicted to decay to $\pi\pi_1$*. Dark nπ* transitions, involving the lone pair (LP) on N3 or on the carbonyl, are not affected by 5-methlyation, but, as their energy gap with respect to $\pi\pi_1$* increases, their population is less likely than what happens for C [36]. A more recent CASPT2/MM PES provides a picture very similar to that described earlier, with $\pi\pi_1$* reaching its minimum, after passing through a "planar" plateau, and the S_1/S_0 CI (reached by C4 puckering) placed +0.35 eV above the minimum [38]. The same study confirms that the $\pi\pi_2$* state decays to $\pi\pi_1$* near the FC region; the dark states are destabilized in water and, therefore, not significantly populated. This deactivation picture is fully consistent with TR spectral data, accounting for the main stimulated emission and excited state absorption experimental features. In addition to these static calculations, the photoactivated dynamics of 5mC has also been studied by quantum dynamics simulations in the gas phase and compared with that of C [68]. On the one hand, when exciting the $\pi\pi_1$*, methylation decreases the population transfer to the dark states; on the other, when exciting $\pi\pi_2$*, methylation leads to a ~10% decrease in the long-time limit transfer to $\pi\pi_1$* and, at the same time, an increase in the dark state population.

In addition to the photophysical changes induced by methylation described earlier, 5mC can also favor photochemical reactions. As an example, photoisomerization reactions are easier in 5mC than in C, occurring even spontaneously and with faster half-lifetimes [59]. Important attention has also been paid to the photodimerization of 5mC, since CPD is the major lesion induced by UV light in DNA [69, 70]. In this respect, experimental studies show that methylation favors the formation of CPDs [43, 71]. Molecular dynamics and QM calculations explained that such effect may be due to methylation favoring puckering combinations and stacking arrangements that are favorable to CPD formation [56, 71]. CASPT2 studies also pointed to more favorable CPD paths by reduced energy barriers leading to this reaction [72]. A general discussion of the effects on the larger reactivity in CPD formation induced by 5-methylation can be found in Martinez-Fernandez et al. [56]. This result is explained by the interplay of several electronic and structural factors: the red-shift of the absorption spectrum, the longer excited state lifetime, the changes in the sugar puckering, which modify the interbase stacking geometry, and an increase in the rigidity of the duplexes containing 5mC, which makes it easier for the excitation energy transfer toward 5mC.

8.2.1.2 5-Formyl-Cytosine

5fC (see Scheme 8.1) is another naturally occurring nucleobase, which can be produced in the methylation/demethylation process of 5mC and which has been discovered to be an epigenetic regulator of different cellular processes, such as transcription regulation or chromatin remodeling [29–33].

The absorption spectrum of 5fC has been characterized both in the gas phase and in water by different electronic methods. Most of the methods (see later) agree that the lowest energy excited state S_1 is a dark nπ^* state involving the LP of the formyl substituent. S_2 is a $\pi\pi^*$ state and corresponds to the lowest energy excited state, responsible for the first experimental absorption band. According to Mato et al. [73] in the gas phase, the S_2 vertical absorption energy varies from 4.2 (MR-MP2/6-31G+(d)) to 4.8 eV (EOM-CCSD/6-31G+(d)). Gas-phase MS-CASPT2/CASSCF(18,13)/6-31G(d) S_2 energy is also in this range, i.e. 4.55 eV [74]. This energy ordering of the states is predicted to be the same in polar solvents as water or acetonitrile, by studies including solvent effect or at the polarizable continuum model (PCM) model or by mixed continuum/discrete models [74, 75] PCM/TD-CAM-B3LYP calculations well reproduce the effect of the different substituent on the C absorption spectrum, with 5-formylation leading to a larger red-shift than that induced by methylation [74]. A more recent CASPT2/MM study instead predicts that the lowest energy bright state corresponds to S_1 for 5fC (vertical excitation energy at 4.36 eV) as well, which is slightly more stable than the lowest energy dark state [38].

The lowest PES (Figure 8.1) has been mapped in the gas phase (by CASPT2) [74, 76] and in water by both CASPT2/MM [38] (or PCM) [76] and TD-CAMB3LYP/PCM [74] (including four water molecules of the first solvation shell, Figure 8.1) calculations. These studies agree on some general features. An effective crossing between the lowest energy $\pi\pi^*$ and nπ^* states is predicted (in the FC region according to CAPT2/MM) [38], and the population on the nπ^* state decays to a stable “nonemissive” minimum, corresponding to the lowest energy singlet excited state minimum. This minimum could act as a funnel towards the population transfer to triplet manifold [38, 74, 76]. Indeed, surface hopping dynamics at the TD-B3LYP level of theory in the gas phase predict that 40% of the population is in the triplet states after 2 ps [75]. This large ISC yield has also been proven experimentally [39]. For what concerns the population of the bright states, these studies provide the passage through a planar plateau. Then, the facility to deactivate to the GS is more debated, with some studies pointing to barrierless mechanisms (CASPT2/MM [38] and CASPT2 in the gas phase [76]) and others predicting some barriers (TD-CAMB3LYP/PCM [74] and CASPT2/PCM [76]). Alternatively, it has been predicted that the population can decay to a close-lying stable minimum, with partial charge transfer (CT) character from the cytosine to the formyl moiety [74]. This minimum has been proposed to act as another important doorway state to the triplet manifold [74]. This CT minimum (whose presence is consistent with the experimental TR Infrared (IR) spectra) is not present in acetonitrile, explaining the very different steady-state emission spectra measured in the two solvents [74]. Notwithstanding these differences, these studies indicate that the fluorescence QY of 5fC should be very small, and, at the same time, the triplet QY large. Both predictions are confirmed by the experimental results, with the fluorescence QY (especially in water) $\leq 10^{-4}$ and the triplet QY > 0.7 in both solvents [74].

The photophysics of 5fC has also been studied when included in the $d(G5fC)_9 \bullet d(G5fC)_9$ duplex DNA duplex, by combining QM calculations with TR experiments [37]. In the duplex, in addition to the excited states localized on 5fC ($\pi\pi^*$ and $n\pi^*$), other states with intra- and inter-CT character between adjacent bases appear in the FC region. Geometry optimization of the lowest energy $\pi\pi^*$ localized on the 5fC moiety indicates that population can be transferred, in addition to the monomer-like 5fC $n\pi^*$ state described earlier, to a state with a significant intrastrand 5fC→G CT character. The cytosine-to-formyl CT state is disfavored with respect to the free base, as the out-of-plane motion of the formyl substituent necessary to reach such a minimum is hampered in the duplex, and the population of the dark $n\pi^*$ state localized on 5fC is reduced, suggesting a lower triplet QY compared to the monomeric species. On the other hand, computational characterization of the triplet electronic states indicates that the lowest energy triplet state has 5fC→G CT character, but that the population of the triplet localized on 5fC is also possible. This picture is fully consistent with the experimental indications. The triplet QY is <0.1, and the spectral signature of two different triplet states is visible; one is similar to that found for 5fC monomer, and the other, the most long-living one, that can be assigned to the intrastrand CT state [37].

8.2.1.3 5-Hydroxymethyl-Cytosine

5hmC is produced in vivo by oxidation of 5mC by TET proteins, and it has been shown to be particularly abundant in the brain, neurons, and brain stem cells. It is the second most frequent epigenetic base, after 5mC, being sometimes labeled as the “sixth” DNA base [31, 77].

The experimental absorption spectrum of 5hmC is more similar to that of 5mC than that of 5fC. Indeed, at the TD-DFT//B3LYP/6-31G(d) level of theory, the excitation energy to the lowest energy bright state is almost identical to the one reported for 5mC [57]. This is also the case at CASPT2/MM level of theory, with an energy difference of ~0.02 eV between 5mC and 5hmC [38].

The excited state PES of 5mC and 5hmC is also similar (see Figure 8.1). Photoexcitation of the first bright state of 5hmC at the CASPT2/MM level leads to a plateau and subsequently to a minimum that is separated by a significant energy barrier of +0.3 eV (+0.35 in 5mC) from the S_1/S_0 deactivation funnel [38]. Similar barriers were also predicted at the MR-MP2 level [73]. In the gas phase, the PES of 5hmC explored by means of CASPT2//CAS(14,10)/6-31G(d)-LIICs provides also large barriers, larger than for 5mC, for accessing the Bend-CI [57]. In this study, it was concluded that ISC pathways are more favorable in the case of 5hmC than in 5mC due to the more accessible S_1/T_1 funnels.

8.2.1.4 5-Carboxyl-Cytosine

5caC is the final product of the oxidative reactions of 5mC, produced by oxidation of 5fC by TET proteins [30, 78] and then can be converted back to C. It has been found in different contexts and, in addition to the methylation/demethylation processes, it can also regulate other cellular processes [32, 79].

5caC has been studied less in detail compared with the other C epigenetic derivatives. According to CASPT2/MM calculations, it seems to have an intermediate impact on the absorption spectrum, i.e. smaller than 5fC (4.36 eV) but larger than 5mC/5hmC (4.54/4.56 eV). The energy of the first bright state is 4.46 eV [38].

As depicted in Figure 8.1, the CASPT2/MM PES is characterized by a relaxation toward a plateau and a direct pathway toward the CI [38]. In the gas phase, this relaxation should be less direct due to the location of barriers but still the major path since ISC is not as favored as nonradiative decay to S_0 [76].

8.3 Methylated Purines

Another important epigenetic process is the methylation of purines, in different positions [80–82], with the most important one being 6-methyl-adenine (6mA), where the amino group bears a methyl substituent (Scheme 8.2) [83], which is the most common covalent modification in prokaryotes [84, 85]. They have also been shown to be involved in the damage response [86], spurring the interest toward the elucidation of their photophysical behavior [87, 88]. The number of computational studies of the photoactivated dynamics of methylated purines is still limited.

PCM/TD-PBE0 calculations have been used [87] to help the interpretation of TR spectra of 7-methyl-guanine (7mG), which affects the structural features of the genome of several viruses [89]. Calculations reproduce the effect of the pH on the steady-state absorption spectra of 7mG, leading to the transition from cationic to zwitterionic (deprotonated at N1) forms. Moreover, a stable minimum is found on the PES of the lowest energy excited state, giving rise to an increase in the fluorescence lifetime with respect to G [87].

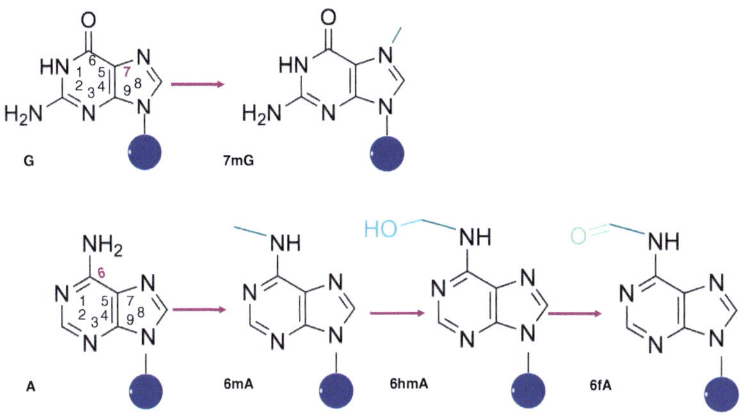

Scheme 8.2 Guanine, adenine (atom labeling included), and some of their methyl derivatives.

6-Hydroxymethyl-adenosine (6hmA) and N6-formyladenosine (6fA), two intermediates in the demethylation of N6-methyladenosine (6mA, see Scheme 8.2), have been studied by TD-DFT calculations [90], with the ωB97XD3 or the B3LYP functionals. The lowest energy singlet and triplet excited states were characterized in the FC region, and particular attention was devoted to define the possible pathways for triplet population, by computing the ISC in the main stationary points. Experiments show, indeed, that for these two compounds, a triplet QY > 0.25 [90]. For 6hmA, calculations show the presence of both locally excited and CT (from the hydroxyl substituent toward the adenine) excited states. This latter state could be the doorway state toward triplet generation. For 5fC, as discussed earlier, the best candidate for triplet generation is instead an nπ* state involving the LP of the formyl group [74].

8.3.1 Natural Derivatives

8.3.1.1 8-Oxoguanine

Guanine is the canonical base with the lowest oxidation potential. It is not surprising, therefore, that 8-hydroxy-2′-deoxyguanine (8oxoG, see Scheme 8.3) is one of the oxidative products most found in DNA, either following UV absorption or the reaction with endogenous or exogenous oxidants [45–47]. 8oxoG is indeed considered a marker of cellular oxidative stress [91]. Moreover, as discussed in more detail later, it can be involved in fast electron transfer reactions with neighboring bases, mediating the repair of CPDs [47]. Consequently, many studies, both experimental and computational, have been devoted to investigating the photoactivated behavior of 8oxoG and sequences containing this base.

Several computational studies of photoexcited 8oxoG in the gas phase exist in the literature, concerning both the nucleobase [40, 92] and the 8oxodG nucleoside [93]. XMCQDPT/CASSCF calculations both in the gas phase and in solution indicate that the two lowest energy excited states resemble those of G: S_1 (L_a according to the Platt labeling) can be described as a HOMO → LUMO excitation, less intense

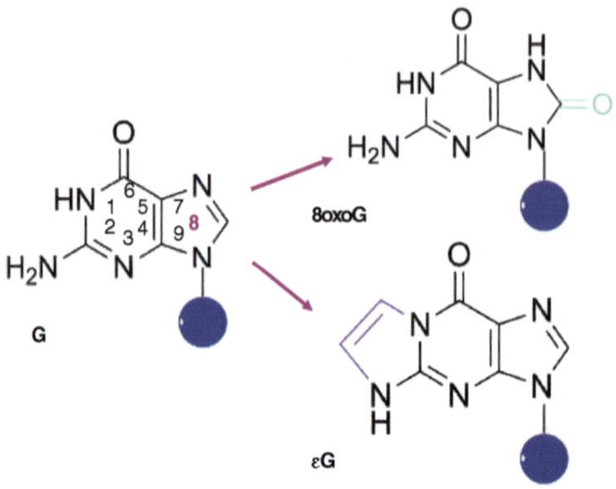

Scheme 8.3 Guanine and its natural derivatives.

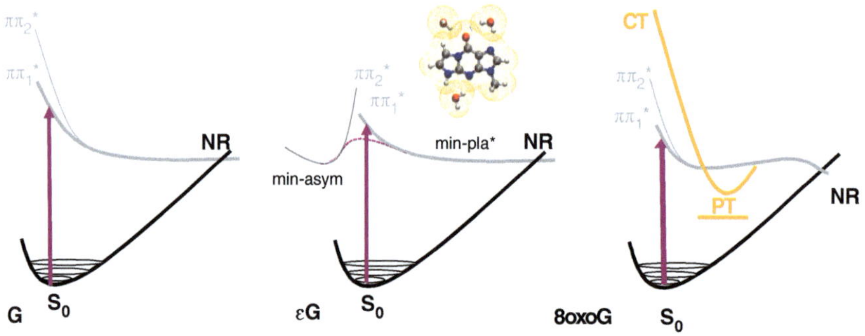

Figure 8.2 Scheme of the main probable excited state mechanisms extracted from calculations for G derivatives.

than S_2 (L_b), with a predominant HOMO → LUMO+1 contribution. The excited state decay routes are predicted to be similar to that found for guanine (Figure 8.2), as detailed later [40, 92].

The main photophysical decay paths of 8oxodG in the gas phase have been mapped at the ADC(2) level by Tuna and Domcke [93]. This study confirms the conclusions obtained on 8oxoG and suggests the possibility that an additional decay channel can be operative for *syn* conformers, which exhibit an intramolecular hydrogen bond between the 5′-O–H group of the sugar and the N3 atom of the base. In addition to the excited states localized on 8oxoG, it is shown that a higher lying CT state, involving the electron transfer from the sugar to the base, is strongly stabilized by intramolecular proton transfer (PT) (5′-O–H→N3). A CI is located on this path, which is slightly less stable than those associated with L_a state (C2-puckered, the lowest energy one, and C6-puckered) and could be accessible without any energy barrier from the L_b state. The nonradiative decay path following excited state intramolecular proton transfer (ESIPT) is instead not accessible for 8oxodG anion.

In water, 8oxoG (with a methyl group mimicking the sugar) has been studied at the PCM/TD-DFT level, adopting PBE0 or M052X functionals, and treating solvent effect with a mixed continuum discrete model (Figure 8.2, five water molecules of the first solvation shell are included in the computational model) [94]. This study provides a description of the FC region similar to that obtained in the gas phase and well reproduces the difference between the experimental absorption spectra of 8oxoG and G. The oxo-substituent leads to a red-shift of the L_a transition (~0.3 eV), whereas the position of the L_b is almost unaffected (its intensity slightly decreases), increasing the L_a/L_b energy gap by ~2000 cm^{-1}, close to the value extracted from the experimental value. By optimizing the excited states in water with the same computational approach, a very similar picture to that found for G is obtained [95]. L_b optimization indicates a very effective decay to L_a. For L_a, it is found that a very steep path connects the FC point with an energy plateau, where the purine planarity is not significantly distorted. After passing this "planar" plateau, the system is predicted to decay to a crossing region with the S_0, by out-of-plane motion of the C2–NH_2 group. This picture provides a rationale for the experimental findings. Emission from the planar plateau is the main responsibility of the experimental emission spectrum, as the computed Stokes shift for this region, i.e. 4000–5000 cm^{-1}, is very similar to the experimental value (5650 cm^{-1}). Moreover, the existence of a barrierless non-radiative deactivation route explains the sub-picosecond excited state lifetime, only slightly longer than that found for G [94].

This decay mechanism has been more recently confirmed by a surface hopping dynamical study, integrating CASSCF and CASPST2 calculations, which provides sub-picosecond GS recovery [96].

QM calculations have investigated the photophysics of 8oxoG in the 8oxodGpdA dinucleotide d(OA) in water [44]. PCM/TD-M052X calculations show that the lowest energy excited state is localized on 8oxodG base, whereas the lowest energy excited states localized on dA (A-L_a and A-L_b) lie more than 0.5 eV higher in energy. A state with well-defined 8oxoG$^+$→A$^-$ CT character (OA-CT) is close in energy to these local bright states, its relative stability depending on the inclusion of eight water molecules of the first solvation shell and on the adopted solvation model (State Specific (SS)-PCM or Linear Response (LR)-PCM). At the SS-PCM level, the CT state corresponds to S_2; it is within 0.1 eV of S_1 and appears strongly vibronically coupled with A-L_a. Geometry optimization of S_1 leads to a barrierless decay to the S_0, by following the same deactivation path described for the 8oxodG, via the C2-puckered crossing region. Optimizing A-L_a, the decay to OA-CT is predicted. Finally, optimization of OA-CT leads to a stable minimum with strong CT character. These findings enable the interpretation of the results of the TR experiments on this dinucleotide. TR IR spectra indicate indeed that, in addition to a fast component, decaying on a few picosecond timescale (which can be associated with the monomer-like decay paths), an excited state surviving for ~60 ps is present. The population of this long living state is larger following the excitation in the blue of the absorption band; it is 40% and 10% when exciting at 265 nm or 295 nm, respectively. Based on the computational results, the long living state has been assigned to OA-CT. Indeed, the computed difference IR spectrum between those of the OA-CT and the S_0 minima is very similar to the spectral signature of the long-living state. Moreover, when excited at 265 nm, the bright states, mainly localized on the dA base, are populated and are strongly coupled with the more stable OA-CT state.

QM calculations have been profitably used to interpret the TR experimental results obtained on 8oxodGpdA$^-$ dinucleotide d(O$^-$A), containing the deprotonated 8oxoG$^-$ anion [97]. The lowest energy excited state in the FC region is an 8oxoG$^{\cdot}$→A$^-$ CT state, involving electron transfer from 8oxoG$^-$ to A, leading to 8oxoG radical and an A$^-$ anion. Indeed, TR IR experiments indicate the formation of such a state in less than 1 ps, with a close to 100% yield, as confirmed by the close similarity between the experimental TR and computed difference IR spectra for the CT state. Back electron transfer is observed in 9 ps, i.e. much faster than the charge recombination observed in the neutral dOA dinucleotide (~60 ps). The computed reorganization energies for the different species enable interpreting all these results in the framework of the Marcus Theory [97].

The photophysics of deprotonated 8oxoG$^-$ has been investigated by QM calculations when hydrogen-bonded with A and C [98]. ADC(2) calculations indicate that in both complexes, the lowest energy excited state in the FC region is a CT state, involving the transfer of an electron from 8oxoG$^-$ to its hydrogen-bonded partner. Geometry optimizations of this state, independently of the partner, reveal that interbase PT reactions can lead, without any energy barrier, to low-lying S_1/S_0 CI. This outcome suggests that interbase hydrogen bond would dramatically decrease the excited state lifetime of 8oxoG$^-$. In a previous TD-DFT study of the neutral 8oxoG·A and 8oxoG·C pairs, a barrierless PT deactivation path was found only for the latter species, suggesting a longer excited state lifetime for 8oxoG·A pair [99]. It is, however, noted that within a duplex, the competition between intrastrand and interstrand CT states should be considered, with the former being characterized by a larger coupling between bright and CT states [9, 100].

8.3.2 Etheno G/C

As anticipated in the Introduction, etheno adducts (ε-adducts) are an important class of "natural" altered nucleosides, which have been isolated in the DNA of vertebrates and in human tissues [48–52]. Their presence is ubiquitarian, as they can be due to both exogenous (as carcinogenic substances like vinyl chloride) and endogenous (as the aldehydes produced during inflammation-induced lipid peroxidation) oxidants [53, 54]. ε-Adducts can be mutagenic and, at the same time, act as biomarkers to assess oxidative stress and cancer risk, motivating the study of their photophysical behavior, by means of QM calculations [50, 101–104].

εG (Scheme 8.3) has been studied by integrating CASPT2 (in the gas phase) and TD-M052X calculations (both in the gas phase and in water), including solvent effect with a mixed continuum explicit model [41]. When compared with G, the etheno ring increases the size of the conjugated π system, leading to a red-shift of the lowest energy absorption band with respect to G. The two lowest energy ππ* bright excited states (S_1 and S_2) are strongly vibronically coupled, with transitions involving MOs delocalized over three rings. As depicted in Figure 8.2, S_1 decays toward a planar stationary point, S_1-min-pla*, with small structural shifts with respect to the FC point and, therefore, a very small fluorescence Stokes shift. However, S_1-min-pla* is not a real minimum of the S_1 PES, as a barrierless path leads to a CI with the S_0, which is characterized by the pyramidalization of the carbonyl group. According to the calculations made in a study [41], S_2 population is transferred to S_1, likely on an ultrafast timescale, where it can decay either to S_1-min-pla* or to another minimum, S_1-min-asym, where the LUMO is mainly localized on the etheno ring. S_1-min-asym, only slightly less stable than S_1-min-pla*, is bright, with an emission energy significantly red-shifted with respect to that of S_1-min-pla*. The energy barrier between S_1-min-asym and S_1-min-pla* is not very large, and this latter minimum can act as a doorway toward the CI with the S_0.

The picture provided by QM calculations gives a general framework to explain the main experimental findings: (i) the steady-state and the TR fluorescence spectra, showing the presence of two minima, whose emission energy is consistent with that predicted by calculations; (ii) the peculiar behavior of TR experimental anisotropies, with the process [absorption on S_2] →[emission from S_1-min-pla*] being characterized by negative anisotropies; and (iii) the larger, but not much larger (about sixfold) fluorescence lifetime and QY measured for εG than for G. The nonradiative deactivation route to the S_0 is barrierless as for G, but part of the excited state population can be trapped in S_1-min-asym.

At difference of εG, QM calculations indicate that etheno substituent does not qualitatively alter the photophysics of εC (Figure 8.1) in water with respect the parent canonical nucleobase (C) [42]. εC has been recently studied using the same computational approach described for εG: CASPT2 and TD-M052X calculations (in the gas phase) and PCM-TD-M052X calculations (in water), including also two explicit water molecules, applied to a computational model where the presence of the ring was mimicked by a methyl group [42]. The two lowest energy excited states in the FC region are two bright $\pi\pi^*$ excitations, with major HOMO → LUMO (for S_1) and HOMO → LUMO+1 (for S_2) contributions, as it is predicted for C. However, due to the contribution of the etheno ring to the frontier orbitals, S_1 and S_2 are closer than in C and the latter state is less intense, explaining why, in the experimental absorption spectrum of εC, the band associated with S_2 is not visible, at difference of what happens in C. Geometry optimization of S_2 predicts the decay to S_1, whereas optimizing S_1 a planar minimum is reached. Both CASPT2 and PCM/TD-DFT calculations predict the existence of a crossing region with the GS accessible from the minimum of S_1, characterized by the out-of-plane motion of the C6 atom and especially of the C5–H5 group with respect to the molecular plane. This is the so-called ethylenic CI, which is predicted to be the most important nonradiative decay channel for C and for the other pyrimidine bases. According to PCM-TD-M052X, the ethylenic crossing region is more stable than the S_1 minimum, but it can be reached only by overcoming an energy barrier of ~0.3 eV. CASPT2 predicts that the ethylenic CI is only 0.1 eV less stable than the S_1 minimum, but the path is barrierless. When applied to εCH^+ protonated derivative (stable at acidic pH), the same computational approach (which reproduces the effect of the pH on the steady-state absorption and emission spectra of εC) [42] provides a description of the S_1 PES like that of εC, but access to the ethylenic crossing region is more difficult (the energy barrier is ~0.5 eV). The QM calculations provide a general picture of the $\varepsilon C/\varepsilon CH^+$ photophysics where the main experimental TR results can be nicely accommodated. The excited state lifetime and the fluorescence QY of εC are three times larger than those of C, and the increase in the lifetime is even larger for $εCH^+$. The access to the ethylenic crossing region for εC is indeed not as easy as predicted for C (the energy barrier ≤ 0.1 eV), and protonation makes this nonradiative decay path more difficult. On the other hand, the rather small increase in the excited state lifetime of εC with respect to C is explained by the accessibility to the ethylenic crossing region also in the presence of the etheno substitution. Thus, this outcome confirms the importance of this decay channel for C, as alternative deactivation routes, e.g. involving the C4–NH_2 group, would have been blocked by etheno substitution. Finally, the rather low value of the time-zero fluorescence anisotropies (<0.3) is instead explained by the contribution of the population S_2 decaying, and emitting, from S_1 on the ultrafast timescale.

8.4 Concluding Remarks

In this chapter, we have concisely reviewed the results of the most recent computational studies tackling the description of the excited state behavior of the noncanonical "natural" nucleobases, i.e. those that are always found, though in small amount, when sequencing the DNA or the RNA of living cells. As discussed in the introduction, many of these modified bases are produced by epigenetic processes, and either have an epigenetic function (in some cases sufficiently well understood) or can be considered "simply" reaction intermediates. Other modifications do not have any specific biological role but are due to the oxidative processes, endogenous or exogenous, unavoidable in living organisms. In the past years, these compounds have been the focus of intense research activities, which has also been extended to the study of their photophysics and photochemistry. Notwithstanding the relatively small absolute occurrence ratio of epigenetic bases, there are key sequences whose photophysical/photochemical behavior is strongly affected by epigenetic bases. Just to make an example, GC sequences are particularly rich in 5mC and 5fC. Moreover, they could act as a mutational hot spot in DNA. Let us consider, for example, 5fC. Its natural abundance is below 0.02, but, considering the large number of bases in the nucleosome, and that even a single mutation can be carcinogenic, the finding that photoexcitation of "free" 5fC leads to the almost quantitative population of triplet states increases the biological relevance of studying its photophysics in DNA.

In addition to their intrinsic importance, moreover, elucidating the photodynamics of modified base can give very useful, indirect insights into the mechanisms active in canonical nucleobases. Let us consider, for example, cytosine. For this molecule, both ethylene-like and sofa-type (involving the out-of-plane motion of the C4–NH_2 group) have been proposed to be involved in the main nonradiative deactivation route [10]. The finding that in etheno-C (for which the sofa-like deactivation route is blocked) the excited state lifetime is only marginally longer [42] than that of C strongly supports the importance of the ethylene-like CI for cytosine. The same indication is provided by the lengthening of the excited state lifetime and an increase in the fluorescence QY induced by 5-methylation [36]. In addition to their relevance for our understanding of oxidative damage in NA, the study of modified bases can give more general information on the effects ruling the photochemical behavior of heterocycle molecules (as nucleobases), with interesting implications also for methodological advances. For example, investigating the factors underlying the different behavior of 5mC and C can give useful insights, on the one hand, on substituent effect, and, on the other, provides an important check of the reliability of different computational methods in capturing this effect.

There are many important features of the photophysics and photochemistry of modified natural nucleobases that would require more in-depth experimental and computational scrutiny, and we hope that this concise review can contribute to pave the route to new, always more accurate, studies.

Acknowledgments

This research project is part of the activities of the National Center for Gene Therapy and Drugs Based on RNA Technology, funded by the European Union—Next Generation EU, Project CN00000041, CUP B93D21010860004. RI also thanks CNR, program "Progetti di Ricerca @cnr," project UCATG4, NUTRAGE funded by FOE-2021 DBA.AD005.225 and MUR 'PRIN 2022 2022K3AY2K' for financial support L. M-F. acknowledges the grant PID2023-151719NA-I00 funded by MICIU/AEI/10.13039/501100011033 and FEDER, UE.

References

1 Improta, R. and Douki, T. (ed.) (2021). *DNA Photodamage: From Light Absorption to Cellular Responses and Skin Cancer*. The Royal Society of Chemistry. https://doi.org/10.1039/9781839165580-FP001.
2 Taylor, J.S. (1994). Unraveling the molecular pathway from sunlight to skin cancer. *Accounts of Chemical Research* 27: 76–82.
3 Crespo-Hernández, C.E., Cohen, B., Hare, P.M., and Kohler, B. (2004). Ultrafast excited-state dynamics in nucleic acids. *Chemical Reviews* 104: 1977–2020.
4 Middleton, C.T., de La Harpe, K., Su, C. et al. (2009). DNA excited-state dynamics: from single bases to the double helix. *Annual Review of Physical Chemistry* 60: 217–239.
5 Gustavsson, T., Improta, R., and Markovitsi, D. (2010). DNA/RNA: building blocks of life under UV irradiation. *Journal of Physical Chemistry Letters* 1: 2025–2030.
6 Gustavsson, T. and Markovitsi, D. (2021). Fundamentals of the intrinsic DNA fluorescence. *Accounts of Chemical Research* 54: 1226–1235.
7 Schreier, W.J., Gilch, P., and Zinth, W. (2015). Early events of DNA photodamage. *Annual Review of Physical Chemistry* 66: 497–519.
8 Matsika, S. and Marcus, A.H. (ed.) (2024). *Nucleic Acid Photophysics and Photochemistry*. Springer Cham. https://doi.org/10.1007/978-3-031-68807-2.

9 Martínez Fernández, L., Santoro, F., and Improta, R. (2022). Nucleic acids as a playground for the computational study of the photophysics and photochemistry of multichromophore assemblies. *Accounts of Chemical Research* 55: 2077–2087.

10 Improta, R., Santoro, F., and Blancafort, L. (2016). Quantum mechanical studies on the photophysics and the photochemistry of nucleic acids and nucleobases. *Chemical Reviews* 116: 3540–3593.

11 Martínez-Fernández, L. and Francés-Monerris, A. (2023). DNA photostability. In: *Theoretical and Computational Photochemistry* (ed. C. García-Iriepa and M. Marazzi), 311–336. Elsevier. https://doi.org/10.1016/B978-0-323-91738-4.00001-4.

12 Martínez-Fernández, L. and Corral, I. (2024). Excited states reactivity of natural and unnatural DNA nucleobases. In: *Comprehensive Computational Chemistry*, 1e (ed. M. Yáñez and R.J. Boyd), 306–329. Oxford: Elsevier. https://doi.org/10.1016/B978-0-12-821978-2.00128-8.

13 Barbatti, M., Borin, A.C., and Ullrich, S. (2015). Photoinduced processes in nucleic acids. In: *Photoinduced Phenomena in Nucleic Acids I: Nucleobases in the Gas Phase and in Solvents* (ed. M. Barbatti, A.C. Borin, and S. Ullrich), 1–32. Cham: Springer International Publishing. https://doi.org/10.1007/128_2014_569.

14 Giussani, A., Segarra-Martí, J., Roca-Sanjuán, D., and Merchán, M. (2015). Excitation of nucleobases from a computational perspective I: Reaction paths. In: *Photoinduced Phenomena in Nucleic Acids I: Nucleobases in the Gas Phase and in Solvents* (ed. M. Barbatti, A.C. Borin, and S. Ullrich), 57–97. Cham: Springer International Publishing. https://doi.org/10.1007/128_2013_501.

15 Holliday, R. and Pugh, J.E. (1975). DNA modification mechanisms and gene activity during development. *Science* 187: 226–232.

16 Johnson, T.B. and Coghill, R.D. (1925). Researches on pyrimidines. C111. The discovery of 5-METHYL-CYTOSINE in tuberculinic acid, the nucleic acid of the tubercle BACILLUS1. *Journal of the American Chemical Society* 47: 2838–2844.

17 Allis, C.D. and Jenuwein, T. (2016). The molecular hallmarks of epigenetic control. *Nature Reviews Genetics* 17: 487–500.

18 Carter, B. and Zhao, K. (2021). The epigenetic basis of cellular heterogeneity. *Nature Reviews Genetics* 22: 235–250.

19 Cannistraro, V.J. and Taylor, J.-S. (2009). Acceleration of 5-methylcytosine deamination in cyclobutane dimers by G and its implications for UV-induced C-to-T mutation hotspots. *Journal of Molecular Biology* 392: 1145–1157.

20 Denissenko, M.F., Chen, J.X., Tang, M.-S., and Pfeifer, G.P. (1997). Cytosine methylation determines hot spots of DNA damage in the human P53 gene. *Proceedings of the National Academy of Sciences* 94: 3893–3898.

21 Drouin, R. and Therrien, J.-P. (1997). UVB-induced cyclobutane pyrimidine dimer frequency correlates with skin cancer mutational hotspots in p53. *Photochemistry and Photobiology* 66: 719–726.

22 Greenblatt, M.S., Bennett, W.P., Hollstein, M., and Harris, C.C. (1994). Mutations in the p53 tumor suppressor gene: clues to cancer etiology and molecular pathogenesis. *Cancer Research* 54: 4855–4875.

23 Ikehata, H. and Ono, T. (2011). The mechanisms of UV mutagenesis. *Journal of Radiation Research* 52: 115–125.

24 Jin, S.-G., Xiong, W., Wu, X. et al. (2015). The DNA methylation landscape of human melanoma. *Genomics* 106: 322–330.

25 Lister, R. and Ecker, J.R. (2009). Finding the fifth base: genome-wide sequencing of cytosine methylation. *Genome Research* 19: 959–966.

26 Rideout, W.M., Coetzee, G.A., Olumi, A.F., and Jones, P.A. (1990). 5-Methylcytosine as an endogenous mutagen in the human LDL receptor and p53 genes. *Science* 249: 1288–1290.

27 Robertson, K.D. and Jones, P.A. (2000). DNA methylation: past, present and future directions. *Carcinogenesis* 21: 461–467.

28 Smith, Z.D. and Meissner, A. (2013). DNA methylation: roles in mammalian development. *Nature Reviews Genetics* 14: 204–220.

29 Ding, G., Chen, B., McCarrey, J.R., and Huang, H. (2022). The epigenetic mechanisms underlying gamete origin of adult-onset diseases. *Science Bulletin* 67: 1724–1727.

30 Ito, S., Shen, L., Dai, Q. et al. (2011). Tet proteins can convert 5-methylcytosine to 5-formylcytosine and 5-carboxylcytosine. *Science* 333: 1300–1303.

31 Iurlaro, M., Ficz, G., Oxley, D. et al. (2013). A screen for hydroxymethylcytosine and formylcytosine binding proteins suggests functions in transcription and chromatin regulation. *Genome Biology* 14: R119.

32 Kellinger, M.W., Song, C.-X., Chong, J. et al. (2012). 5-formylcytosine and 5-carboxylcytosine reduce the rate and substrate specificity of RNA polymerase II transcription. *Nature Structural & Molecular Biology* 19: 831–833.

33 Nakano, S., Suzuki, T., Kawarada, L. et al. (2016). NSUN3 methylase initiates 5-formylcytidine biogenesis in human mitochondrial tRNAMet. *Nature Chemical Biology* 12: 546–551.

34 Jones, P.A. (2012). Functions of DNA methylation: islands, start sites, gene bodies and beyond. *Nature Reviews Genetics* 13: 484–492.

35 Cooper, D.N. and Youssoufian, H. (1988). The CpG dinucleotide and human genetic disease. *Human Genetics* 78: 151–155.

36 Martínez-Fernández, L., Pepino, A.J., Segarra-Martí, J. et al. (2017). Photophysics of deoxycytidine and 5-methyldeoxycytidine in solution: a comprehensive picture by quantum mechanical calculations and femtosecond fluorescence spectroscopy. *Journal of the American Chemical Society* 139: 7780–7791.

37 Wang, X., Martínez-Fernández, L., Zhang, Y. et al. (2024). Ultrafast formation of a delocalized triplet-excited state in an epigenetically modified DNA duplex under direct UV excitation. *Journal of the American Chemical Society* 146: 1839–1848.

38 Kabaciński, P., Romanelli, M., Ponkkonen, E. et al. (2021). Unified description of ultrafast excited state decay processes in epigenetic deoxycytidine derivatives. *The Journal of Physical Chemistry Letters* 12: 11070–11077.

39 Wang, X., Yu, Y., Zhou, Z. et al. (2019). Ultrafast intersystem crossing in epigenetic DNA nucleoside 2′-deoxy-5-formylcytidine. *The Journal of Physical Chemistry B* 123: 5782–5790.

40 Lu, Z., Beckstead, A.A., Kohler, B., and Matsika, S. (2015). Excited state relaxation of neutral and basic 8-oxoguanine. *The Journal of Physical Chemistry B* 119: 8293–8301.

41 Lizondo-Aranda, P., Gustavsson, T., Martínez-Fernández, L. et al. (2024). The excited state dynamics of a mutagenic guanosine etheno adduct investigated by femtosecond fluorescence spectroscopy and quantum mechanical calculations. *Chemistry – A European Journal* 30: e202401835.

42 Lizondo-Aranda, P., Martínez-Fernández, L., Miranda, M.A. et al. (2022). The excited state dynamics of a mutagenic cytidine etheno adduct investigated by combining time-resolved spectroscopy and quantum mechanical calculations. *The Journal of Physical Chemistry Letters* 13: 251–257.

43 Banyasz, A., Esposito, L., Douki, T. et al. (2016). Effect of C5-methylation of cytosine on the UV-induced reactivity of duplex DNA: conformational and electronic factors. *The Journal of Physical Chemistry B* 120: 4232–4242.

44 Zhang, Y., Dood, J., Beckstead, A.A. et al. (2014). Efficient UV-induced charge separation and recombination in an 8-oxoguanine-containing dinucleotide. *Proceedings of the National Academy of Sciences* 111: 11612–11617.

45 Cadet, J., Douki, T., and Ravanat, J.-L. (2008). Oxidatively generated damage to the guanine moiety of DNA: mechanistic aspects and formation in cells. *Accounts of Chemical Research* 41: 1075–1083.

46 Kanvah, S., Joseph, J., Schuster, G.B. et al. (2010). Oxidation of DNA: damage to nucleobases. *Accounts of Chemical Research* 43: 280–287.

47 Nguyen, K.V. and Burrows, C.J. (2011). A prebiotic role for 8-oxoguanosine as a flavin mimic in pyrimidine dimer photorepair. *Journal of the American Chemical Society* 133: 14586–14589.

48 Chung, F.-L., Chen, H.-J.C., and Nath, R.G. (1996). Lipid peroxidation as a potential endogenous source for the formation of exocyclic DNA adducts. *Carcinogenesis* 17: 2105–2111.

49 Fernando, R.C., Nair, J., Barbin, A. et al. (1996). Detection of 1, N6 -ethenodeoxyadenosine and 3, N4 -ethenodeoxycytidine by immunoaffinity/ 32 P-post-labelling in liver and lung DNA of mice treated with ethylcarbamate (urethane) or its metabolites. *Carcinogenesis* 17: 1711–1718.

50 Yu, Y., Cui, Y., Niedernhofer, L.J., and Wang, Y. (2016). Occurrence, biological consequences, and human health relevance of oxidative stress-induced DNA damage. *Chemical Research in Toxicology* 29: 2008–2039.

51 Yu, Y., Guerrero, C.R., Liu, S. et al. (2016). Comprehensive assessment of oxidatively induced modifications of DNA in a rat model of human Wilson's disease. *Molecular & Cellular Proteomics* 15: 810–817.

52 Garcia, C.C.M., Angeli, J.P.F., Freitas, F.P. et al. (2011). [13C2]- Acetaldehyde promotes unequivocal formation of 1,N2-propano-2′-deoxyguanosine in human cells. *Journal of the American Chemical Society* 133: 9140–9143.

53 Garcia, C.C.M., Freitas, F.P., Di Mascio, P., and Medeiros, M.H.G. (2010). Ultra-sensitive simultaneous quantification of 1,N2-etheno-2′-deoxyguanosine and 1,N2-propano-2′-deoxyguanosine in DNA by an online liquid chromatography electrospray tandem mass spectrometry assay. *Chemical Research in Toxicology* 23: 1851–1851.

54 Barbin, A., Ohgaki, H., Nakamura, J. et al. (2003). Endogenous deoxyribonucleic acid (DNA) damage in human tissues: a comparison of ethenobases with aldehydic DNA lesions. *Cancer Epidemiology, Biomarkers & Prevention* 12: 1241–1247.

55 Matsika, S. (2015). Modified nucleobases. In: *Photoinduced Phenomena in Nucleic Acids I: Nucleobases in the Gas Phase and in Solvents* (ed. M. Barbatti, A.C. Borin, and S. Ullrich), 209–243. Cham: Springer International Publishing. https://doi.org/10.1007/128_2014_532.

56 Martinez-Fernandez, L., Banyasz, A., Esposito, L. et al. (2017). UV-induced damage to DNA: effect of cytosine methylation on pyrimidine dimerization. *Signal Transduction and Targeted Therapy* 2: 17021.

57 Ai, Y., Xing, J., Zhang, A. et al. (2018). Computational study on the excited-state decay of 5-methylcytosine and 5-hydroxymethylcytosine: the common form of DNA methylation and its oxidation product. *The Journal of Physical Chemistry B* 122: 10424–10434.

58 Ho, J.-W. and Cheng, P.-Y. (2022). Ultrafast excited-state dynamics of gas-phase 5-methylcytosine tautomers. *Journal of the Chinese Chemical Society* 69: 107–119.

59 Lapinski, L., Reva, I., Rostkowska, H. et al. (2014). Near-IR-induced, UV-induced, and spontaneous isomerizations in 5-methylcytosine and 5-fluorocytosine. *The Journal of Physical Chemistry B* 118: 2831–2841.

60 Feyer, V., Plekan, O., Kivimäki, A. et al. (2011). Comprehensive core-level study of the effects of isomerism, halogenation, and methylation on the tautomeric equilibrium of cytosine. *The Journal of Physical Chemistry A* 115: 7722–7733.

61 Segarra-Martí, J., Tran, T., and Bearpark, M.J. (2019). Computing the ultrafast and radiationless electronic excited state decay of cytosine and 5-methyl-cytosine cations: uncovering the role of dynamic electron correlation. *ChemPhotoChem* 3: 856–865.

62 You, Y.-H. and Pfeifer, G.P. (2001). Similarities in sunlight-induced mutational spectra of CpG-methylated transgenes and the p53 gene in skin cancer point to an important role of 5-methylcytosine residues in solar UV mutagenesis11Edited by J. Miller. *Journal of Molecular Biology* 305: 389–399.

63 Nakayama, A., Yamazaki, S., and Taketsugu, T. (2014). Quantum chemical investigations on the nonradiative deactivation pathways of cytosine derivatives. *The Journal of Physical Chemistry A* 118: 9429–9437.

64 Trachsel, M.A., Lobsiger, S., and Leutwyler, S. (2012). Out-of-plane low-frequency vibrations and nonradiative decay in the $1\pi\pi^*$ state of jet-cooled 5-methylcytosine. *The Journal of Physical Chemistry B* 116: 11081–11091.

65 Martínez-Fernández, L., Pepino, A.J., Segarra-Martí, J. et al. (2016). Computing the absorption and emission spectra of 5-methylcytidine in different solvents: a test-case for different solvation models. *Journal of Chemical Theory and Computation* 12: 4430–4439.

66 Lobsiger, S., Trachsel, M.A., Frey, H.-M., and Leutwyler, S. (2013). Excited-state structure and dynamics of keto–amino cytosine: the $1\pi\pi^*$ state is nonplanar and its radiationless decay is not ultrafast. *The Journal of Physical Chemistry B* 117: 6106–6115.

67 Hoehn, S.J., Krul, S.E., Pogharian, M.M. et al. (2023). Photochemical stability of 5-methylcytidine relative to cytidine: photophysical insight for mRNA therapeutic applications. *The Journal of Physical Chemistry Letters* 14: 10856–10862.

68 Yaghoubi Jouybari, M., Liu, Y., Improta, R., and Santoro, F. (2020). Quantum dynamics of the $\pi\pi^*/n\pi^*$ decay of the epigenetic nucleobase 1,5-dimethyl-cytosine in the gas phase. *Physical Chemistry Chemical Physics* 22: 26525–26535.

69 Cadet, J. and Douki, T. (2018). Formation of UV-induced DNA damage contributing to skin cancer development. *Photochemical & Photobiological Sciences* 17: 1816–1841.

70 Mouret, S., Baudouin, C., Charveron, M. et al. (2006). Cyclobutane pyrimidine dimers are predominant DNA lesions in whole human skin exposed to UVA radiation. *Proceedings of the National Academy of Sciences* 103: 13765–13770.

71 Esposito, L., Banyasz, A., Douki, T. et al. (2014). Effect of C5-methylation of cytosine on the photoreactivity of DNA: a joint experimental and computational study of TCG trinucleotides. *Journal of the American Chemical Society* 136: 10838–10841.

72 González-Ramírez, I., Roca-Sanjuán, D., Climent, T. et al. (2011). On the photoproduction of DNA/RNA cyclobutane pyrimidine dimers. *Theoretical Chemistry Accounts* 128: 705–711.

73 Mato, J., Keipert, K., and Gordon, M.S. (2017). Excited state properties of 5-formylcytosine and 5-hydroxymethylcytosine. *Molecular Physics* 115: 2721–2730.

74 Wang, X., Martínez-Fernández, L., Zhang, Y. et al. (2021). Solvent-dependent stabilization of a charge transfer state is the key to ultrafast triplet state formation in an epigenetic DNA nucleoside. *Chemistry – A European Journal* 27: 10932–10940.

75 Francés-Monerris, A., Lineros-Rosa, M., Miranda, M.A. et al. (2020). Photoinduced intersystem crossing in DNA oxidative lesions and epigenetic intermediates. *Chemical Communications* 56: 4404–4407.

76 Xing, J., Ai, Y., Liu, Y. et al. (2018). Theoretical studies on the photophysics and photochemistry of 5-formylcytosine and 5-carboxylcytosine: the oxidative products of epigenetic modification of cytosine in DNA. *The Journal of Physical Chemistry B* 122: 2704–2714.

77 Rusk, N. (2012). The sixth base and counting. *Nature Methods* 9: 646–646.

78 Klungland, A. and Robertson, A.B. (2017). Oxidized C5-methyl cytosine bases in DNA: 5-hydroxymethylcytosine; 5-formylcytosine; and 5-carboxycytosine. *Free Radical Biology and Medicine* 107: 62–68.

79 Wang, L., Zhou, Y., Xu, L. et al. (2015). Molecular basis for 5-carboxycytosine recognition by RNA polymerase II elongation complex. *Nature* 523: 621–625.

80 Suzuki, M.M. and Bird, A. (2008). DNA methylation landscapes: provocative insights from epigenomics. *Nature Reviews Genetics* 9: 465–476.

81 Motorin, Y. and Helm, M. (2011). RNA nucleotide methylation. *Wiley Interdisciplinary Reviews. RNA* 2: 611–631.

82 Motorin, Y.A.-O.X. and Helm, M.A.-O. (2022). RNA nucleotide methylation: 2021 update. *Wiley Interdisciplinary Reviews. RNA* 13: e1691.

83 Ji, P., Wang, X., Xie, N., and Li, Y. (2018). N6-methyladenosine in RNA and DNA: an epitranscriptomic and epigenetic player implicated in determination of stem cell fate. *Stem Cells International* 2018: 3256524.

84 Wang, X., Zhao, B.S., Roundtree, I.A. et al. (2015). N6-methyladenosine modulates messenger RNA translation efficiency. *Cell* 161: 1388–1399.

85 Luo, G.-Z., Blanco, M.A., Greer, E.L. et al. (2015). DNA N6-methyladenine: a new epigenetic mark in eukaryotes? *Nature Reviews Molecular Cell Biology* 16: 705–710.

86 Xiang, Y., Laurent, B., Hsu, C.-H. et al. (2017). RNA m6A methylation regulates the ultraviolet-induced DNA damage response. *Nature* 543: 573–576.

87 Wang, D., Wang, X., Jiang, Y. et al. (2022). Excited state dynamics of methylated guanosine derivatives revealed by femtosecond time-resolved spectroscopy. *Photochemistry and Photobiology* 98: 1008–1016.

88 Wang, D., Jia, M., He, X. et al. (2024). Methylation induces a low-energy emissive state in N6-methyladenine containing dinucleotides. *ChemPhotoChem* 8: e202300235.

89 Liu, L., Dong, H., Chen, H. et al. (2010). Flavivirus RNA cap methyltransferase: structure, function, and inhibition. *Frontiers in Biology* 5: 286–303.

90 Wang, D., Wang, X., Jin, P. et al. (2023). Distinct doorway states lead to triplet excited state generation in epigenetic RNA nucleosides. *Physical Chemistry Chemical Physics* 25: 15153–15161.

91 Fleming, A.M. and Burrows, C.J. (2025). Why the ROS matters: one-electron oxidants focus DNA damage and repair on G-quadruplexes for gene regulation. *DNA Repair* 145: 103789.

92 Lu, Z., Beckstead, A.A., Kohler, B., and Matsika, S. (2016). Correction to "Excited state relaxation of neutral and basic 8-oxoguanine". *The Journal of Physical Chemistry B* 120: 597–597.

93 Tuna, D. and Domcke, W. (2016). Excited-state deactivation in 8-oxo-deoxyguanosine: comparison between anionic and neutral forms. *Physical Chemistry Chemical Physics* 18: 947–955.

94 Changenet-Barret, P., Gustavsson, T., Improta, R., and Markovitsi, D. (2015). Ultrafast excited-state deactivation of 8-hydroxy-2′-deoxyguanosine studied by femtosecond fluorescence spectroscopy and quantum-chemical calculations. *The Journal of Physical Chemistry A* 119: 6131–6139.

95 Karunakaran, V., Kleinermanns, K., Improta, R., and Kovalenko, S.A. (2009). Photoinduced dynamics of guanosine monophosphate in water from broad-band transient absorption spectroscopy and quantum-chemical calculations. *Journal of the American Chemical Society* 131: 5839–5850.

96 Zhao, L., Zhou, P., Liu, X. et al. (2021). Theoretical studies of the ultrafast deactivation mechanism of 8-oxo-guanine on the S1 and S2 electronic states in gas phase. *Spectrochimica Acta Part A: Molecular and Biomolecular Spectroscopy* 244: 118884.

97 Zhang, Y., Dood, J., Beckstead, A.A. et al. (2015). Photoinduced electron transfer in DNA: charge shift dynamics between 8-oxo-guanine anion and adenine. *The Journal of Physical Chemistry B* 119: 7491–7502.

98 Wu, X., Karsili, T.N.V., and Domcke, W. (2017). Role of electron-driven proton-transfer processes in the ultrafast deactivation of photoexcited anionic 8-oxoguanine-adenine and 8-oxoguanine-cytosine base pairs. *Molecules* 22: 135.

99 Kumar, A. and Sevilla, M.D. (2013). Excited state proton-coupled electron transfer in 8-oxoG-C and 8-oxoG-A base pairs: a time dependent density functional theory (TD-DFT) study. *Photochemical & Photobiological Sciences* 12: 1328–1340.

100 Fernández, L.M. and Improta, R. (2024). Computational studies on photoinduced charge transfer processes in nucleic acids: from Watson–Crick dimers to quadruple helices. In: *Nucleic Acid Photophysics and Photochemistry* (ed. S. Matsika and A.H. Marcus), 29–50. Cham: Springer Nature Switzerland. https://doi.org/10.1007/978-3-031-68807-2_2.

101 Tudek, B., Zdżalik-Bielecka, D., Tudek, A. et al. (2017). Lipid peroxidation in face of DNA damage, DNA repair and other cellular processes. *Free Radical Biology and Medicine* 107: 77–89.

102 Hecht, S.S. (2017). Oral cell DNA adducts as potential biomarkers for lung cancer susceptibility in cigarette smokers. *Chemical Research in Toxicology* 30: 367–375.

103 Chen, H., Krishnamachari, S., Guo, J. et al. (2019). Quantitation of lipid peroxidation product DNA adducts in human prostate by tandem mass spectrometry: a method that mitigates artifacts. *Chemical Research in Toxicology* 32: 1850–1862.

104 Ohnishi, I., Iwashita, Y., Matsushita, Y. et al. (2021). Mass spectrometric profiling of DNA adducts in the human stomach associated with damage from environmental factors. *Genes and Environment* 43: 12.

Index

Excited State Dynamics in DNA and RNA: Experimental and Computational Approaches on Triplet State, First Edition. Edited by Jinquan Chen.